To Chris

Putting Africa First

The Making of African Innovation Systems

Edited by

Mammo Muchie
Middlesex University Business School, United Kingdom

Peter Gammeltoft
Copenhagen Business School, Denmark

Bengt-Åke Lundvall
Aalborg University, Denmark

Aalborg University Press
Aalborg, Denmark

Putting Africa First: The Making of African Innovations Systems

ISBN 87-7307-699-6

Published by
Aalborg University Press
Niels Jernes Vej 6B
DK-9220 Aalborg Ø
Phone: 96357140
Fax: 96350076
E-mail: aauf@forlag.auc.dk
http://www.forlag.auc.dk

Cover design
Peter Gammeltoft

Layout
Peter Gammeltoft

Printed in Denmark by
J. Tengstedt offset-digitaltryk

Preface

Christopher Freeman

The expression 'national systems of innovation' was introduced by Bengt-Åke Lundvall to emphasise the interdependence between technical and institutional change and a great deal of work has been done to illustrate this interdependence since he and his colleagues published their seminal book in 1992. Originally, most of this work was in Europe or in North America and dealt with nation states in these parts of the world. More recently, and again stimulated mainly through the initiative of Aalborg University Centre and the DRUID Programme, the research frontier moved forward in two main directions. On the one hand, researchers in Asia and Latin America began to participate on a significant scale; on the other hand, there was growing recognition of the need to examine not merely national systems (NSI) but also regional systems (RSI), sectoral systems (SSI) and even continental systems. A big gap in all these developments was the lack of much research relating to Africa. Now, thanks again in part to the initiative of AUC and Lundvall, this gap is beginning to be filled.

For many reasons, the work on Africa is especially important. No continent has a more complex pattern of national boundaries or of ethnic, religious and tribal sub-systems, interacting with sectoral systems. The heritage of colonial exploitation, colonial divisions and colonial rule and the struggles for independence have left behind an extraordinary legacy, now also overlaid with the operations of multi-national corporations and other international organisations as well. To trace and understand these multiple interacting cultural and institutional sub-systems is indeed a huge challenge.

Yet no continent is more in need of the potential benefits, which science and technology can bring, than the African continent. The direst poverty, the cruellest burdens of disease, of armed conflict, and of environmental disasters still ravage Africa. Yet technology cannot yield its potential benefits without appropriate cooperative social initiatives organised by Africans for Africans. Therefore to understand the patterns of existing innovation systems, and their limitations and to devise ways to deliver much greater benefits to all the peoples of Africa is a fundamental need for the continent.

Personally, I have only very briefly visited a few African countries and I am not competent to propose any substantial ideas or conclusions but I was delighted to see the variety and scope of the chapters which have been prepared for this book. As an external observer with the warmest feelings of respect for Africa, it seems to me that the problems of the continent need a combination of interdependent studies at all levels of society. Some of these relate to the villages, towns, districts and industries of both large and small countries, some relate to national systems and still others to the entire continent. The Pan African ideal surely has some part to play in mobilising, energising and inspiring all these other institutions and initiatives. The heritage of existing knowledge, for example in botany, in medicine, in care for the environment, can only be harnessed in numerous local grassroots projects and institutions, but these need a continent-wide interchange of ideas.

The applications of wider international advances in science and technology demand greater resources and higher level institutions. The protection of the environment, the resolution of problems of indebtedness, the avoidance of abuse by powerful corporations and the defence of African interests in technology and science more generally may require initiatives and institutions on a continental scale as is increasingly the case in other continents too. At first, only a part of a continent may lead the way, as is the case with both India and China. Baskaran and Muchie's chapter on India and Africa is of particular interest in this connection.

I congratulate Mammo Muchie and his colleagues on their splendid initiative and wish the book every possible success: especially at this time of global economic instability, this work is needed more than ever to protect science and education from inept and misdirected programmes of 'structural adjustment' and to 'put the last first'. Then the stone that the builders rejected may indeed become the cornerstone of the arch.

Chris Freeman
SPRU,
The Freeman Centre,
University of Sussex,
Falmer, Brighton,
BN1 9QE, UK

Contents

Abbreviations

AU	African Union
CDF	Comprehensive Development Framework
FDI	foreign direct investment
GNP	gross national product
ICT	information and communication technology
IDRC	International Development Research Centre
IMF	International Monetary Fund
KIS	knowledge-intensive service
LDC	less developed country
MC	Maghreb countries
MENA	Middle East and North Africa
MSTQ	metrology, standards, testing, and quality assurance
MNC	multinational corporation
NIS	national innovation system
NSI	national system of innovation
NEPAD	New Partnership for Africa's Development
NGO	non-governmental organisation
NIE	newly industrialised economy
OAU	Organisation of African Unity
R&D	research and development
RCA	revealed comparative advantage
RPED	Regional Programme on Enterprise Development
RSI	regional system of innovation
SIDO	Small Industries Development Organisation
SME	small and medium-sized enterprise
SSA	Sub-Saharan Africa
SSI	sectoral system of innovation
TC	technological capability
TCB	technology capability building
TNC	transnational corporation
TILC	total innovation and learning culture
TRIM	trade-related investment measure
TRIPS	trade-related aspects of intellectual property rights
S&T	science and technology
SAP	structural adjustment programme
UNCTAD	United Nations Conference on Trade and Development
UNIDO	United Nations Industrial Development Organisation
WTO	World Trade Organisation

Contributors

ALICE H. AMSDEN, Barton L. Weller Professor of Political Economy, Massachusetts Institute of Technnology, Cambridge, USA.

LOU ANNE BARCLAY, Researcher, Institute for New Technologies, United Nations University (UNU/INTECH), Maastricht, the Netherlands.

ANGATHEVAR BASKARAN, Lecturer, Middlesex University Business School, London.

PERNILLE BERTELSEN, Associate Professor, Department of Development and Planning, Aalborg University, Denmark.

MONA DAHMS, Associate Professor, Department of Development and Planning, Aalborg University, Denmark.

ABDELKADER DJEFLAT, University of Lille, France.

CHRISTOPHER FREEMAN, Emeritus Professor of Science Policy, SPRU – Science and Technology Policy Research, University of Sussex, UK.

PETER GAMMELTOFT, Assistant Professor, Department of Intercultural Communication and Management, Copenhagen Business School, Denmark.

SHULIN GU, Visiting Professor, School of Economics and Technology, Tsinghua University, China.

ANDREW JAMISON, Professor of Technology and Society, Aalborg University, Denmark.

BJÖRN JOHNSON, Senior Associate Professor, Department of Business Studies, Aalborg University, Denmark.

JOHN KUADA, Associate Professor of International Management, Aalborg University, Denmark.

SANJAYA LALL, Professor of Development Economics, International Development Centre, Queen Elizabeth House, Oxford, UK.

LOBNA ABDEL-LATIF, Professor of Economics, Cairo University, Egypt.

BENGT-ÅKE LUNDVALL, Professor of Economics, Department of Business Studies, Aalborg University, Denmark.

SUNIL MANI, Researcher, Institute for New Technologies, United Nations University (UNU/INTECH), Maastricht, the Netherlands.

GILLIAN MARCELLE, Consultant to telecommunication companies in Africa, based in Johannesburg, South Africa.

MAMMO MUCHIE, Middlesex University Business School, UK.

JENS MÜLLER, Senior Associate Professor, Department of Development and Planning, Aalborg University, Denmark.

LYNN K. MYTELKA, Director, Institute for New Technologies, United Nations University (UNU/INTECH), Maastricht, the Netherlands.

CONTRIBUTORS

BANJI OYELARAN-OYEYINKA, Researcher, Institute for New Technologies, United Nations University (UNU/INTECH), Maastricht, the Netherlands.

CARLO PIETROBELLI, Professor of International Economics, University of Rome, Italy.

MARIO SCERRI, Dean, Faculty of Economic and Management Sciences, Technikon North-West, South Africa.

OLAV JULL SØRENSEN, Professor of International Business, Department of Business Studies, Aalborg University.

MARK TOMLINSON, Research Coordinator, Department of Economics, National University of Ireland, Galway.

SAMUEL WANGWE, Principal Research Associate, Economic and Social Research Foundation (ESRF), Dar es Salam, Tanzania.

Foreword

The idea behind this book came out of conversations between Bengt-Åke Lundvall and Mammo Muchie at Aalborg University in 1999. Mammo was attracted by the idea of applying the national innovation system concept on Africa because he was occupied by the needs to create a unification-nation building process in Africa. Bengt-Åke Lundvall was in the process of working out the Globelics concept (www.globelics.org). Building the first parts of the global network of scholars working on innovation systems together with African colleagues was an attractive idea.

We concluded that it might be a good idea to gather scholars from Africa and elsewhere that worked on innovation and innovation systems at an international conference 'African Systems of Innovation and Competence Building' and the conference took place March 2001 in Aalborg.

Some time has passed since the conference took place and the idea to publish this book became a realistic option only when Peter Gammeltoft joined the team of editors in the Summer 2003. Had it not been for his management and editorial effort it would not have been possible to finalise the project. We also thank Søren Koed who as a student assistant took on advanced tasks with speed and flexibility, and Ina Drejer and Jan Vang who took time out to go over an early version of the manuscript.

The conference was funded by Aalborg University and by a grant from the Danish Social Science Research Council. We want to take this opportunity to thank those who helped us raise the funding, especially Institute for History and International and Social Studies, Department for Business Studies, the Social Science Faculty, and the Chancellor of Aalborg University, Sven Caspersen.

One reason that Mammo and Bengt-Åke found together in this project was that both of us belong to the global circle of Christopher Freeman's intellectual 'god children'. A statement from him opened the conference in Aalborg and another opens this book. We dedicate the book to him with the utmost respect for the way he combines academic integrity with compassion for the poor countries and their people.

Cape Town and Aalborg September 2003

Mammo Muchie   Bengt-Åke Lundvall

Introduction

Bengt-Åke Lundvall, Mammo Muchie, and Peter Gammeltoft

Introduction

When Muchie and Lundvall initiated the international conference 'African Systems of Innovation and Competence Building' March 2001 in Aalborg the intention was to re-examine the problems of African development by mobilising the scholarship on innovation systems. The major objective was to respond to academic and analytical questions: what lessons could be drawn from applying the innovation system concept on Africa? The indirect objective was to address more political and controversial questions: could the introduction of the innovation system concept help to break the ideological stalemate where Africa appears as the hopeless victim doomed to permanent poverty and could it also help to bolster the idea of a Pan African renaissance that Mandela and other African leaders have tried to build recently? Could it open up a new vision 'where the last becomes the first'?

We managed to bring together leading experts from various parts of the world and later on a number of further contributions have been added. The chapters in this volume relate to different aspects of African innovation systems and they contribute to a revitalised agenda for African development. The contributions span a wide set of issues relevant for African innovation systems. Among them are general contributions to the understanding of innovation and economic development (Johnson and Lundvall; Jamison; Mytelka; Wangwe; Oyelaran-Oyeinka and Barclay) as well as sector studies (Sørensen; Tomlinson; Marcelle; Lall and Pietrobelli). The geographical level of analysis goes from the mobilisation of local knowledge (Dahms; Kuada) and regional innovation systems (Djeflat; Scerri) to national (Mani; Latif) and Africa-wide (Muchie; Baskaran and Muchie; Shulin Gu; Gammeltoft) studies. Some chapters have their focus on science-based activities (Amsden) while others focus on activities rooted in informal and tacit knowledge (Müller and Bertelsen).

The concept 'national system of innovation' can be traced to the work of Friedrich List. In his book the National System of Political Economy, List criticised what he called 'the cosmopolitan' approach of Adam Smith for being too focused on competition and resource allocation to the neglect of productive forces (List 1841). His analysis took into account a wide set of national institutions including those engaged in education and training as well as tangible infrastructures such as railroads (Freeman 1995). In relation to the current challenges for Africa and the Pan African response to them, it is interesting to note that List married his analysis of national production systems to a political agenda for nation-building in what was to become

1

Germany. In this introduction we will relate the basic elements in the national innovation system concept to the African reality.

Why Innovation?

In this volume innovation refers to the creation, diffusion and use of new ideas applied in the economy. The applications can take the form of new production processes, new products, new forms of organisation and new markets.

If a less developed economy has the ambition to grow and create more and better paid jobs for its citizens, continuous innovation, in this broad sense, must be part of the solution. Hard work, investment and more efficient use of resources are important elements in any development strategy. But in order to keep the momentum of economic development, technical and organisational innovations are important.

First, it might help to move into rapidly growing sectors characterised by the production and use of advanced technology. The entrance into the production and exports of information technology in Taiwan, South Korea and Singapore has been one reason for the rapid economic growth in these countries. This is an option that is not easy to exploit in most of the African states. One prerequisite is access to a highly trained labour force another is a reliable infrastructure. It should also be taken into account that this route to wealth may be less attractive than originally assumed in the height of the New Economy era. The experience from South East Asian economies is that there is instability built into high tech specialisation.

Second, the capability to use the new technologies in old sectors is highly important. This capability can be built through experimentation and learning. Without adaptation of organisational forms and human competence, the introduction of complex technologies in production processes may actually do more harm than leaving it aside? A major challenge is to develop technical training in the interface between schools and industry in such a way that the formation of theoretical and practical skills supports each other.

Third, product innovation is not something that takes place only in science-based sectors such as electronics and biotechnology. Developing gradually higher quality and more attractive agricultural products, textiles and machinery is one of the most important ways to establish economic growth and employment growth. Sector specific and technical institutes with a close connection to producers may be crucial for developing the new qualities.

It should be noted that in an open economy 'not to innovate' means that domestic producers will bear the negative impact of innovations made by others. Most African countries are quite open in terms of trade and have export and import shares of GNP over 20 per cent. Domestic producers are constantly confronted with competition from innovating producers abroad. This is true also for those countries producing agricultural products for the world markets – there is a permanent drive toward more attractive and cheap products also in this field. Therefore, in most markets, not to innovate means continuously to lose market share and income.

It is also helpful to note that in the African context, the most important innovation and competence building activities are not a specialty reserved for engineers and scientists. Diffusing good practices in toiling the land, transporting the products to the market and training farmers to use new crops and technologies are the elements of innovation processes that may have the most important effects on economic growth and well being.

Why Systems?

At the end of the Second World War, technology policy was based on a linear model of innovation. It was assumed that if a firm or a country invested in scientific research, this would create the basis for new technologies, and when these were transformed into new products, the result would be a strong international competitiveness and economic growth.

Especially in the sixties and on-wards, empirical research on technical change and innovation demonstrated that this was not what actually takes place in innovation processes. Innovating firms may draw upon science but then it is usually quite old research results. Crucial for innovation is the *interaction* with many other organisations and firms. Firms do not innovate alone – they innovate in an interaction with customers, suppliers and even, sometimes, with competitors.

The research also demonstrated that the quality of linkages and relationships between firms and knowledge institutions, not only universities but also technical institutes and schools, was important for innovation. The single firm operates in a local and national environment and this environment forms an innovation system that may be more or less supportive to its innovative activities.

The system approach helps to keep an eye on the needs of users in the system and on the importance of having good feedback links from users to producers. In the context of Africa users may refer to households utilising new infrastructure and technologies, including the Internet (Dahms). It may also refer to private firms interacting with knowledge institutions such as schools, technological institutes and universities.

In most African economies, the most important linkages are those between agriculture on the one hand and the rest of the production system. They are crucial when it comes to develop better methods and products. This involves building and establishing links to 'land universities' with research and training addressing the needs of agriculture. It also implies establishing links to the manufacturing industries that produce the crucial means of production for agriculture. To diffuse good practice within agriculture is of course also important, but again, this might take place through consultants and producers addressing many users in primary agriculture.

Why National?

It might appear paradoxical that the concept 'national' system of innovation has become increasingly popular among policy makers and in the business community in a period when the focus in upon globalisation. It reflects that while competition becomes more global, some of the factors that lie behind competitiveness remain local and national. The most important localised factor is specialised competence and learning. Competing exclusively on low wages and hard work has become increasingly difficult and strong positions based on such advantages will be gradually undermined if not supported by competence building.

The focus on the national level makes explicit what has been implicit in strategies aiming at strengthening 'the international competitiveness of the national economy' and it broadens the understanding of what such strategies need to include. Especially for less developed economies it offers a concept that might be used to mobilise agents around an agenda of action that is forward looking and optimistic rather than backward looking and defensive. Its historical roots demonstrate that it is a concept strongly connected to 'catching-up' with world leaders.

But not least in the context of Africa, analysis and action at the national level cannot stand alone. Weak and split nation states and weak political institutions at the national level call for combining alternative perspectives. Building capabilities in local communities, combining the forces of economies located close to each other in economic integration and not least drawing upon Pan African resources and idealism are as important as building institutions at the level of nation states.

Why 'Innovation and Competence Building Systems'?

Our attempt to relate the concept National System of Innovation (NSI) to a largely agrarian Africa may at first sight appear as an impossible exercise. Africa South of the Sahara has about 70 per cent of the population in rural areas. Agriculture employs 45 per cent of the 600 hundred million people. Industry in Africa is still largely waiting to be organised. According to the World Bank Sector Analysis (2002), the size and structure of firms in the African manufacturing sector have continued to show a lack of development toward a dynamic industrial economy (World Bank, RPED 2002). The studies of the Bank's Regional Programme on Enterprise Development claim that the growth rate of exports of African countries continues to be either negative or well below those of other developing countries.

It is thus far from obvious that it is meaningful to apply the NSI-concept developed mainly as a tool to analyse industrial and post-industrial economies (Freeman 1987; Lundvall 1992; Nelson 1993) to the circumstances in Africa. When we are dealing with Africa we are speaking of a continent regarded by the international financial institutions as 'the last in the world'. It is from such an unpromising condition that African systems of innovation have to emerge.

The modern revival of the NSI-concept some 12-15 years ago gave rise to different more or less broad definitions of innovation systems. The US-approach (Nelson 1988) linked the concept mainly to high-tech industries and put the interaction between R&D-departments in firms, the university system and national technology policy at the centre of the analysis. Freeman (1987) introduced a broader perspective that took into account national specificities in the organisation of firms – he emphasized for instance how Japanese firms increasingly used 'the factory as a laboratory'. The Aalborg approach (Lundvall 1985; Andersen and Lundvall 1988) also saw national systems of innovation as rooted in national production systems.

In order to justify our endeavour we need to broaden and enrich the NSI-concept so that it becomes a useful tool for promoting structural transformation in Africa. The title of the international conference in Aalborg in 2002 'African Systems of Innovation and Competence Building' was chosen to signal such a need to broaden the innovation system-approach (Lundvall 2002). In Africa science-based activities still play a miniscule role in economic and social development. The most advanced economy in Africa is South Africa which has a population of 40.5 million people. '[South Africa] produced only a total of 1088 doctoral and master's graduates in science and technology in 1993. In the same year, South Africa produced a meagre total of 4264 master's and doctoral graduates overall in the fields of natural sciences and engineering, health sciences, and social sciences and humanities…[In the same year] the total enrolment of natural science and engineering master's and doctoral students in South African universities was only 5602' (FRD 1996). In reality, this will be the case for the near future as well. One of the indicators at our disposal shows that so-called high technology products form between 0 per cent and 10 per cent of exports with most countries close to zero and only South Africa close to the upper limit. Building effective universities of good quality is certainly becoming increasingly important (Mytelka) but what will be at least as important for economic development is to enhance the competence among the majority of those working in agriculture, manufacturing and services.

Some of the chapters in this volume analyse R&D-systems, others link innovation and technological capabilities to the production system while some introduce broader perspectives that include competence building in the informal economy and in service sectors. Combining these insights in a holistic systemic framework that includes all kinds of innovation and competence building helps us to understand what are the major problems and opportunities for African innovation systems.

Knowledge and Learning

In order to understand the concept of innovation system and why competence building needs to be integrated with innovation, it is necessary to understand the role of knowledge and learning in the economy. In Lundvall (1992: 1) it was proposed that 'the most fundamental resource in the modern economy is knowledge and, accordingly, the most important process is learning'. Over the last decade the attempt to get a better understanding of the knowledge-based economy and the learning economy has created a more satisfactory theoretical foundation for innovation systems.

It has been argued that what is new in the present phase of development is not the use of knowledge in production, but rather the speed of learning and forgetting. We have coined the concept 'the learning economy' to capture these characteristics and argued that the success of individuals, organisations, regions and countries today will reflect, not what specific knowledge they have at a specific moment, but rather the capability to learn (Lundvall and Johnson 1994).

The understanding has been developed using the basic distinctions between information and knowledge, between 'knowing about the world' and 'knowing how to change the world' and between knowledge that is explicit and codified versus knowledge that remains implicit and tacit. In this context, it was proposed to make a distinction between four different kinds of knowledge. *Know-what* refers to access to information, *know-why* to understanding causal relationships, *know-how* to capability to do things and *know-who* to the access to knowledge and capabilities of others. While it is easier to make the kind of knowledge belonging to the first two categories explicit, the crucial elements of know-how and know-who to a much higher degree remain tacit knowledge.

What makes a difference in economic terms between success and failure for a firm or a region is knowledge that is not easily appropriated by others. This will typically be knowledge with tacit elements or explicit knowledge protected by intellectual property right instruments. The last category has certainly been growing in importance over the last decade, not least in pharmaceuticals and biotechnology. In more and more areas the richest countries initiate restrictive intellectual property regimes and more strict sanctions are imposed on those who try to break the rules.

It is legitimate to ask if Africa can get away from the bottom of the ladder respecting fully the existing rules set by the established players of the world economy. Historically every country that has succeeded to industrialise has broken one rule or another. The USA has done it. As report from the Office of Technology Assessment (OTA) puts it: when 'the United States was still a relatively young and developing country [...] it refused to respect international intellectual property rights on the grounds that it was freely entitled to foreign works to further its social and economic development' (OTA 1986: 228). In the context of the establishment of the German state, the rules set by the British Empire were broken. It might be argued that the African countries – 'being the last in the world economy' – would have good moral grounds to break some of the rules when these get in the way for development. Even better would it be if Africa, acting as one player on the world economic scene, could negotiate a change in the rules so they did not hinder learning from abroad.

Innovation and Allocation

The advice to African governments coming from international organisations such as the World Bank and IMF is strongly focused on macroeconomic stability and on 'structural reform' aiming at establishing well-defined private property rights and well functioning markets. This type of advice is based on a neo-classical perspective where the most important issue is assumed to be the efficient allocation of resources. A shift of focus to innovation and learning changes the policy agenda as well as the agenda for 'structural reform' changes.

The distinction between knowledge about the world and know-how are especially helpful when it comes to contrast the theoretical micro-foundations of innovation systems with those of standard economics. If agents are allowed to learn, at all, in a neo-classical model, learning is either understood as getting access to more or more precise information about the world or it is a black-box phenomenon as in growth models assuming 'learning by doing'. The very fundamental fact that agents – individuals as well as firms – *are more or less competent* in what they are doing and that they may learn how to become more competent is abstracted from in order to keep the analysis simple so that it can be based upon 'representative firms' and agents. This abstraction is absolutely fatal in an economy where the capability to learn tends to become the most important factor behind the economic success of people, organisations and regions (Lundvall and Johnson 1994).

Table 1 illustrates how the analytical framework connected to innovation systems relates to mainstream economic theory. The theoretical core of standard economic theory is about rational agents making choices between well-defined (but possibly risky) alternatives and the focus of the analysis is on the allocation of scarce resources. What is proposed here is a double shift in focus that can be illustrated by the following table.

Table 1 Four Different Perspectives in Economic Analysis

	Allocation	Innovation
Choice making	Standard neoclassical	Management of innovation
Learning	Austrian economics	Innovation systems

The table illustrates that learning as well as innovation, in principle, can be analysed in analytical frameworks closer to the mainstream neoclassical economics. It is possible (but not logically satisfactory) to apply the principles of rational choice to the analysis of innovation. It may, for instance, be assumed that 'management of innovation' is aiming at funds getting allocated to alternative R&D-projects according to the private rate of return, taking into account the risk that the projects do not succeed.[1]

Austrian economics (Hayek and Kirzner) has the focus on allocation of scarce resources in common with neoclassical economics. But Hayek presents the market as a dynamic learning process where the allocation of scarce commodities is brought closer to the ideal of general equilibrium without ever reaching this state.

The analysis of innovation systems moves the focus toward the combination of innovation and learning. Innovation processes may be seen as a process of joint production where one output is innovation and the other a change in the competence of the involved agents. Applying the perspective of 'national innovation system' in a country is basically a way to re-assess its structure, institutions and organisations – including the public sector. While the focus of traditional economics is to evaluate their capacity to utilise existing resources the innovation system perspective helps to evaluate their capacity to create new resources and to build new competence in the economy.

In an economy where innovation and learning are of fundamental importance for economic growth the institutional recommendations and the program for structural reform derived from neoclassical thought are misleading. A key issue is to establish institutions that support learning (and forgetting) in relation to all markets, including financial markets and labour markets. From the perspective of innovation systems the very idea of 'pure markets' as the ideal form tends to evaporate as an illusion. From a neo-classical perspective, 'market failure' appears as exceptions. A combination of 'market failure' and 'system failure' offers a rich agenda for intelligent public policy making when we see the economy as an innovation system.

Building and Integrating Innovation Systems as an Element in African Renaissance?

This book is particularly timely since the African Union and the New Partnership for Africa's Development (NEPAD) have been formed at the turn of the century in Africa. There is some optimism in the air and agreement that Africa must undergo a transformation of its social, political and economic structure. For the first time African leaders have accepted 'sovereignty with responsibility' permitting peer reviews of their conduct and possible intervention and sanction if they carry out activities that put large classes of their citizens to danger or use unlawful seizure of power. A new post cold war desire to respect citizens and tame arbitrary authority seems now to have a chance to become a reality. The strong focus upon Africa's many disasters should not hide that there are also signs of a positive development (Muchie 2003).

One of the key expressions of the new positive trend is the New Partnership for Africa's Development (Gammeltoft). NEPAD aims at establishing a new type of interaction between Africa on the one hand and the Northern industrialised regions and multilateral financial institutions on the other. One goal is to stimulate African GDP growth so that it reaches to 7 per cent per annum. While this objective is laudable, this ambition to create a new framework of interaction to alter radically Africa's current rate of growth is strongly linked to building and integrating African systems of innovation.

This volume, amongst other things, asks the difficult question of whether the national system of innovation concept can be appropriated from its earlier applications in industrial economy contexts and usefully applied to Africa. The contributors have divergent views. But they have in common an intellectual willingness to put the innovation system as an issue on the African agenda. It is a first step and there is a need to go further in the analysis of how to build systems of innovation and competence building in Africa. The intellectual and political endeavour is to find paths for transforming Africa much like Friedrich List broke the orthodoxy of Smithian and Ricardian comparative advantage in order to create a German national system of political economy. The African system of political economy is waiting to be written.

One of the most useful aspects of the concept of national system of innovation is that even with globalisation, nations and states matter. The concept may be seen as a framework for policy learning when it comes to policies aiming at promoting innovation and competence building. Such learning processes may take place at the national level but they would certainly benefit from being organised also at the African level. This could become an integrated element in the current trend by Africans to accelerate the African Union and NEPAD initiatives. African leaders have recognised that 'Development is a process of empowerment and self-reliance. Africans must not be wards of benevolent guardians; rather they must be architects of their own sustained upliftment' (OAU-NEPAD 2001). Even more radical would be the implementation of a long-term vision of a Pan African innovation system that could complement and strengthen AU/NEPAD.

Notes

[1] Arrow has pointed out the obvious that innovation is a phenomenon not ideal for that kind of analysis because innovation has as its most fundamental characteristic that it gives rise to something that is not known in advance – and it is not possible to apply the principles of rational choice if the choice set is not defined in advance. But it is still the case that, for instance, new growth theory operates with models that combine on-going innovation with assumptions of rational choice.

References

Andersen, Esben S. and Bengt-Åke Lundvall (1988), 'Small National Innovation Systems Facing Technological Revolutions: An Analytical Framework', in Christopher Freeman and Bengt-Åke Lundvall, *Small Countries Facing the Technological Revolution*, London: Pinter Publishers.

Freeman, Christopher (1987), *Technology Policy and Economic Performance: Lessons from Japan*, London: Pinter Publishers.

Freeman, Christopher (1995), 'The National Innovation Systems in Historical Perspective', *Cambridge Journal of Economics*, Vol. 19, No. 1.

Foundation for Research and Development (FRD) (1996), *South Africa Science and Technology Indicators*, South Africa: Pretoria.

List, Friedrich (1841), *Das Nationale System der Politischen Ökonomie*, Basel: Kyklos, (translated and published under the title: *The National System of Political Economy by Longmans*, London: Green and Co.).

Lundvall, Bengt-Åke (1985), *Product Innovation and User-Producer Interaction*, Aalborg: Aalborg University Press.

Lundvall, Bengt-Åke (2002), *Innovation, Growth and Social Cohesion: The Danish Model*, London: Edward Elgar Publishers.

Lundvall, Bengt-Åke (ed.) (1992), *National Innovation Systems: Towards a Theory of Innovation and Interactive Learning*, London: Pinter Publishers.

Lundvall, Bengt-Åke and Björn Johnson (1994), 'The Learning Economy', *Journal of Industry Studies*, Vol. 1, No. 2: 23-42.

Muchie, Mammo (2003), *The Making of Africa-Nation: Pan-Africanism and the African Renaissance*, London: Adonis-Abbey Publishers.

Nelson, Richard R. (1988), 'Institutions Supporting Technical Change in the United States', in Giovanni Dosi, Christopher Freeman, Richard R. Nelson, Gerald Silverberg, and Luc Soete (eds.), *Technology and Economic Theory*, London: Pinter Publishers.

Nelson, Richard R. (ed.) (1993), *National Innovation Systems: A Comparative Analysis*, Oxford, Oxford University Press.

OAU (2001), *OAU-NEPAD Document*, Addis Abeba.

Office of Technology Assessment (OTA) (1986), *Intellectual Property Rights in an Age of Electronic Information*, Washington, D.C.: US Government Printing Office.

World Bank, Regional Programme on Enterprise Development (RPED) (2002), *Nigeria Firm Survey*, Paper No. 118, April, World Bank.

PART I

Conceptual Specification and Framework Setting

1

National Systems of Innovation and Economic

Development

Björn Johnson and Bengt-Åke Lundvall

Introduction

In the first section of this chapter we give some background on how the concept 'national system of innovation' has developed. In the second section we look at how it may be seen as a response to as well as be enriched by recent new ideas on analysing economic development. Finally we reflect upon what specific adaptations are necessary in order to make the concept relevant for countries in Africa. One main conclusion is that it is necessary to *broaden the innovation system concept* so that it includes all institutions and structures that promote or hinder competence building in all parts of the economy. Another conclusion is that there is a need to understand how to *build* national innovation systems. This is especially true when the focus is on African innovation systems where some of the most fundamental prerequisites for establishing coherent national innovation systems are yet to be established. On this background we end up by sketching a research agenda aiming at understanding how to establish and develop innovation systems in Africa.

The National System of Innovation – a Concept with Roots Far Back in History

The history and development of the innovation system concept indicates that it can be useful for analysing less developed economies. Some of the basic ideas behind the concept 'national systems of innovation' go back to Friedrich List (List 1841) and they were developed as the basis for a German 'catching-up' strategy. His concept 'national systems of production' took into account a wide set of national institutions including those engaged in education and training as well as infrastructures such as networks for transportation of people and commodities (Freeman 1995).

List's analysis focused on the development of productive forces rather than on allocation issues. He was critical and polemic to the 'cosmopolitan' approach of Adam Smith, where free trade was assumed always to be to the advantage of the weak as well as the strong national economies. Referring to the 'national production system', List pointed to the need to build national infrastructure and institutions in order to promote the accumulation of 'mental capital' and use it to spur economic development rather than just to sit back and trust 'the invisible hand' to solve all problems. It was a perspective and a strategy for the 'catching-up' economy of early 19th century Germany.

The first written contribution that used the concept 'national system of innovation' is, to the best of our knowledge, an unpublished paper by Christopher Freeman from 1982 that he worked out for the OECD expert group on Science, Technology and Competitiveness (Freeman 1982: 18). The paper, titled 'Technological Infrastructure and International Competitiveness', was written very much in the spirit of Friedrich List, pointing out the importance of an active role for government in promoting technological infrastructure. It also discusses in critical terms under what circumstances free trade will promote economic development.

It is also interesting to note that while the modern version of the concept of national systems of innovation was developed mainly in rich countries (Edquist 1997; Lundvall 1992; Nelson 1993) some of the most important elements going into the combined concept actually came from the literature on development issues in the third world. For instance the Aalborg version (Andersen and Lundvall 1988) got some of its inspiration from Hirschman (1958) and Stewart (1977).

Also the idea that institutions matter for the performance of the economy that is central in the innovation system approach (Johnson 1988) was originally more generally accepted for 'less developed countries' than for full blown market economies where the market was assumed to solve most problems in an institution-less world. As we shall see below the importance of institutions for economic development has got new attention recently for example in the contributions of IMF.

To apply the NSI-concept to developing countries may therefore be seen as a kind of 're-export'. Gunnar Myrdal's ideas, inspired by Veblen and developed in 'Asian Drama' (1968), of positive and negative feedback, of cumulative causation and of virtuous and vicious circles are all easily reconciled with the idea of innovation system. But the sum of all these connections between the NSI-concept and economic development does not mean that it is directly applicable on for instance the African states. Especially, it would be problematic to apply a narrow definition of the innovation system to African economies.

Different Definitions of the 'National Innovation System'

It is obvious that different authors mean different things when referring to a national system of innovation. The major differences have to do with focus and breadth of definition in relation to sectors, institutions and markets.

Authors from the US with a background in studying science and technology policy, tend to focus on 'the innovation system in the narrow sense'. The NSI-concept is seen as a broadening of earlier analyses of national science systems (see for instance the definition given in Mowery and Oxley 1995: 80). The focus is upon the systemic relationships between R&D-efforts in firms, S&T-organisations, including universities, and public policy. The relationships at the centre of the analysis are the ones between knowledge institutions and firms and the focus is on high tech-sectors. This narrow approach is not so different from the 'triple helix' concept were universities, government and business are seen as the three important poles in a dynamic interaction (Etzkowitz and Leydesdorff 2000).

The Freeman and the 'Aalborg-version' of the national innovation system-approach (Freeman 1987; Freeman and Lundvall 1988) aims at understanding 'the innovation system in the broad sense'. The definition of 'innovation' is broader. Innovation is seen as a continuous cumulative process involving not only radical and incremental innovation but also the diffusion, absorption and use of innovation. Second, a wider set of sources of innovation is taken into account. Innovation is seen as reflecting, besides science and R&D, interactive learning taking place in connection with on-going activities in procurement, production and sales.

To a certain degree, these differences in focus reflect the national origin of the analysts. In small countries such as Denmark, as in developing countries, it is obvious that the competence base most critical for innovation in the economy as a whole is not scientific knowledge. Incremental innovation, 'absorptive capacity' and economic performance will typically reflect the skills and motivation of employees as well as manifold inter and intra organisation relationships and characteristics. Science-based sectors may be rapidly growing but their shares of total employment and exports remain relatively small. In the US, aggregate economic growth is more directly connected with the expansion of science-based sectors. In these sectors big US-firms have an international lead and they introduce radical innovation in areas where the interaction with science is crucial for success.

Common Characteristics of Innovation System Approaches

While there are competing conceptions regarding what constitutes the core elements of an innovation system, it might still be useful to see what the different definitions have in common.

A first common characteristic is the assumption that national systems differ in terms of specialisation in production, trade and knowledge (Archibugi and Pianta 1992). This is not controversial – for instance neoclassical trade theory would lead us to a similar assumption. One important difference from neoclassical theory, however, is that among NSI-analysts the focus is upon the co-evolution between what countries do and what people and firms in these countries know how to do well. This implies that both the production structure and the knowledge structure will change only slowly and that such change involves learning as well as structural change.

A second common assumption behind the different approaches to innovation systems is that elements of knowledge important for economic performance are localised and not easily moved from one place to another. In a fictive neoclassical world where knowledge equalled information and where society was populated with perfectly rational agents, each with unlimited access to information, national (or local) innovation systems would be an unnecessary construct. A common assumption behind the innovation system perspective is that knowledge is something more than information and that it includes tacit elements (Polanyi 1966).

A third assumption that makes it understandable why knowledge is localised is that important elements of knowledge are embodied in the minds and bodies of agents, in routines of firms and not least in relationships between people and organisations (Dosi 1999).

A fourth assumption central to the idea of innovation systems is a focus on interaction and relationships. The relationships may be seen as carriers of knowledge and interaction as processes where new knowledge is produced and learnt. This assumption reflects the stylised fact that neither firms and knowledge institutions nor people innovate alone. Perhaps the most basic characteristic of the innovation system approach is that it is 'interactionist'.[1]

Sometimes characteristics of interaction and relationships have been named 'institutions' referring to its sociological sense – institutions are seen as informal and formal norms and rules regulating how people interact (Johnson 1992). In a terminology emanating from evolutionary economics and the management literature 'routines' are regarded as more or less standardised procedures followed by economic agents and organisations when they act and when they interact with each other (Dosi 1999).

This is a major dimension in which different national systems approaches tend to be in agreement with each other. While neoclassical theory in its ambition to become a general theory imposes one general rule of behaviour (utility and profit maximisation) on all agents, independently of time and space, the institutional approach recognises that the history and context makes a difference when it comes to how agents interact and learn.[2]

National Systems in the Context of the Learning Economy

In the public debate knowledge is increasingly seen as the crucial factor for the development of both society and the economy. In a growing number of publications from the World Bank and OECD it is emphasised that we currently operate in 'a knowledge-based economy'. For several reasons we prefer the term 'the learning economy' in characterising the current phase of socio-economic development (Lundvall and Johnson 1994).

The basic assumption is that, working together, technological developments, globalisation and political processes of deregulation have led to an *acceleration* of technical and economic change. Consequently, access to any given knowledge base is less important for the economic success of firms and individuals, than their ability to rapidly acquire new competences as they get confronted with new types of problems. New knowledge is created at an increasing rate, but the quantity of relevant knowledge is also being reduced as knowledge becomes obsolete at a faster pace than before.

16

This often implies 'de-learning' of old competencies which otherwise may delay or block the development of new competences.

This acceleration in economic change is reflected in different indicators. The life cycle of new products becomes shorter. New production processes are diffused more rapidly than before. For employees, work tasks change character, and many will change employer more often. In a report from the Danish Ministry of Education, a German study is cited, maintaining that it only takes one year from the exam, before half of what a computer engineer has learnt has become obsolete. The 'halving times' of what has been learnt in the education system are longer for other specific professions but on average, it is argued, it is about 8 years (Ministry of Education 1998: 56f.).

As illustrated by the example, the most dramatic acceleration in the rate of change is something that characterises the most technology intensive elements of rich countries. But 'the gales of destruction' do not spare the least developed parts of the world. On the contrary, underdevelopment in African states reflects their lack of specific learning capabilities necessary in the modern economy and that the institutional set ups do not promote relevant learning. Access to the most dynamic global networks of learning is also crucial in the globalising learning economy (Archibugi and Lundvall 2000).

In the Learning Economy Social Capital is Important for the Performance of the Innovation System

In the learning economy there is a strong connection between intellectual capital and social capital (Putnam 1993; Woolcock 1998). Since specific competencies always build upon elements of tacit knowledge the learning capability of a system will depend crucially on elements of social capital. The learning capability will be high in systems where citizens are used to collaborate in civic organisations and networks, trust each other and are open to an interaction with wider communities. Systems where citizens are focused on narrow individual or family interests and where experts are unwilling to co-operate with experts with a different background will have increasing competitiveness problems in the globalising learning economy (Lundvall and Borras 1998).

This issue points to the importance of social capital, not least for development strategies. It is certainly a good idea to invest in education and training and in building scientific and technological capabilities. This is a necessary element in a successful development strategy. But such investments may give very limited returns if they do not have some basis in terms of social capital to build upon. In the context of the learning economy, the role of education institutions in supporting social capital may be even more fundamental than its role in creating intellectual capital.

New Tendencies in Development Thinking

One way to specify the need for adaptation of the innovation system approach to the needs of developing countries is to look at three recent tendencies in development thinking:

First, there is an increasing focus on capabilities rather than resource endowments as the main instruments and values in development. This tendency can be exemplified by work by Amartya Sen.

Second, you can observe a new focus on knowledge as the perhaps most crucial resource driving development. This can be illustrated by several recent reports from the World Bank.

Third, there is a tendency to underline the primary importance of institutions as the 'root causes' of development dwarfing the importance of all other factors such as geography and policies. The IMF has recently underlined this view.

In this section of the chapter we will demonstrate how these three dimensions may be integrated into the NSI-approach and vice versa, how this results in a broadening of the approach that makes it more relevant for development studies. Integrating them helps us to form a new perspective on development, which is already implicit in the broad version of the national innovation system approach.

The three tendencies mentioned above are quite interesting since each of them represents a broadening of the more narrow economistic views on economic development. At the same time the tendencies are myopic when seen from the perspective of 'innovation systems in the context of the learning economy'.

The capability approach introduces and underlines several capabilities (or 'freedoms' in the language of Sen) but more or less disregards learning capabilities. The World Bank focuses on knowledge and its diffusion from the North to the South but seems to underestimate the role of learning and innovation processes in the North and particularly in the South for knowledge creation and utilisation. In a related manner the new interest in institutions of the IMF gives little attention to the institutions supporting learning and innovation. Bringing the three perspectives into the theoretical framework of innovation systems in the learning economy confirms that institutions are crucial for economic development. But what comes out, as the most significant institutions, will be different when focus is moved toward innovation and competence building.

A Capability Based Approach

Amartya Sen (1999) presents a capability-based approach where development is seen as an expansion of the substantive freedoms that people enjoy. Substantive freedoms are defined as the capabilities people have to live the kind of lives they have reason to value. They include things like being able to avoid starvation and under nourishment, diseases and premature mortality. It also includes the freedoms of being literate, able to participate in public life and in political processes, having ability and possibility to work and to influence one's work conditions, having entrepreneurial freedom and possibilities to take economic decisions of different kinds. Enhancement of freedoms like these is seen as both the ends and means of development. Freedoms constitute development and freedoms are the driving forces behind development.

This way of looking at development refers to the capabilities people have to act and to choose a life they value, rather than to their level of income and possession of wealth. Poverty, for example, is in this perspective more a deprivation of basic capabilities than just low income. Human capabilities rather than resource endowments are the fundamental factors of development.

The capability approach includes an institutional dimension. Individual capabilities clearly depend on the institutional set-up of society, on its political, social and economic arrangements. To build, sustain and improve such freedoms/capabilities processes of organisational and institutional change are required. Simple accumulation of wealth will not do it. When one tries to create institutions, suitable for development, one needs to focus not only on how to 'break down the barriers of caste, colour, religion, ethnic origin, culture, language, etc.' (Myrdal 1968) but also on how to liberate and improve human capabilities. Development depends on interconnected instrumental freedoms rooted in political, social and economic institutions.

Another aspect of Sen's approach is that from the instrumental point of view the different freedoms – political freedoms, economic facilities, social opportunities, learning opportunities and so on – are linked and feed upon each other. Political participation depends on education and trust; education and training depends on income and social security; economic facilities depend on health care, education and participation, etc. At a more basic level this has to do with the systemic character of the institutional set-up, which is an important aspect of Sen's way of thinking about development.

The Missing Capability

Sen's approach represents a systemic, broad and dynamic view of development that fits well into a system of innovation approach. It is noteworthy however that learning and innovation capabilities generally do not seem to be included in his capability based approach to development.

A similar 'omission' seems to be common in approaches which have a focus on information and knowledge. Expressions like 'information divides', 'technology divides' and 'knowledge divides' between North and South have become common and accepted by dominating policy actors such as the World Bank. This is an important shift from earlier positions. But as an aspect of a capability based development concept it may be more important to identify and analyse the learning divide between North and South. Knowledge may indeed be viewed as a central resource and development factor, but the capability to create, spread and utilise it is even more important. The learning divide more than the technology divide may, thus, be the crucial factor in the North/South relationship which development policies have to take into account (Arocena and Sutz 2000). This should not be misunderstood. There is an enormous amount of different kinds of learning in the South. What we primarily refer to here, however, is learning related to technological and organisational capabilities strong enough to establish and defend competitive positions on the world market.

The developmental value of learning capabilities is closely related to the connection between learning and innovation. In economic terms development depends on technical and organisational change brought about by continued processes of innovation. Innovations introduce technical and organisational knowledge into the economy. They are important means in the process of development. We can think of them as 'learning results' contributing to the removal of 'un-freedoms' like ignorance, lack of learning opportunities and lack of economic opportunities and we can think of them as contributing to the enhancement of substantive freedoms like the capability to work, communicate, learn and to participate democratically in political processes.

We can also think of learning processes as forming the preconditions for innovation; learning does not always result in innovation, but without learning there would be no knowledge to introduce into the economy as innovation. Technological capabilities of firms, for example, develop over time as a result, not only of firm specific learning but also of different kinds of interaction, co-operative as well as competitive, between firms and other organisations. Capability building involves interactive learning by individuals and organisations taking part in processes of innovation of different kinds.

The learning capability is thus one of the most important of the human capabilities. It does not only have an instrumental role in development but also, under certain conditions, substantive value. When learning takes place in such a way that it enhances the capability of individuals and collectives to master and co-exist with their environment it contributes directly to human well-being. To be able to participate in learning and innovation at the work place may be seen as 'a good thing' contributing to a feeling of belonging and significance. This is also the case for possibilities for education and participation in democratic processes.

Institutions and Development

We have seen that Sen's capability based approach regards and underlines the importance of a broad spectrum of interrelated institutions. We have also mentioned that there is a new tendency of focusing on institutions as perhaps the most important development factor. This tendency is interesting and useful. One may wonder, however, if the relatively narrow spectrum of institutions, which are in focus really can explain so much of the development process as it is claimed.

According to the World Bank (2002: 8), institutions have three main objectives:

• They channel information about market conditions, goods, and participants.

- They define and enforce property rights and contracts, determining who gets what and when.
- They increase (or decrease) competition in markets.

Within this framework, transaction costs that limit market opportunities are seen as emanating from institutional problems related to especially inadequate information, incomplete definition and enforcement of property rights, and barriers to entry for new participants. Institutions that help manage risks of market exchange and raise returns increase market opportunities and support development.

Admittedly, there is a tendency to draw a broader range of institutions into the picture, for example in the World Development Report from 2003, but the main focus remains on transactions costs and markets.

In recent publications also IMF strongly emphasises the importance of institutions for growth. World Economic Outlook from April 2003 (IMF 2003) for example concludes that if the quality of institutions in sub-Saharan Africa were to 'improve to the levels in developing Asia', per capita income would rise by 80 per cent and if its institutions 'rose to world average levels' the average per capita economic growth rates would become 2 per cent higher. The question of how to close these 'institutional gaps' (Johnson and Lundvall 1992) are not discussed very much by the IMF, however. Like the World Bank, IMF focuses on a narrow range of market supporting institutions related to the security of property rights, good governance and measures to restrict corruption.

We are not arguing here that well functioning markets are not important for development, far from it. The point, however, is that the crucial question of how institutions may support learning and innovation is not raised (except of course for the role of the formal school system, which belongs to the 'established' growth factors). The impact on learning and innovation of for example labour market institutions, financial institutions, economic policy regimes and a host of norms supporting (or undermining) a learning culture are not analysed.

Innovation, Innovation Systems and Development

As a way of summarising these critical observations on some new tendencies in development thinking one might say that even if both institutions and knowledge seem to be moving into the centre of the stage the question of how they interact and co-evolve shaping learning and innovation and driving the process of development is not explicitly raised.

It seems to us that the broad definition of national systems of innovation may be useful in this context. It fits well with both Sen's focus on capabilities and the focus on institutions since a broad spectrum of socially based inter-linked capabilities is necessary for efficient innovation processes or for well performing innovation systems. The concept of innovation systems may be a tool for understanding the relations between different kinds of capabilities and between the constitutive and instrumental aspects of freedoms in developing countries. We can think of them as contributing to the removal of un-freedoms like ignorance, lack of economic opportunities and poverty and we can think of them as contributing to the enhancement of substantive freedoms. The viewpoint taken here is that improving learning and innovation capabilities is not only a question of more resources for education and research (more and better schools and universities) but also of reshaping a broad set of institutions so that they support interactive learning and innovation in all parts of society including the individual families, communities, firms and organisations.

The introduction of knowledge into the economy (and into the society at large) cannot be realised just by giving citizens access to flows of information through IT-networks. It requires active and interactive learning by individuals and organisations taking part in processes of innovation of different kinds. The efficiency of these learning activities and, hence, the performance of the innovation systems depends of economic, political and social infrastructures and institutions. It also depends on past experiences as they are reflected in the tangible and intangible aspects of the structure of production and on values and policies.

We have argued here that there are good reasons for using a broad concept of innovation system in connection with development analysis especially when focusing on countries in the South. A narrow innovation system concept focusing on the research and development system and on high tech and science-based innovations does not make sense in the South. There are several reasons for this.

In a relatively 'complete' national system of innovation it may be less problematic to analyse a specific subsystem. If there are adequate knowledge infrastructures and intellectual property rights and if there are good networking capabilities and high levels of trust, there is also a suitable basis for an efficient research and development system. It may then be quite possible to analyse the details of this subsystem without worrying too much about the rest of the innovation system. But this is typically not the case in the South, which makes a broad approach preferable.

Another reason is that the need and opportunity to build on local knowledge and traditional knowledge may be relatively bigger in the South than in the North. This would underline the importance of tacit knowledge and it would also draw attention to the need not to loose important parts of largely not codified and undocumented local competencies. Local knowledge is easily forgotten when economies in a quick tempo are opened up to international competition and societies accordingly restructured. A broad concept of innovation systems helps to see the importance of different kinds of knowledge and the ways they complement each other.

Weaknesses of the Innovation System Approach

When applied to countries in the South it is important to be aware of some weaknesses of the innovation system approach, as it has been used so far. Some of these have directly to do with the fact that it has mostly been applied to the North. It has been used mainly as an ex-post rather than as an ex-ante concept. It has been used to describe and compare relatively strong and diversified systems with well developed institutional and infrastructure support of innovation activities. It has not, to the same extent, been applied to system building. When applied to the South the focus ought to be shifted in the direction of system construction and system promotion. Furthermore, the relationships between globalisation and national and local systems need to be further researched. It is important to know more about how globalisation processes might affect the possibilities to build and support both national and local systems of innovation in developing countries.

Another important weakness of the system of innovation approach is that it is still lacking in its treatment of the power aspects of development. The focus on interactive learning – a process in which agents communicate and even cooperate in the creation and utilisation of new economically useful knowledge – may lead to an underestimation of the conflicts over income and power, which are also connected to the innovation process. Interactive learning and innovation immediately sounds like a purely positive sum game, in which everybody may gain. In fact, there is little learning without forgetting. Increasing rates of learning and innovation may lead not only to increasing productivity but also to increasing polarisation in terms of incomes and employment. It may be more common in the South than in the North that interactive learning possibilities are blocked and existing competences destroyed for political reasons related to the distribution of power.

It is, thus, clear that the innovation system approach needs to be adapted to the situation in developing countries if it is to be allied to system building. It seems also clear, however, that the holistic and systemic character of the approach and its focus on production based tacit knowledge and on learning by doing, using and interacting should make it possible to implement such adaptations.

National System of Innovation – a Legitimate Tool for Analysing and Promoting African Development?

To apply the concept of a national innovation system to nation states in Africa may be seen as something of a provocation. The 54 states of Africa are very different – for instance there are great differences between the Maghreb states north of Sahara, South Africa and the countries in between. But in most of the African countries many people suffer from fundamental problems having to do with insecurity, poverty and diseases. In a situation where half the population on the African continent live on less than $1 a day it might seem irresponsible to focus on fancy concepts such as innovation and competence building.

To a certain degree we accept this view. There is an important connection between solving the fundamental problems of poverty, bad health and insecurity on the one hand and promoting competence building on the other. To neglect these problems would be dangerous. There is even a risk that for policy makers the use of new fancy development concepts becomes a cover up for not tackling the fundamental problems. In order to be able to build competence and innovate, it is important to establish institutions that enhance order, trust and predictability in the life of individuals and in the workings of firms and other organisations.

On the other hand there is little doubt that the long term effort to promote economic development needs to be oriented towards competence building and innovation also in what may appear to be a dismal situation. To mobilise existing technology and knowledge when building new production capacity is necessary in order to make production competitive in the global economy. To constantly and gradually upgrade technological capabilities is necessary in order to remain competitive. This implies a need for mobilising people in processes of education and life-long learning. The intelligent use of information technology offers new opportunities in all these respects but it will only take place if the infrastructure is built.

The concepts 'virtuous and vicious circles' (as cumulative self-reinforcing processes) – developed many years ago by Gunnar Myrdal (1968) in his work Asian Drama – perhaps captures what is at stake. Building institutions to create order and stable living conditions is necessary to give people the opportunity and incentives to engage in learning new competences. But such institutions cannot be built without engaging people in competence building and learning. We believe that new and more fruitful approaches to economic development may be stimulated by introducing a 'double focus' on living conditions and on competence building.

Also in a more strict sense doubts may be raised about the very existence of national systems of innovation on the African continent. It is true that innovation in the sense of radical innovation in products and processes is a scarce phenomenon in Africa especially if we hereby imply innovations new to the world market. The organisations focusing on innovation are not strongly interlinked into a national 'system'. On the contrary they are often established by copying similar institutions from the rich countries and not well integrated with each other and with the socio-economic environment. This is re-enforced by the fact that many states in Africa are heterogeneous colonial constructs and therefore the very meaning of what is 'national' is less clear than in most European nation states.

These are reasons why we need to rethink the national innovation system concept in different contexts in order to make it relevant for African states. First we need to include all sectors – agriculture, services as well as manufacturing and low tech as well as high tech sectors – in the analysis. Second we need to include all aspects of innovation including diffusion, imitation and use of new technologies together with all forms of competence building – spanning from learning at the job by workers and farmers to the formal training of scientific personnel. Third we need to change the focus from the reproduction of the system to its construction. And, fourth, doing so we need to take into account the wider living conditions and how they affect learning and innovation.

Concluding Remarks

With the caveats made above, we still believe that the concept national system of innovation may be useful as an analytical tool and as a tool for promoting well-being in countries in Africa. At the same time we recognise the need to adapt and further develop the concept so that it becomes more adequate for the situation in, for instance, African countries.

There are at least five issues that are crucial for building innovation systems in Africa and that need to be addressed in future research:

- The role of nation states.
- Prerequisites for building education systems.
- Building social capital and good governance.
- Finding ways to utilise local knowledge systems.
- Finding strategies for appropriate insertion in the world economy.

It is extremely important to understand the historical and current role of nation states in Africa. The historical background and the current functioning of nation states indicate that they are quite different from the nation states that helped to bring together resources and stimulate capability building in Europe in the 19th century. This raises three other questions. Is the current drive toward Pan African solutions – such as the building of an African Union in the image of EU – a rational response to the weakness of nation states or is it a distraction from problems that seem to be unsolvable for those who are in charge? Or could an alternative strategy be to start to further strengthen the integration of some smaller supranational regional blocks of states in the North, West and South East of Africa that have already some economic linkages among them? And thirdly, is there a potential for supporting and improving local and regional innovation systems as growth poles at the sub-national level?

A second major issue has to do with how poverty alleviation, the education system and the labour market may be made to interact in the different African states. Historically a fundamental step in building the foundation of national innovation systems has been to create a national, first, primary and, later, secondary school system. But this can only be done with some success if children, young people and parents can 'afford' the school. Poverty and diseases are important hindrances for school attendance. And if there are no job openings for the educated after they have finished school the incentives to attend school are diminished. Is it possible to locate good practices in Africa making the three spheres work together?

The international discourse on bad governance in the less developed countries is normally quite narrowly focused about the introduction of market institutions, private ownership rights and on corruption. Even if the discourse has a certain neo-colonial ring to it, African problems of tribalism and corruption need to be taken seriously. Especially in the current broader context when learning is a key to economic development all incidents of unequal treatment of minorities, undeserved privileges for the governing and not meeting contractual or informal obligations to the governed tend to undermine the development potential. It depreciates social capital and thereby the learning capability of society. Especially it is crucial that all institutions directly involved in the creation, distribution and use of knowledge are seen as incorruptible. To locate good practices that work in African states is an important task for research.

Finally, both the local and the global context for African competence building needs to be taken into account. Africa is rich in competences and skills and some of them may be used to produce attractive goods for the world market. There is a need to map, develop and utilise such skills. But in order to eradicate poverty and to engage in social and economic development it is not sufficient to produce on the basis of such skills. For Africans, as for all catch up countries, a key to development must be 'borrowing' and adapting technologies that the technological lead countries control today. The combination of reverse engineering, licensing, sending scholars abroad, inviting foreign firms and experts and engaging in international scientific collaboration may be difficult to achieve; but all these elements need to be considered in building the national innovation system. To develop national strategies that make it possible to select technologies and institutions from abroad that support the national system is a big challenge when building national systems of innovation.[3]

We have repeatedly referred to the need to locate and map 'good practices' (or well-functioning institutions) already in use in African countries. Seen in this light the NEPAD-initiative (Gammeltoft 2003) where African countries establish a procedure of national review of good practice among themselves could become an important instrument for policy learning in Africa. Especially if the reviews give proper attention to institutions that promote innovation and competence building and build into the procedures comparison across national systems they may become most helpful in helping to construct national and regional innovation systems that are both viable and dynamic.

Finally, when it comes to building national innovation systems ordinary state-led science, technology and innovation policy might not be sufficient to overcome the obstacles for economic development. A broader social mobilisation might be necessary in order to overcome barriers to socio-economic development. But this opens up for a much broader discussion than we can bring in this chapter.

Notes

[1] Actually the NSI-approach has much in common with the methodological perspectives of the social psychological pragmatist school of Chicago and not least with the ideas of George Herbert Mead (Mjøset 2002).

[2] Concepts such as institutions and routines are useful in a theoretical context but they are somewhat elusive when it comes to empirical and historical studies. It is easier to track the history of R&D-departments, universities and professional training of engineers than it is to capture changes in how people interact and communicate. But the aim for a full-blown analysis of innovation systems remains to understand how international differences in both the tangible technological infrastructure and behaviour affect innovation outcomes.

[3] It is an interesting issue if learning from Asia and Latin America rather than just from Western Europe and the US could be an option. It might be easier to learn when the technological differences are less dramatic.

References

Andersen, Esben S. and Bengt-Åke Lundvall (1988), 'Small National Innovation Systems Facing Technological Revolutions: An Analytical Framework', in Christopher Freeman and Bengt-Åke Lundvall, *Small Countries Facing the Technological Revolution*, London: Pinter Publishers.

Archibugi, Daniele and Bengt-Åke Lundvall (eds.) (2000), *The Globalising Learning Economy: Major Socio-Economic Trends and European Innovation Policy*, Oxford: Oxford University Press.

Archibugi, Daniele and Mario Pianta (1992), *The Technological Specialization of Advanced Countries*, Dordrecht: Kluwer Academic Publishers.

Arocena, Rodrigo and Judith Sutz (2000), *Interactive Learning Spaces and Development Policies in Latin America*, DRUID working papers, No. 00-13.

Dosi, Giovanni (1999), 'Some Notes on National Systems of Innovation and Production and Their Implication for Economic Analysis', in Daniele Archibugi, Jeremy Howells, and Jonathan Michie (eds.), *Innovation Policy in a Global Economy*, Cambridge: Cambridge University Press.

Edquist, Charles (ed.) (1997), *Systems of Innovation: Technologies, Institutions and Organizations*, London: Pinter Publishers.

Etzkowitz, Henry and Loet Leydesdorff (2000), 'The Dynamics of Innovation: From National Systems and 'Mode 2' to a Triple Helix of University-Industry-Government Relations', Introduction to the special 'Triple Helix' issue, *Research Policy*, Vol. 29, No. 2: 109-123.

Freeman, Christopher (1982), *Technological Infrastructure and International Competitiveness*, unpublished paper for the OECD Expert Group on Science, Technology, and Competitiveness.

Freeman, Christopher (1987), *Technology Policy and Economic Performance: Lessons from Japan*, London: Pinter Publishers.

Freeman, Christopher (1995), 'The National Innovation Systems in Historical Perspective', *Cambridge Journal of Economics*, Vol. 19, No. 1.

Freeman, Christopher and Bengt-Åke Lundvall (eds.) (1988), *Small Countries Facing the Technological Revolution*, London: Pinter Publishers.

Gammeltoft, Peter (2003), 'National Innovation Systems, Development and the NEPAD', in this volume.

Hirschman, Albert O. (1958), *The Strategy of Economic Development*, Clinton: Yale University Press.

IMF (2003), *World Economic Outlook: Growth and Institutions*, April, Washington: International Monetary Fund.

Johnson, Björn (1992), 'Institutional learning', in Bengt-Åke Lundvall (ed.), *National Innovation Systems: Towards a Theory of Innovation and Interactive Learning*, London: Pinter Publishers.

Johnson, Björn and Bengt-Åke Lundvall (1992), 'Closing the Institutional Gap?', *Revue D'Economie Industrielle*, No. 59.

List, Friedrich (1841), *Das Nationale System der Politischen Ökonomie*, Basel: Kyklos, (translated and published under the title: The National System of Political Economy' by Longmans, Green and Co., London 1841).

Lundvall, Bengt-Åke (ed.) (1992), *National Innovation Systems: Towards a Theory of Innovation and Interactive Learning*, London: Pinter Publishers.

Lundvall, Bengt-Åke and Björn Johnson (1994), 'The Learning Economy', *Journal of Industry Studies*, Vol. 1, No. 2: 23-42.

Lundvall, Bengt-Åke and Susana Borras (1998), *The Globalising Learning Economy: Implications for Innovation Policy*, DG XII-TSER, Bruxelles: The European Commission.

Ministry of Education (1998), *National kompetenceudvikling*, København: Undervisningsministeriet

Mjøset, Lars (2002), *An Essay on the Foundations of Comparative Historical Social Science*, ARENA Working Paper, No. 22, Oslo: ARENA.

Mowery, David C. and Joanne E. Oxley (1995), 'Inward Technology Transfer and Competitiveness: the Role of National Innovation Systems', *Cambridge Journal of Economics*, Vol. 19, No. 1.

Myrdal, Gunnar (1968), *Asian Drama: An Inquiry into the Poverty of Nations*, New York: Penguin Books.

Nelson, Richard R. (ed.) (1993), *National Innovation Systems: A Comparative Analysis*, Oxford: Oxford University Press.

Polanyi, Michael (1966), *The Tacit Dimension*, London: Routledge and Kegan.

Sen, Amartya (1999), *Development as Freedom*, Oxford: Oxford University Press.

Stewart, Frances (1977), *Technology and Underdevelopment*, London: Macmillan.

Woolcock, Michael (1998), 'Social Capital and Economic Development: Toward a Theoretical Synthesis and Policy Framework', *Theory and Society*, No. 2, Vol. 27: 151-207.

World Bank (2002), *World Development Report 2002: Building Institutions for Markets*, New York: Oxford University Press.

World Bank (2003), Sustainable Development in a Dynamic World: Transferring Institutions, Growth, and Quality of Life, New York: Oxford University Press.

2

The Dynamics of Catching Up

The Relevance of an Innovation System Approach in Africa

Lynn K. Mytelka

Introduction

This chapter briefly traces the emergence of two bodies of literature each of which has a bearing on the dynamics of catching up in Africa today. The first, growing out of the identification, by Alfred Marshall, of the support afforded by Italy's industrial districts to the growth of small and medium-sized enterprises, focuses on externalities provided by the geographical cluster as a stimulus to technological change. The second, emerging from the work of Joseph Schumpeter, draws attention to the importance of interaction for learning and innovation and the systemic properties of this process.

The purpose of this chapter is to bring these two bodies of literature together and to explore some of the factors involved in transforming clusters into dynamic systems of innovation in a developing country context. It provides a conceptual framework for analysing the dynamics of catching up from this perspective with a particular focus on the role that policy can and must play despite the growing openness of African economies.

Understanding Technological Change and Development: An Innovation System Perspective

In suggesting that an innovation system focus is an appropriate framework for policy-making in Africa, this chapter directly counters two of the beliefs still common in development circles. The first is the traditional approach to building local technological and productive capabilities in developing countries. It regards developing countries as 'technology users' and emphasises North–South technology transfer. From this perspective, learning remains imitative as opposed to innovative and catching up is an incremental process from a low wage, low skill base in which tertiary education and local research capacity play only a minor role at best.

The second is the notion that innovation is something that only takes place in countries like Japan or the United States, in large multinational corporations or in what are regarded as the high tech industries. Indeed, much of the conventional literature continues to associate innovation with the kind of activity by firms that takes place at the technological frontier or what Schumpeter has called invention. A narrow definition that equates innovation with invention of this sort, however, denies the importance of:

- building upon indigenous knowledge;
- exercising creativity in the development of new products, processes, management routines or organisational structures that correspond to local conditions and needs;
- creating the local linkages that support the modification of production processes to bring costs down, increase efficiency and ensure environmental sustainability;
- mastering imported technology in order to transform it in new ways;
- developing policies that stimulate and support a continuous process of learning and innovation.

In the mechanically-based technologies of the past, production of goods such as textiles, consumer electrical and electronic products or automobiles, in their mature phase, focused on standardisation, mass production and incremental innovation along the established technological trajectory. This opened opportunities for catching up by developing countries, which had build up production capabilities. Catching up in these technologies required the capabilities to modify and adapt products and processes thus moving the firm up market to higher value added products. The process of innovation in these industries can best be understood as one in which '[...] firms master and implement the design and production of goods and services that are new to them, irrespective of whether or not they are new to their competitors – domestic or foreign' (Ernst, Mytelka, and Ganiatsos 1998: 12-13).

The last quarter of the twentieth century, have been marked by new waves of technological change. Information and communications technologies (ICTs) were the first of these. With the advent of the semi-conductor, waves of technological change transformed traditional telecommunications processes, led to the creation of knowledge-based products in a wide variety of different industries and launched a new era in information processing. Information and communication technologies have generated many benefits, but these have not been evenly distributed around the globe. Alongside the literature on the ICT revolution, there thus came its sequel that of the digital divide.

Despite their generic and transformatory nature, ICTs, in many respects were a transitional technology and thus only a foretaste of what may be in store for developing countries. They emerged within the paradigm of earlier mechanically-based industrial revolutions and only now are beginning to incorporate the science-based nano-level technologies and products that are central to new wave technologies. Their earlier introduction was incremental and this created opportunities for those developing countries with financial, organisational and knowledge resources and capabilities, to begin a process of catching up.

In the second wave of technological change, that growing out of biotechnology and its application to pharmaceuticals and agriculture, a strong science base was coupled with intensive patenting. The belief that development pathways of the past will be those trodden in the future must thus be tempered by this new reality. In a knowledge-based economy, the locus of knowledge creation and the forms through which knowledge is appropriated will increasingly shape opportunities for learning, for innovation and thus for growth and development.

A third wave of technological change, involving nano-materials and hydrogen as an energy source, will soon be upon us. Just how soon is a matter of conjecture since the extensive systems embeddedness of these new wave technologies, in social, economic and technological terms, have until recently, slowed their application, particularly in the transport sector. But that is changing rapidly and it is in this context that the role of the South as a 'technology user' is becoming increasingly problematic. Looking back to the impact of ICTs and biotechnology, it raises the spectre of a cumulative and path dependent growth in *inequalities* between North and South in the future.

Innovation in industries based on new wave technologies will require new policy thinking. Tertiary education, so long neglected in Africa, will be needed from the outset. So, too, will the building of local research capacity. Though innovation will still involve hands on technicians and engineers as the agents of innovative changes in products, processes, management routines, and marketing strategies, both their problem-solving ability and the kinds of modifications in products and processes that will keep them competitive over time, will depend upon linkages to local research and development facilities and to actors in fields with complementary technologies. Their ability to play an innovative role, is thus more systemically embedded. An innovation system framework has particular relevance to the analysis of both sorts of innovation and the embeddedness of the innovation process.

An innovation system is defined as a network of economic agents, together with the institutions and policies that influence their innovative behaviour and performance, (Lundvall 1992; Nelson 1993; Nelson and Winter 1982). A number of features differentiate an innovation system from both earlier equilibrium-based models of economic systems and the notion of action and reaction (feedback) contained in more mechanical versions of systems thinking brought into the social sciences (Clark, Stokes, and Mugabe 1995; Nelson and Winter 1982). An innovation system, for example, is conceptualised as an evolutionary system in which enterprises in interaction with each other and supported by institutions[1] and organisations such as industry associations, R&D, innovation and productivity centres, standard setting bodies, universities and vocational training centres, information gathering and analysis services, and banking and other financing mechanisms play a key role in bringing new products, new processes and new forms of organisation into economic use. Figure 1 graphically represents an innovation system. In this figure 'organisations' such as universities, public sector research bodies, science councils and firms, that are the traditional focus of science and technology studies are distinguished from 'institutions' which are understood as 'sets of common habits, routines, established practices, rules or laws that regulate the relations and interactions between individuals and groups' (Edquist 1997: 7), that '[...] prescribe behavioural roles, constrain activity and shape expectations' (Storper 1998: 24).

Figure 1 Innovation Systems

Source: Mytelka (2000: 17).

The utility of this distinction is fivefold. First, it lies in the fact that simply having potentially critical actors co-located within a geographical space, does not necessarily predict to their interaction. Actor competences, habits and practices with respect to three of the key elements that underlie an innovation process – linkages, investment and learning – are also important in determining the nature and extensiveness of their interactions (Mytelka 2000). The innovation system approach, moreover, acknowledges the role of policies, whether tacit or explicit, in setting the parameters within which these actors make decisions about learning and innovation.

Second it builds awareness of the extent to which habits, practices and institutions are learned behaviour patterns, marked by the historical specificities of a particular system and moment in time. As such, their relevance may diminish as conditions change. Learning and unlearning on the part of firms and policymakers are thus essential to the evolution of a system in response to new challenges.

Third, it redirects attention towards the flows of knowledge and information that are at the heart of an innovation system. Although these may, on occasion, move along a linear path from the 'supply' of research to products in the market, more often they are multidirectional and link a wider set of actors than those located along the value chain. Which actors other than, suppliers and clients, will be critical to a given innovation process cannot always be known *a priori* and they are likely to be sector specific. So, while it is important to have an overview of the 'national' system of innovation, sector specificity – in industrial structure and technological terms – and the particular habits and practices of actors in that sector will be major factors in shaping policy dynamics and policy impacts. Continuous monitoring of policy dynamics generated by the interaction between policies and the varied habits and practices of actors in the system will be of importance in fine-tuning policies for maximum impact. Adaptive policy-making is part of what makes an innovation system, a learning system.[2]

Fourth the innovation system approach brings the demand-side within the system. Demand flows are amongst the signals that shape the focus of research, the decision as to which technologies from among the range of the possible will be developed and the speed of diffusion. Demand is not solely articulated at arms length through the market, but may take place as Lundvall has shown, through a variety of non-market mediated collaborative relationships between individual users and producers of innovation (Lundvall 1988: 35). In still broader systems terms demand may be intermediated by policies. In the case of pharmaceuticals, policies directed at the local health care system have a powerful impact on the structure of demand and hence the opportunities or constraints on the local supply of research, development and production of (bio-)pharmaceutical products.

Fifth, from a policy perspective the innovation system approach draws attention to the behaviour of local actors with respect to three key elements in the innovation process: learning, linkage and investment (Mytelka 2000: 18). This is particularly important in developing countries where over time actors have developed a set of habits and practices with regard to these three underlying processes that are often inimical to innovation. SMEs, for example, are risk adverse, lack the linkages needed for learning and the finance to support a continuous process of innovation. Yet as Schumpeter acknowledged in his *Theory of Economic Development*, for entrepreneurs to become the driving force in a process of innovation, they must be able to convince banks to provide the credit with which to finance innovation (Christensen 1992: 147). High transaction costs and risks, however, have meant that banks are reluctant to lend to SMEs. Under these conditions, care must be taken to tailor, time and sequence policies aimed at stimulating innovative behaviour to the habits and practices of local actors.[3]

Building Innovation Systems in a Changing Global Environment

Underlying the 'system of innovation' approach was a resurgence of interest in innovation and a reconceptualisation of the firm as a learning organisation embedded within a broader institutional context (Freeman 1988; Freeman and Perez 1988;

Lundvall 1988; Nelson and Winter 1982). That context is pre-eminently national and domestic policies have conventionally been viewed as a critical means to orient the behaviour of national actors towards innovation. Today, however, the institutional set-up at the global level has become a powerful force that shapes the parameters within which actors make critical decision with respect to innovation. As Figure 1 illustrates, these include the profound shaping effects of transnational corporations on the structure of markets and the pace and direction of technological change as well as the set of international agreements dealing with trade, investment and intellectual property.

Since the 1970s, production has become increasingly more knowledge-intensive as investments in intangibles such as R&D, software, design, engineering, training, marketing and management have come to play a greater role in the production of goods and services. Much of this involves tacit rather than codified knowledge and mastery requires a conscious effort at learning by doing, by using and by interacting (Mytelka 1987, 1999).

Gradually the knowledge-intensity of production extended beyond the so-called high technology sectors to reshape a broad spectrum of traditional industries from the shrimp and salmon fisheries in the Philippines, Norway and Chile, the forestry and flower enterprises in Kenya and Colombia, to the furniture, textile and clothing firms of Denmark, Italy, Taiwan and Thailand.[4] Indeed where linkages were established to a wider set of knowledge inputs and the local knowledge base was deepened to include scientific and engineering skills, design, search and quality control capabilities, these traditional industries have shown a remarkable robustness in the growth of output and exports. Most developing countries have not kept up in building these knowledge-based capabilities and this is particularly true for the least developed countries, many of which are located in Africa.

Within the context of more knowledge-intensive production, firms began to compete not only on price but also on the basis of their ability to innovate. The entrenchment of an innovation-based mode of competition radically reduced product life cycles. One out of every two pharmaceutical products, for example, is less than ten years old. In information technology, generations of semiconductor chips or software succeed each other in less than 18 months. In more traditional industries such as textiles and clothing and even eyeglass frames, design changes have turned commodities into fashion goods.

As product life cycles shortened, pressures rose for market opening, which would enable innovative firms from the developed countries to amortise the rising costs of R&D and other knowledge-intensive investments over a larger number of markets. With liberalisation and deregulation in the late 1980s and into the 1990s, innovation-based competition diffused rapidly around the globe and powerful competitors were introduced into the local markets of developing countries. This put additional pressure on local firms to engage in a process of continuous innovation for which they were ill prepared. As a result, innovative firms from the developed countries have enjoyed a large measure of natural protection against potential competitors. The high level and variety of accumulated knowledge and scientific and engineering skills and the scales of production and marketing that are needed to reverse engineer, replicate and commercialise new products and processes, make it exceedingly difficult for firms in the least developed or in most other developing countries, to breach these natural defences.

To this must be added a variety of strategies employed by large firms to internalise new knowledge or appropriate it for lengthy periods of time. In consumer-oriented knowledge-intensive industries, such as pharmaceuticals and agro-industry, a systematic process of market segmentation through the establishment of brand name loyalty via advertising and trade marking became common. Broader patent protection, extended over longer durations was sought and in the TRIPs agreement was obtained.[5] In these and other knowledge-intensive industries a wave of mergers and acquisitions internalised new knowledge and turned it into a proprietary asset. By the end of the 1990s, mergers and acquisitions accounted for well over 50 per cent of global foreign direct investment (UNCTAD 1998b, 2000).

The failure to deal with the global in the context of national systems of innovation has led many to believe that national governments and within nation-states, provincial and municipal governments have only a limited role to play in fostering dynamic innovation systems. This, however, is not the case. Domestic policies can matter and international institutions are not immutable.

International trade agreements, for example, do limit the use of tariffs, non-tariff barriers and subsidies available to support productive activity as it emerges and moves towards competitiveness. But they do not eliminate these all together, as the European Union's regional, innovation and competition policies have shown. Anti-dumping charges and the creation of new sanitary and phytosanitary standards have become strategic arms in the struggle for market access. As negotiations within the WTO continue on subsidies, agriculture, Trade-Related Intellectual Property systems (TRIPs), Trade-Related Investment Measures (TRIMs) and other subjects with a high impact on opportunities for strong technological capability building, new thinking will also be needed to shape rules for international trade that allow small and medium-sized local enterprises the time and support that are required to build and sustain competitiveness in a more open trading system.

Intellectual property rights that affect the life and scope of patent grants have significantly reduced opportunities for catching up through imitation and close follower strategies. As ever-higher proportions of R&D around the world take place within the broad networks of transnational corporations, transnational corporations (TNCs) are in a position to shape the direction and pace of technological change. Reduced possibilities for compulsory licensing further slow the diffusion of new technology. The TRIPs agreement, even in its present form, however, does not prevent the growth of publicly funded R&D, of its ownership by public entities such as research institutions and universities nor does it impede public access to publicly funded R&D. Innovation policy, including a stronger role for publicly funded R&D can widen choice, deepen technological capacity building and provide the tools for technological foresight and impact assessment. The TRIPs agreement itself will soon be subject to review and numerous proposals for revision have already been made.

International investment agreements offer considerable protection to foreign investors. But just as they affect the rights of foreign firms, so, too, can they deal with the responsibilities of these enterprises. The TRIMS agreement affects some of the earlier requirements negotiated with TNCs. The obligation to meet local content requirements can no longer be imposed. Some have argued that under a possible Millennium round, one might carve out certain TRIMS on the basis of developmental considerations. Others have emphasised the need to work around this constraint by developing programs for investor targeting and by turning investment promotion agencies into local development agencies that stimulate the building and strengthening of clusters of suppliers around anchor investments. Incentives put in place to attract foreign firms can be designed to achieve broader regional and local developmental objectives, such as the creation of clusters that enhance local context and linkages.

Clusters and Innovation Systems

As global competition intensified, researchers working on development issues in Asia, Latin America and to a lesser extent in Africa exhibited a growing interest in clusters, understood mainly in terms of spatial agglomerations of enterprises and related supplier and service industries. Elsewhere I have distinguished between 'spontaneous' clusters and 'constructed' clusters, where the latter are agglomerations that emerged not spontaneously through an incremental process of co-location but rather through the actions of public policy.[6]

Industrial parks and Export Processing Zones (EPZs) were the earliest forms of constructed cluster. Their focus was on production and exports and the vehicle to achieve these objectives has traditionally been through the attraction of foreign firms. Little attention was paid to learning and technological capability building and still less to innovation as critical development goals in either the creation of EPZs or the analysis of their achievements and failures, though competitive conditions were changing, making the sustainability of export growth problematic without more conscious attention to technological upgrading. Not until the late 1990s do we find a few studies that begin to take learning and innovation into consideration in their analysis.[7]

Spontaneous clusters had also emerged in many developing countries and these were now 'discovered' as their products became competitive in export markets. All were based on traditional industries such as surgical instruments in Sialkot, Pakistan (Nadvi 1998), ceramic tiles in Santa Catarina, Brazil (Meyer-Stamer, Maggi, and Seibel 2001) and cotton knitwear in Tirrupur, India (Cawthorne 1995). Modelled upon the work of sociologists, geographers and economists in Europe, where clustering of small enterprises appeared to have withstood the pressures for economies of scale, concentration and a more ruthless form of capitalism, these development economists saw in the externalities provided by clusters and the collaboration amongst actors within the cluster, a means to stimulate industrialisation and technological upgrading (Nadvi and Schmitz 1997; Schmitz 1997).

For SMEs in particular, clustering is believed to offer unique opportunities to engage in the wide array of domestic linkages between users and producers and between the knowledge producing sector (universities and R&D institutes) and the goods and services producing sectors of an economy that stimulate learning and innovation (Nadvi and Schmitz 1997; UNCTAD 1998a). Stable *vertical relationships* between users and producers, for example, can reduce the costs related to information and communication, the risks associated with the introduction of new products and the time needed to move an innovation from the laboratory or design table to market (Ernst, Mytelka, and Ganiatsos 1998; Lundvall 1988, 1992). *Horizontal collaboration* between same-sector small and medium-sized enterprises can also yield '*collective efficiencies*' (Schmitz 1997) in the form of reduced transaction costs, accelerated innovation through more rapid problem solving and greater market access. Still other studies have pointed to the *positive* externalities generated by agglomerations – the availability of skilled labour, of certain kinds of infrastructure, of innovation-generating informal exchanges and learning made possible through the adoption of conventions (Maskell 1996; Maskell and Malmberg 1999; Storper 1998).

Initially, spontaneous clusters that had emerged in Africa, attracted less attention since they were peopled, for the most part, by actors from the informal sector and their potential for growth and transformation appeared limited. More recent studies, such as those of the metalworking cluster in the Suami Magazine in Kumasi, Ghana (Powell 1995), the autoparts cluster in Nwewi, Nigeria (Oyelaren-Oyeyinka 1997), and the fish processing cluster in Uganda (ongoing research at UNU/INTECH) and a variety of clusters in Kenya and South Africa (McCormick 1999) show the emergence of considerable adaptive behaviour by local actors and some organisational coordination in technological upgrading, though they have not been transformed into innovative systems. These studies reveal the need for new policy thinking to support both the growth of clusters and their transformation into dynamic innovative system.

The Policy Dimension in the Design of Innovation Systems

From a 'national' system of innovation perspective, it is assumed that enterprises are more attuned to conditions and actors in the domestic market both in their perception of problems and in the way they will deal with them. It is also assumed that actors with whom interaction will be needed in order to respond to a problem are present within the system, have the necessary competence and have habits and practices conducive to the formation of linkages, to learning and to innovation more broadly.

These assumptions can be challenged in a developing country context. There, both the firms and other actors with whom learning through interaction might take place, have developed habits and practices that are not conducive to innovation and the domestic presence of many critical actors is not assured. The establishment of TNCs and the development of long distance linkages have stretched the concept of 'national' when applied to an innovation system. In developed and developing countries, the varied application of the innovation system approach to economic sectors and to geographical spaces has also led to the confusion of 'clusters' with innovation systems. A static bias that does not do justice to the dynamics of industrial, technological and institutional change at the global level within which national and local systems of innovation are embedded is also common to most applications of the system of innovation concept.

Despite these limitations, the system of innovation approach has the potential to become a powerful tool for national and local policy-making. It provides a new way to organise knowledge as an input into policy-making, a means to analyse the support structures and policies needed for innovation and a framework for situating the local in the context of dynamic processes of change at the global level. This is particularly important in the design of policies that support learning and innovation.

Cluster studies have traditionally fallen short here. Though they tend to focus on specific sectors, they rarely situate these sectors within the broader policy environment. Policy dynamics, that are the interface between policies and the habits and practices of the actors in a sector, are lost in this process and so, too, is the need to monitor the impact of policies in a continuous fashion.

In contrast, reconceptualising sectors as innovation systems and embedding them within the broader national system of innovation opens a multitude of new opportunities to identify and strengthen knowledge bases that are common to several sectors, to build linkages across these sector-based systems and to identify critical actors and the nature of their interactions in local sector-based systems using detailed comparative analysis. The latter, however, is only the first step in what Edquist (2001: 233) described as the need to go beyond the analogous firm-level 'benchmarking' process now so common. The typical benchmarking exercise leads to a static view of the change process as one of 'catching up'. It overly focuses attention on the pathway of the 'lead' firm and the policies of the country in which this lead firm has emerged. It fails to recognise that catching up sometimes means running in a new direction. Work underway at UNU/INTECH on (bio-)pharmaceutical innovation systems in Cuba, Egypt, Ghana, India and Taiwan, illustrates the multiplicity of pathways and policies to building an innovation system and such possibilities need to be further explored in the context of other developing countries and other sector-based innovation systems.

By situating sector-based innovation systems studies within the global context, the innovation system approach provides a useful methodological framework for analysing the strengths and weaknesses of the actors in an innovation system – their competences and the nature and extent of the interactions between them – and for competence building at all levels – individual and organisational, as well as across all actors in the system. It also alerts us to the need to focus on a broader range of competencies than the more usual set of capabilities found in the development and innovation literature. These include openness, experimentation, coping with uncertainty, dealing with change, questioning established truths, building trust, and working within collaborative partnerships both across ministries as well as among firms and between firms and universities or research institutes.

Competencies such as these are not as amenable to standard training processes in which codified knowledge is transferred or through traditional apprenticeship practices where existing bodies of tacit knowledge are passed along. They cannot simply be acquired from outside or imitated by rote, because the very ability to build such competencies requires that they be internalised by the individual or organisation and subjected to continuous scrutiny, feedback and change. All actors in an innovation system must thus become learning organisations and traditional transfers of technology models are limitative.

In building innovation systems, therefore, policies must not only be directed towards ensuring the presence of critical actors and building a number of new competences, but to the different purposes that such competences and organisations serve under contemporary conditions from those of the past. Four sets of actors and competences are of critical importance in the transformation of clusters into innovation systems and illustrate these differences.

- Creating, strengthening, and networking knowledge institutions such as universities or research organisations, productive enterprises in agriculture, other natural resource sectors, manufacturing and services and the policymaking processes for greater openness in information and knowledge flows, experimentation and dealing with change.
- Strengthening the ability of producing firms and their network components such as suppliers, contractors to deal with uncertainty; risk-taking and collaborative partnerships. Intermediaries who work with such networks will have to be trained in facilitating the development of such networks and of new financing instruments tailored to local conditions.
- Ensuring the kind of interactivity with technical and non-technical institutions providing metrological, marketing, financing and information services, and support that enables actors to problem solve, take a longer-term perspective and engage in strategic planning.
- Building trust–based relationship and channels of communication between policymakers, enterprises and civil society more generally.

Notes

[1] Formal definitions of 'institutions' stress the 'persistent and connected set of rules, formal and informal, that prescribe behavioural roles, constrain activity and shape expectations […] they […] give order to expectations and allow actors to coordinate under conditions of uncertainty' (Storper 1998: 24). See also (Edquist and Johnson 1997).

[2] For Metcalfe, at issue, then, '[…] is how well policymakers learn and adapt in the light of experience' (Metcalfe 1997: 275).

[3] Elsewhere I have discussed the nature of policy dynamics that result from the interaction between policies and the traditional habits and practices of actors in greater detail. Case studies in that volume provide evidence for the ability of government to successfully tailor and sequence policies designed to induce change under different local conditions (Mytelka 1999).

[4] In fish farming, for example, pond technologies are based on advanced materials and incorporate complex design knowledge. Monitoring depends upon computer imaging and pattern recognition technologies. Feeding and health systems involve the use of robotics, pharmaceutical inputs and knowledge of nutrition which is increasingly related to biotechnology. New techniques for preservation, storage and packaging are based on freezing technology, hermetics, bacteriology and microbiology as well as engineering and informatics (Smith 1999: 10, 19).

[5] See (Mytelka 2000) for a lengthier treatment of this subject.

[6] For a full exposition of the classification scheme used here and its subsequent application in Brazil see (Mytelka and Farinelli 2000, 2003).

[7] In Africa, the best studied from this perspective is the EPZ that housed much of the early clothing cluster in Mauritius (Lall and Wignaraja 1998; Wignaraja and O'Neil 1999).

References

Cawthorne, Pamela M. (1995), 'Of Networks and Markets: The Rise and Rise of a South Indian Town, the Example of Tiruppur's Cotton Knitwear Industry', *World Development*, Vol. 23, No. 1: 43-56.

Clark, Norman, Kathryn Stokes, and John Mugabe (2002), 'Biotechnology and Development: Threats and Promises for the 21st Century', *Futures*, Vol. 34: 785-806.

Christensen, Jesper L. (1992), 'The Role of Finance in National Systems of Innovation', in Bengt-Åke Lundvall (ed.), *National Systems of Innovation: Towards a Theory of Innovation and Interactive Learning*, London: Pinter Publishers: 146-168.

Edquist, Charles (2001), 'Innovation Policy: a Systemic Approach', in Daniele Archibugi and Bengt-Åke Lundvall (eds.), *The Globalizing Learning Economy*, Oxford: Oxford University Press: 219-238.

Edquist, Charles and Björn Johnson (1997), 'Institutions and Organizations in Systems of Innovation', in Charles Edquist (ed.), *Systems of Innovation: Technologies, Institutions and Organizations*, London: Pinter Publishers.

Ernst, Dieter, Lynn K. Mytelka, and Tom Ganiatsos (1998), 'Technological Capabilities in the Context of Export-Led Growth: a Conceptual Framework', in Dieter Ernst, Lynn K. Mytelka, and Tom Ganiatsos (eds.), *Technological Capabilities and Export Success in Asia*, London: Routledge: 5-45.

Freeman, Christopher (1988), 'Japan: a New National System of Innovation?', in Giovanni Dosi, Christopher Freeman, Richard R. Nelson, Gerald Silverberg, and Luc Soete (eds.), *Technical Change and Economic Theory*, London: Pinter Publishers: 349-369.

Freeman, Christopher (1992), 'Formal Scientific and Technical Institutions in the National Systems of Innovation', in Bengt-Åke Lundvall, *National Systems of Innovation: Towards a Theory of Innovation and Interactive Learning*, London: Pinter Publishers.

Freeman, Christopher and Carlota Perez (1988), 'Structural Crises of Adjustment, Business Cycles and Investment Behaviour', in Giovanni Dosi, Christopher Freeman, Richard R. Nelson, Gerald Silverberg, and Luc Soete (eds.), *Technical Change and Economic Theory*, London: Pinter Publishers: 38-66.

Lall, Sanjaya and Geneshan Wignaraja (1998), *Mauritius Dynamising Export Competitiveness*, London: Commonwealth Secretariat.

Lundvall, Bengt-Åke (1988), 'Innovation as an Interactive Process: From User-Producer Interaction to the National System of Innovation', in Giovanni Dosi, Christopher Freeman, Richard R. Nelson, Gerald Silverberg, and Luc Soete (eds.), *Technical change and Economic Theory*, London: Pinter Publishers: 349-369.

Lundvall, Bengt-Åke (ed.) (1992), *National Systems of Innovation: Towards a Theory of Innovation and Interactive Learning*, London: Pinter Publishers.

Maskell, Peter (1996), *Localised Low Tech Learning*, paper presented at the 28th International Geographical Congress, Haag, 4-10 August.

Maskell, Peter and Anders Malmberg (1999), 'Localised Learning and Industrial Competitiveness', *Cambridge Journal of Economics*, Vol. 23, No. 2: 167-185.

Meyer-Stamer, Jörg, Claudio Maggi, and Silene Seibel (2001), *Improving Upon Nature: Creating Competitive Advantage in Ceramic tile Clusters in Italy, Spain, and Brazil*, INEF Report, Germany, Institut für Entwicklung und Frieden, Gerhard-Mercator-Universitat, Duisburg.

McCormick, Dorothy (1999), 'African Enterprise Clusters and Industrialization: Theory and Reality', *World Development*, Vol. 27, No. 9: 1531-1551.

Mytelka, Lynn K. (1987), 'The Evolution of Knowledge Production Strategies within Multinational Firms', in James Caporaso (ed.), *A Changing International Division of Labour*, Boulder: Lynne Reiner: 43-70.

Mytelka, Lynn K. (1999), 'Competition, Innovation and Competitiveness: A Framework for Analysis', in Lynn K. Mytelka (ed.), *Competition, Innovation and Competitiveness in Developing Countries*, Paris: OECD Development Centre: 15-32.

Mytelka, Lynn K. (2000), 'Local Systems of Innovation in a Globalized World Economy', *Industry and Innovation*, Vol. 7, No.1: 15-32.

Mytelka, Lynn K. and Fulvia Farinelli (2000), *Local Clusters, Innovation Systems and Sustained Competitiveness*, UNU/INTECH Discussion Paper, No. 2000-5, October, http://www.intech.unu.edu.

Nadvi, Khalid (1998), 'Knowing Me, Knowing You: Social Networks in the Surgical Instrument Cluster of Sialkot, Pakistan', in *IDS Discussion Paper, No. 364*.

Nadvi, Khalid and Hubert Schmitz (1997), *SME Responses to Global Challenges: Case Studies of Private and Public Initiatives*, paper presented at the Seminar on New Trends and Challenges in Industrial Policy, October, Vienna: UNIDO.

Nelson, Richard R. (ed.) (1993), *National Innovation Systems: A Comparative Analysis*, Oxford: Oxford University Press.

Nelson, Richard R. and Sidney G. Winter (1982), *An Evolutionary Theory of Economic Change*, Cambridge: Harvard University Press.

Oyelaran-Oyeyinka, Banji (1997), *Nnewi: An Emergent Industrial Cluster in Nigeria*, Ibadan: Technopol Publishers.

Powell, John (1995), *The Survival of the Fitter: Lives of Some African Engineers*, London: Intermediate Technology Publications.

Schmitz, Hubert (1997), *Collective Efficiency and Increasing Returns*, Sussex: IDS Working Paper, No. 50 (March).

Storper, Michael (1998), 'Industrial Policy for Latecomers: Products, Conventions, and Learning', in Michael Storper, Stavros B. Thomadakis, and Lena Tsipouri (eds.), *Latecomers in the Global Economy*, London: Routledge: 13-39.

UNCTAD (1998a), *Promoting and Sustaining SMEs Clusters and Networks for Development*, Paper prepared for an Expert Meeting on Clustering and Networking for SME Development, Geneva, 2-4 September, No. TD/B/COM.3/EM.5/2.

UNCTAD (1998b), *World Investment Report 1998, Trends and Determinants*, Geneva: United Nations.

UNCTAD (2002), *World Investment Report 2002*, Transnational Corporations and Export Competitiveness, Geneva: United Nations.

Wignaraja, Ganeshan and Sue O'Neil (1999), SME Exports and Public Policies in Mauritius, Commonwealth Trade and Enterprise Paper Number 1, London: Commonwealth Secretariat.

3

Re-thinking Africa's Development through the National

Innovation System

Mammo Muchie

> When in Africa we speak and dream of and work for, a
> rebirth of that continent as a full participant in the affairs
> of the world in the next century, we are deeply conscious
> of how dependent that is on the mobilisation and
> strengthening of the continent's resources of learning.
> (Mandela 1998)

> A paradigm can, for that matter, even insulate the
> community from those socially important problems that
> are not reducible to the puzzle form, because they cannot
> be stated in terms of the conceptual and instrumental tools
> the paradigm supplies.
> (Kuhn 1962: 37)

Introduction[1]

This chapter will discuss why the national innovation system (NSI) in the African context has to be constructed from a basis that goes beyond the existing states. It will be argued that the NSI approach can be used to reinforce the Pan-African framework expressed in the AU/NEPAD continental process. It can be used to orient the Pan-African institutions and to interlink and diversify the continent's social economic structure and create new products and processes for the world economy.

The chapter will argue that the attempt to build NSIs on the basis of the existing fractured economic, political and social circumstances with weak states and entities that hardly can be called nations will reproduce the traditional division of labour that Africa has with the industrialised countries within the world economy. What is needed is systemic coherence at a Pan-African level that promotes internally generated innovative products and processes that will re-articulate Africa's position within the world economy. The chapter will identify the salient components of the NSI conceptual approach and re-articulate them within a Pan-African perspective.

A Brief Overview of the National System of Innovation Literature

The literature on national innovation system (NSI) takes firms as the main actors of innovation and learning. The concept explicitly introduces the wider environment that can constrain or facilitate technological learning by firms and accentuates that the interaction of the firm with institutions, governments, universities, civil society, other firms, and various stakeholders matter a great deal for success or failure. An additional point is that, in spite of globalisation or even because of it, nations remain a legitimate unit of analysis when it comes to describe and explain variations and uneven locations of innovation globally. For the industrialised world, identifying fully formed nations has not been a problem. The NSI conceptual framework has been used in the industrialised countries and the innovation system and policies of specific nations such as the USA, UK, Japan, Germany, and Denmark etc. have been scrutinised in detail (Freeman 1987, 1994b; Lundvall 1992, 2002; Nelson 1993).

In addition, the NSI of the newly industrialising countries in Asia have been analysed as well (Kim and Nelson 2000; Lim 1999). Their NSI were scrutinised in relation to how these countries managed to build and accumulate domestic technological capability that, in turn, facilitated transfer of knowledge, and technology from the external sources to their firms and institutions. The type of links of firms to the education system, the financial system, the state system, the users and other relevant institutional actors and stakeholders have been studied. Such studies have also been undertaken for Latin America (Katz 1987).

In Africa there have been a number of research networks such as the Africa Technology Policy Studies (ATPS), African Centre for Technology Studies (ACTS), the Magtech in North Africa and so on, but there has not been a systematic study that employs the NSI concept to the specific context of Africa inside Africa. In the UN System, UNCTAD and UNIDO and some of the specialised research units like INTECH are doing work employing the NSI framework. The exception is South Africa where the White Paper on Science and Technology Policy in 1996 was produced based on the NSI framework (Kaplan 1999; Scerri 1998).

However while the NSI of South Africa is more developed in certain respects it appears that it also reproduces some of the characteristics of the post-colonial innovation systems in other parts of Africa. The system of innovation exists in a rather peculiar form. The main reason is that South Africa's system of innovation remains heavily constrained by the apartheid legacy. The ANC has the political throne, but it does not control the economic goal to pursue a fully independent social policy that can empower the historically under-privileged black African population. South Africa is still like two nations, where the historically privileged side can boast rightfully of having innovation of world frontier vintage and performance in certain technologies related to aerospace, arms, mining, IT and medicine. The mandate of the black universities was to teach but not to do research. Even the teaching of engineering and natural sciences was carried out as an exception, not as part of the production of the nation's intellectual grain. This stark contrast within the South African system of innovation – excellence to the privileged nation and legally enforced mediocrity to the majority nation – still colours the particular evolution of the NSI in that country. One of the most interesting developments that is taking place at the moment in tertiary education in South Africa is to redress this by a strategy to merge historically disadvantaged universities with white research universities.

Thus the NSI in South Africa has peculiar, lop-sided and bifurcated characteristics. Like every other African country, in South Africa, an integrated and coherent national framework for innovation and learning, where the link between firms and knowledge institutions is stronger is yet to be forged. In this sense South Africa is an ordinary African country with a privilege of one of the nations. This means that South Africa's NSI is not so radically different that it cannot be integrated with the NSI from other parts of Africa. The call by Nelson Mandela and Thabo Mbeki for South Africa to join Africa by spearheading an African renaissance suggests that South African exceptionalism is no barrier for joining the Pan-African national project. Some commentators are sceptical of the new role of South Africa in helping to 'pull the continent out of its enduring crisis' (Swatuk and Black 1997: 15).

The key question thus becomes: what would one mean by 'nation' in the African context including South Africa? Is the 'nation' the existing states that are often pronounced as 'failures' or 'bifurcated' in the case of South Africa? Is it an imagined community waiting to be made? This is the central issue when one thinks of appropriating (thus re-inventing it in the new context) the NSI-concept into the African context. What does the 'National' stand for in the making of Africa's innovation system?

The NSI is more relevant in the African context precisely because the nation along with the innovation, and the systems, are yet to be made. In the case of Africa, a principal value of the NSI concept is that it sets an ambitious but necessary future agenda in these three dimensions. The building of separate NSIs in fifty-four states – most of which have populations of five to ten millions would be incredibly difficult. Thus thought about NSI has to begin not by the affirmation of the fragmented Africa, but by rejecting it. The problem of the boundary for the African nation as unit of analysis is thus up for debate. The Norwegian economist, Erik S. Reinert says that 'The term state is hardly applicable to several African countries...many institutions such as the educational system... have broken down' (Reinert 2003). Not only is the idea of the Africa nation contested, but also the idea of state is problematic.

We are thus dealing with concepts that are highly contested, whose materialisation or lack of it in the African setting will have profound consequences. A country may get its prices wrong, but there is a much heavier price in getting its national system of innovation wrong. It must get its NSI right especially in Africa that has yet to perform its structural transformation on the foundation of continuous technological learning. It is important to put the NSI on the agenda as the situation in Africa is in a state of flux and getting Africa's NSI right will have long-term consequences for building Africa's wealth and self-reliance.

The fact that Africa has yet to make its own NSI may be a blessing in disguise. Consider that, at present Africa suffers from chronic food crises. But, given innovation and learning, Africa should not have been in this predicament. According to the US National Academy of Sciences, 'Africa has more native cereals than any other continent. It has its own species of rice, as well as finger millet, fonio, peral millet, sorghum, teff, guinea millet, and several dozen wild cereals whose grains are eaten. This is a food heritage that has fed people for generations, possibly stretching back to the origins of mankind. It is also a local legacy of genetic wealth upon which a sound food future might be built. But strangely, it has been bypassed in modern times.... Forward thinking scientists are starting to look at the old cereal heritage with unbiased eyes. Peering past the myths, they see waiting in the shadows a storehouse of resources whose qualities offer promise not just to Africa, but the world' (NAS 1996).

What is essentially missing is an innovation and competence building system and that can mine and add value to this endowment in natural resources. Instead of trying to manufacture products that others are already producing, Africa has vast opportunities to create new lines of production and fabricate and introduce new products into the world market. What it needs is a robust system of innovation that links up to the world's leading centres of innovation whilst re-articulating a distinctly African-wide national political economy of production.

What then is the appropriate national unit to organise an African system of innovation? The NSI in the African context can evolve if we develop an imagined community called the Africa unification-nation and bring it closer to becoming a reality. In Africa, the nation in the NSI can be a 'unification-nation'. Historically unification-nations such as Germany and Italy were also latecomers. While they were able to create unified systems of national production systems, they were also authoritarian and their rival relationship with the big powers resulted in two tragic wars. Other weaker unification nations like the former USSR and the Balkans have broken asunder leading to nationalism that has degenerated into ethnic cleansing. When we propose a unification nation for building an African system of national production, the lessons from these historical experiences must be in our memory. The African unification-nation must be participatory and emancipating, founded on tolerance, democratic principles, rule of law, respect of universal human rights very much as it is currently enshrined in the South African constitution. It is thus a democratic and constitutional African unification-nation that we must imagine and translate into reality.

The NSI-framework in the African Context

The making of an African innovation system is both an ambition and a call to create and mobilise an enabling environment in Africa that is simultaneously conducive and responsive to fostering a total innovative and learning culture in the continent. Africa's post-colonial states, universities and industries have not provided the attractive context and enabling environment to make innovations and learning thrive. As stated in the introduction even the most advanced country on the continent (South Africa) does not have an integrated innovation system.

The making of an innovation system embodies both the desire and strategy to introduce well functioning states, well functioning universities and well functioning industries, well-functioning and stable civil society, well-functioning markets – in the framework of a well-functioning Africa-nation by taking a systemic perspective on their co-evolutions. After forty years of colonialism, Africa has only four per cent of the world's share of cognitive resources (Wamahiu and Bunyi 2001). During the last forty years fifty-four states have emerged in Africa. Most of these entities are fragile or weak. Such a fragmented pattern of nation building bequeathed from the post-colonial condition has proved more a hindrance than a facilitator in the pursuit of any meaningful national independent industrial and technology strategy and policy. The post-colonial state suffered from two major constraints: a) inability to manifest an agency different from those external actors who provide aid to sustain it, b) inability to govern society by breaking out of the maze of domestic conflicts, political, economic, social and cultural problems.

Social and economic policies largely fashioned by the ideas and paradigm of neo-classical and proto neo-Keynesian development economics have been inserted in Africa's fractured economic and policy space. The received economic theories and policies did not engage with 'the socially important problems that are not reducible to the puzzle form' to use Thomas Kuhn's formulation. They took fragmentation as the normal order of things and simply produced technicist and economist recipes that did not go anywhere. Far from stimulating structural transformation, the policies became part of Africa's burden. The policies did not do much to reduce Africa's accentuated dependence on external donors, nor did they assist in reducing the internal conflicts and social problems of poverty and inequality.

To make matters worse, in the last twenty years the policy employed has come from the neo-liberal interpretation of neo-classical economics. The result appears to have been more poverty and inequality creation than wealth creation. The outcome from the structural adjustment policies has been a sort of universal 'de-industrialisation' or 'primitivisation' (Reinert 2003: 453).

At present, development economics seems to be in intellectual disarray as a field with many voices from Right to Left making damning criticisms. From the Right a neo-classical economist has written about the poverty of development economics (Lal 1983). Others from a neo-Keynesian perspective, have tried to defend development economics against such attacks (Toye 1987). Marxists have also written about the rise and fall of development theory (Leys 1996). There is a real alarm that development economics has been marginalized in the academy (Chang 2003). There is a need to challenge this orthodoxy and look hard for intellectual resources that may assist Africa to cut roots from the domination of ideas that have brought more poverty than prosperity, except to the few. We need to look for intellectual ideas that may be appropriated to re-orient policy direction to create more wealth rather than poverty in Africa. We think shaping and organising an African NSI provides the necessary, but not sufficient, condition to stimulate and provide a fresh perspective for African development.

The relevant variables that emerge from the NSI framework are:

1. The idea of the *'nation'* as a unit of analysis for innovation and learning.
2. A *'system'* perspective for establishing an institutional framework for continuous technological learning.
3. Embedding a *total* innovation, learning and competence building culture in economy, society, institutions and stakeholders.
4. Creating an *enabling environment* or framework for fostering innovation in economy, polity and society in general and firms, actors and specific institutions in particular.

Let us take each component in the context of re-launching development as structural transformation in Africa.

The Idea of the 'Africa-Unification Nation'

To date, the effort to make nations out of the disparate communities that exist from the casually and arbitrarily carved African states/entities has not been successful. Many of the African states remain insecure and unstable to threats from a constellation of internal and external actors. The governance pattern of the existing states has not been conducive to embedding the total innovation and learning culture. Left to their own devices the governing elites have little incentive to open their societies and create an enabling political, social and cultural environment to foster innovation and learning. It is thus interesting that the African Union has given up on the notion of non-interference and autonomy to sovereignty as limited by the responsibility of those who govern to the governed. Those who sign and ratify the African Union's Constitutive Act admit also to make it a law in their own respective states. As Nelson Mandela put it:

> Africa has the right and the duty to intervene to root tyranny… we must all accept that we cannot abuse the concept of national sovereignty to deny the rest of the continent the right and duty to intervene when behind those sovereign boundaries people are being slaughtered to protect tyranny. (Speech at OAU Summit 1998)

The requirements for creating an enabling environment to foster innovation and learning culture as part of the daily routine of life and existence in Africa are based on the following considerations:

a. The forty year historical record of existing states has not provided the required context to foster innovation, learning and competence building.
b. Whether they acknowledged this failure or not, states have begun to pull sovereignties together within the AU/NEPAD framework.
c. There is a need to pull resources together to create viable institutions that can inter-link and influence the development process in Africa.

d. The more robust the Africa Union becomes, the better the chance to benefit rather than lose from globalisation.
e. The AU can provide the framework for deepening and broadening the concept of Africa as a unification-nation.
f. There has been a new wave of Pan-African resurging after 1994 when Nelson Mandela and Thabo Mbeki coined the coming of South Africa to the common Africa home with their optimistic speeches of building the Africa renaissance.

Before that the Organisation of African Unity (OAU) was so focused on the task of de-colonisation and removing white supremacist rule, that it hardly addressed the idea of an African-unification nation.

Together the AU/NEPAD may be seen as attempts to establish a political and economic environment that will facilitate the diffusion of innovation and technological learning across Africa. Politically the AU's Constitutive Act permits African intervention in other African states when a) an unconstitutional take over of power through military coup de etat, mercenaries, armed dissidents and where an incumbent refuses to hand over power after election defeat. Sanctions such as suspension from AU meetings have been employed to coup makers and those who did not respect popular verdict in elections; b) Any repeat of a Rwandan type of genocide is expected to invite AU intervention; c) Internal instability that has regional or continental consequences (e.g. Liberia) would also invite AU intervention. What is interesting is that such interventions and sanctions are already taking place. Sovereignty is no longer a barrier to AU intervention as in the old OAU days. There is now recognition that many crimes have been committed in Africa under the guise of protecting sovereignty. Democracy and good governance are on the agenda, even though they are unevenly distributed across the continent. Thus the willingness to allow inter African intervention paves the way to the self-erosion of state sovereignty and possible incremental migration of authority to the Pan-African framework the AU provides.

In the economic field the NEPAD framework promotes a peer review mechanism for purposes of disseminating good practice, to build self-reliance in African development strategy, engage in policy learning and identify problems and monitor progress in a constitutional and democratic process.

The ideas and rhetoric coming from the AU/NEPAD suggest that an integrated Africa-wide political and economic unification programme is beginning to move on to the agenda. The constraint is the degree to which the very actors that declared these intentions will have the political will to fulfil this plan. Intention is one thing practice another.

That is precisely the reason why intellectual effort must be invested to align intention with practice and implementation. Intellectuals can engage and provide knowledge from an independent institutional base, regarding how a unified economic and political programme may be sustained. That is why the proposition is made that the NSI concept can accommodate the 'national' as the Africa-unification nation. An alternative beyond the fragments has yet to emerge. I suggest the creation of an Africa-unification nation to provide a logo, a new Pan-African framework for promoting Africa's sustainable development.

It is difficult to see how the African renaissance could succeed without a simultaneous construction of a robust Africa-wide national system of innovation. The attempt to create an African development strategy through the New Partnership for African Development suggests that there is an implicit (if not explicit yet!) recognition to strive for a unification-nation through Pan-Africanism and African renaissance by organising an Africa-national unification system of innovation.

Pan-Africanism or the African Renaissance makes sense when they signal a strategy to promote transformation. They can provide coherence and integration of the fragments. They can serve both to re-articulate the internal maze of relationships and provide an equitable relationship of Africa with the rest of the world economy. The authors of AU/NEPAD are trying to provide to the current wave of transition a Pan-African framework and imprint. Such a unified framing for transformation provides a fresh perspective to Africa's long quest for political and economic integration.

Another important reason why the Africa-unification nation is useful is that it rebuts Washington consensus recipes that has further de-industrialised much of Africa. In the 1980s neo-liberal ideas shaped the so–called Washington Consensus. It imposed 'stabilisation, liberalisation and privatisation' as the chief vehicles for growth and development on vulnerable and poor economies destroying the little industry they had built up. Getting prices right through an unfettered free market driver dominated the intellectual grain for development economics. Policies of structural adjustment minimising the role of the state prevailed. States were expected solely to create enabling environment for property rights and money supply. Economies were expected to seek comparative advantage through open competition, deregulation, privatisation and de-nationalisation. In the 1990s neo-classical economics came up also with the new growth economics and new institutional economics. The World Bank re-introduced the poverty reduction rhetoric back on the agenda from its total eclipse in the 1980s (World Bank 2001).

The NSI approach can provide new tools and cognitive vision to reframe development research within broader Africa. The concept of national system of innovation links dialectically knowledge, technology and research with different levels of space, types of institutions, activities and stakeholders. It is one of the few concepts from the western academy that would not resist discussion of development or transformation in Africa by positing the concept or imagination of the making or creation of an Africa-unification nation. I propose the concept of an Africa national-unification system of innovation to provide the missing link in Africa's transformation. This concept is proposed to structurally transform the African social economy not by fooling with the endless game of poverty alleviation, but by creating wealth directly through a comprehensive and deliberate internalisation of innovation and learning to Africa's existing production arrangements (See Muchie 2003).

Embedding a Total Innovation and Learning Culture (TILC) in Africa

If the NSI conceptual framework is to guide public policy for African development, there is a need to develop a generalised innovation and learning culture in Africa. This culture can evolve when we learn to convert Africa's problems of transformation into challenges for innovation and learning.

With a continental enabling environment, possibilities open up for Africa's numerous communities to self-define and associate with a diverse manifestation of their identities. Innovation and learning will provide the means to empower people in creating a productive engagement with African development. They learn to investigate the challenges of African transformation as a problem of a complex interplay of institutional and technical innovation.

Innovation has been widened beyond its role in accelerating economic growth to include functions such as quality of life improvement, human-well-being development, for ecological solutions and above all changing the metaphysics, assumptions, mentalities and attitudes in society (Freeman 1994a). The promotion of innovation and continuous learning is critical in order to change the prevailing pessimist assumption about Africa's capacities and possibilities. 'The image of the African... has been built up... upon his lowest common denominator' (Sertima 1999: 330). Evidence has been acknowledged by astronomers and engineers from Western research universities that in many major innovations and inventions, documented contributions have been made by Africans: carbon steel was made 2000 years ago in Tanzania, astronomical observations in the Mali of the Dogons have been acknowledged by Carl Segall of Cornell University, language, mathematical systems, architecture, agriculture and cattle-rearing, navigation on inland waterways and open sea, medicine and communication and writing systems – in all these fields technological innovations have taken place in Africa (Sertima 1999). There is thus evidence of a lost science and technology that Africans made important contributions to. It is crucial to use this memory to construct and support innovation and learning culture to confront contemporary challenges in Africa.

Africa has been reported as lacking in innovation and learning. This is the dominant view that prevails. This type of perception has contributed to a pessimistic thought pattern on Africa. We think that how a matter is named and framed ontologically is part of and becomes constitutive of the changes the matter is likely to undergo, and the way knowledge is secured about the dynamics of change including the methodology for making/constructing knowledge. In Africa, the observer's ethical compass, and his defining lens has been as much a problem for Africa as the manifold problems bedevilling the continent. There is a need to select issues that can assist to question matters that often feed negative cognition and practice.

It has become conventional wisdom to make damning judgements from the events and occasions that frequently happen in Africa. Almost anyone who writes on Africa from any academic direction seems to assert the 'failure' of Africa. This is indeed worrying as what is said to fail is Africa, not policy, market, the state, development or any other variable. Analysts do talk about such failures, but they also claim a 'failure of Africa'. This sort of bland assertion is unhelpful and it does not make sense. For example some Afro-pessimists like the French Africanist Bayart have given up any hope for optimism for Africa. They seem to say that the only source from which Africa may generate hope is from evil (Muchie 2001).

Even scholars that wish to renew or rethink African development begin with a disconcerting assertion that 'Africa is mired in developmental crises' (Stein 2003: 153) having paralysed oneself with such an assertion, the writer finds that 'the challenge of reversal of the African malaise is daunting' (ibid.). Even those who wish to rethink Africa's possibilities start with a self-paralysing cognitive framing. Africa's economic performance is widely seen as 'disastrous and prospects for the future are bleak' (Belshaw and Livingstone 2002: 5).

Neither are those writing from the perspective of innovation and development free from betraying a similarly pessimistic thought process. Some researchers write that Sub-Saharan Africa is 'locked in a development tragedy…Sub-Saharan Africa … (is) probably the most technologically backward area of the world today'. (Oyelaran-Oyeyinka, Laditan, and Esubiyi 1996) Another scholar adds his twist to the same view: 'In a world of accelerating technical change, intensifying competition and globalising production, Africa is not only failing to improve its international competitive position – it is falling behind' (Lall 2001). Even I myself went with the paralysing view of 'tragedy, failure or death' – having been contaminated by the prevailing common sense about Africa – to study the pollution crises of a major agro-industrial sector in Kenya: the Leather Industry. The material I found on the ground was impressive (Muchie 1999).

There is a need for the analyst to become free from such tragic invocations. Such assertions are mere expressions of opinion especially when they are used to open a discussion on Africa's possibilities and do not emerge as conclusions from any serious scholarly investigation or project. I shall propose a simple alternative, approach or paradigm to examine Africa's problems and proffer possible lines of productive engagement. I suggest that that we couple innovation and learning within an Africa-unification nation framework with emancipating research in order to build knowledge that will assist new trajectories for making wealth in the interest of eradicating poverty and inequality in Africa as a whole. This perspective should be mobilized as an alternative to the endless see-sawing between neo-liberal structural adjustments, and the World Bank's rhetoric of poverty reduction. The key is to create wealth without sacrificing social cohesion. This requires the creation of systems of innovation and learning that support the making of a sphere for independent public policy formation.

We should change the terms of talking about Africa's problems by fostering the Total Innovation and Learning Culture (TILC) to embed in all aspects of productive activities. This is to encourage Africans to become problem solvers. TILC can assist to paint an upbeat picture on a broader canvass to convey the simple maxim that problems beget solutions provided the agency of the problem solvers is released from all confining encumbrances.

The tendency by writers to essentialise Africa's many problems, and compress Africa's complex situation into evocative symbols of 'tragedy, failure, risk, hopelessness and death' needs contesting. The diffusion of the total innovation and learning culture in Africa can provide and assist in making research acquire a more nuanced cognitive shift and possibly a new and a less condemnatory gaze on Africa and Africans. The total innovation and learning culture can also provide a useful cognitive and epistemological re-orientation away from 'failure-speak' to 'innovation-speak'. The total embedding of an innovation and learning culture can assist also in re-framing Africa's challenges of transformation within a post-pessimistic conceptual foundational principle.

The actual state of innovation is one thing; the potential of innovation to bring about transformation in Africa is another. The actual and the potential, the empirical and the normative can be twinned in analyses, description, explanation, modelling, understanding, interpretation and prediction. Empirical findings and regularities need to be interpreted in the light of policy perspectives that seek new avenues for expanding Africa's transformation possibilities. The question of 'what is the state of existing innovation threshold in unlocking Africa's challenges of transformation' should be connected with how the innovation system can be embedded within the transformation dynamic of the continent? Raising the point of 'is' without the 'can' will make research *passive* while connecting the 'is' and the 'can' will make it *active*.

a) How does innovative activity actually take place?
b) What can we learn from the past inventions and innovations by Africans in Africa?
c) In what way may innovation systems and activities be mobilised to bring about social-economic transformation of Africa without transgressing ecological and social constraints?
d) The actual and the normative direction can be made to feedback on each other to outline the innovation and learning trajectories of Africa's specific transformation within the AU/NEPAD broad framework.

Finally pessimism will not disappear until there is a systematic initiation to sustain transformation by the systematic fostering of a comprehensive and total embedding of a generalised knowledge, innovative and learning culture in stakeholders, institutions, activities, spaces and their systemic connections across Africa's economy, polity and society.

The Role of Systems from the NSI Perspective

It has been suggested that in a system interaction is everything. The important problem is to identify both conceptually and empirically the interactions that are significant and matter more than others. The knowledge system, the production system, the financial system, the legal and human resource base have to be interlinked at the Africa-wide level in order to articulate the African social economy both internally and in relation to the world economy. What are the system drivers and how do they interact? How can systems simultaneously maintain closure under the conditions of openness both down stream to the local and upstream to the international economy? How do the various relevant structures, actors and practices of the African-national level activity influence and impart to the development of a specific identity of the NSI? The influence of the NSI at a continental scale within the AU/NEPAD framework has to be engaged with the various specific institutions, locations, activities and stakeholders. The challenge is to create systemic coherence and integration of the various levels from firms to the educational system, the financial system, civil society in the development of the African NSI. This is not something that can be claimed. It has to be investigated and it is an important field for research in its own right.

A system of innovation confronts varied environments and situations to which it must respond and deal with. In the process the internal coherence of the system of innovation may either be strengthened or weakened. A system of innovation can evolve in a variety of ways depending on the power balance: a) NSI can increase efficient economic performance while sustaining social cohesion, or b) it can increase economic growth whilst creating increased income disparities, or c) it can increase economic growth while maintaining income equality. Lundvall suggests the Danish model of the national system of innovation reflects economic dynamism with social cohesion unlike the Anglo-Saxon system. From a normative stance, an innovation system that increases economic growth while maintaining income equality represents a desirable form if one puts a premium on social cohesion to social discontent, growth of crime and other undesirable social consequences. 'The Danish system of innovation and competence-building is small in global terms but it has certain characteristics that makes it interesting as a model for international institutional learning. Denmark has one of the most egalitarian societies in the world in terms of income distribution and at the same time it has an income level that is almost the highest in the world' (Lundvall 2002: 2).

Assuming threats and opportunities that are similarly faced by different national economic arrangements, the response can vary depending on the difference in innovative systemic capabilities to innovate and to cope with change. National or even local systems have different policies, institutions and styles of sharing the costs and benefits of change. This has an impact on the dynamism of the economic system and the cohesion of the social arrangement.

A well-functioning system of innovation is open to new opportunities and deals with threats that emanate from the environment. A strong system of innovation functions reasonably well both in periods of stability and instability. A weak system of innovation may function in periods of stability, while it may not respond effectively in periods of dynamic instabilities. A learning innovation system copes with new problems while an innovation system that is not quick to adapt to changing circumstances get stuck. Some analysts have explained the failure of the Soviet system as a failure of innovation system. The argument is that the Soviet system became locked into economic practices and institutions that were incongruent with the new era where economies were driven by technologies, knowledge and services. Powerful vested interests resisted the change and the successful innovation of the first wave of Soviet industrialisation was inadequate to keep the Soviet economy delivering the consumer goods that the population wanted. Thus the national system of Soviet political economy broke down. It was fit in the period of extensive industrialisation, but failed to maintain the same momentum in the period of intensive industrialisation. New actors, new technologies and new businesses are required to stimulate the innovation and economic performance of the society. When that fails, crises set in.

In a similar vein some analysts have tried to explain the East Asian crises of 1997 as a failure of their system of innovation. The argument is that there was a successful innovation system for a stable economic period, but when the world economy was transformed with the ICT revolution, the system of innovation was not prepared to cope with the transforming or what some have called 'the learning globalising economy'. Vested interests resisted, new actors could not establish themselves. The ensuing period saw massive dislocation.

In the case of Africa, the system of innovation concept is useful in providing a conceptual tool for connecting the pressures and changes in the world economy, with the mobilisation of an African national–unification project to deal with the structural problems of social-economic transformation and eradicating poverty. African economies are lured to liberalise their economies with the incentives of donor funding. This may stimulate production while aggravating poverty. A specific African system of innovation is necessary to assist in the development of an Africa-unification nation in order to deal with the challenge of bringing about comprehensive transformation in line with Africans' own principles.

A system of innovation can thus be linked to different internal and external problems different economies are facing. In Africa I see it as one of building a system of innovation for national unification and the African-renaissance. In Russia, it is one of dealing with a transition from an overregulated political economic system to one that creates a new balance between social regulation and economic development. In East Asia, it is one of dealing with a learning globalising economy and adjusting to it. In Denmark it is one of keeping an innovation system that seems to be working well. In The Anglo-Saxon world, it is one of finding a system that promotes economic development without increasing further income inequalities. Different systems of innovation exist since nations face different internal and external challenges.

The mutual dependence of NSI and production unit, sector-level or region level system of innovations means that synergies or lack of synergies lead to differentiated technological, business and national economic performance. The fact that knowledge, learning and innovation are interactive means that policy perspectives have to address the problem of building synergies amongst actors, institutions and activities. The key is to promote the appropriate type of interactions: changing negative interactions into positive links, strengthening positive interactions and rejecting negative ones. The Achilles heel of policy decision is making sure that synergies are attained in knowledge, learning, and innovation from the various forms of interactions in order to stimulate technological, business and economic performance.

Africa has been locked into a primary commodity exporting economic structure. The central actors, institutional arrangements and outlooks that constitute a particular economic practice can be represented in the form of a system of innovation. Powerful vested interests continue to keep this unproductive structure to Africa and resist the mobilisation of a new system of innovation. The creation of an African system of national political economy emerges by unfettering and unlocking the primary exporting political economic structure.

The Creation of an Enabling Environment

Innovation and learning are to transformation and development as water is to plants. Continuous technological learning is critical for change. The environment is equally an important factor to continue a regime of technological learning, competence building and capability accumulation. The environment can enable or disable the innovation and learning process. The environment is both internal to the firm undergoing innovation and learning and external deriving from the way the wider political, economic, cultural, legal (regulatory), financial and institutional arrangements function. For continuous technological learning to occur, both the internal and external environments should be enabling rather than disabling. One of the central arguments in creating a national system of innovation is to foster an enabling environment for continuous technological learning. Conversely an enabling environment also influences positively a national system of innovation.

In Africa, the problem of creating an enabling environment is not settled yet. It has been one of the most troubling problems that the post-colonial state confronted immediately after decolonisation. Unfortunately Africa got caught in the Cold War despite efforts to organise Pan-Africanism as an alternative to the USSR and the US competition. Many states became polarised and failed to manifest a united voice. The continent turned into an armed field from which it has not fully extricated itself. Super power rivals fought proxy wars in Africa. The political instability and lack of a predictable and constitutionally anchored power transfer created a security problem in the continent. That in turn affected patterns of trade, investment, technology and economic development to this day.

This engagement with external actors also affected governance. Elites had the option to side with one or the other camp in the Cold War. This allowed the possibility to elites to play opportunist games, often without any scrutiny or sanction.

Consequently, corruption became rife. Repression rather than development was imposed on the population. Home-grown ideas for transforming the continent were also given short thrift. Ideas from outside backed with dollars tempted the elite not to pursue independent social policy. Africa followed the agenda of the International Financial Institutions and abandoned its own such as the long-term structural transformation approved by the OAU in Lagos, known as the Lagos Plan of Action. In 1981 the Berg report on accelerated development in Africa came linked to donor lending. Africa entered the structural adjustment epoch that led to the so-called 'lost-decade'. If Africans do not have their own agenda and are too weak to believe and push the agenda they put their signature to collectively, they will end up following someone else's agenda. As it happens the structural adjustment period was distinguished by the near-total absence of social policy. The consequence is the collapse of infrastructure and learning in much of Africa. For over twenty years there has not been an enabling environment to foster innovation and learning in Africa.

During the current post-cold war phase, we cannot say that the conducive environment has been created. What seems to be happening though is that there is recognition of the centrality of fostering an enabling environment at the continental level. We believe that the national system of innovation framework can be useful for creating a Pan-African enabling environment, as the later will also be useful to strengthening the NSI.

NSI and Pan-Africanism: Towards a Research Agenda

There is a need to orientate current NSI research related to Africa with a Pan-African perspective. The key components that form the NSI need to be re-framed in the context of Pan-African evolution. I shall indicate some of the salient issues that need to be on the research agenda in and for Africa.

- How is learning and innovation undertaken into existing institutions? How might learning and innovation become rooted into the traditional institutions? The importance of TILC becoming embedded within the variety of traditional arrangements and finding different strategies to do it constitutes an important issue for research. For any change to be enduring, it must be rooted in the native institutions to transform them root and branch.
- How do modern institutions relate to traditional ones? How do new institutions or hybrids form? How does the building or shaping of new institutional systems evolve in relation with techno-economic networks with a traditional institutional arrangement?
- Does the connection of existing and new institutional arrangements initiate and bring structural transformation?
- Identification of specific policies and strategies in using TILC to change the norms, values, interests and attitudes of those engaged in the transformation

process on the creation and use of knowledge to make resourceful learners out of sleepy persons and institutions.

- Comparative knowledge base accumulation relating TILC competence or mastery by the core organisations such as industry, farms, trade unions, offices and firms and their supporting networks.

- How does knowledge creation or use become part of the every day culture and common sense for responding to or dealing with any expected or unexpected difficulty and problem coming from within and outside the environment or the interaction of the internal and external environment?

- How do the units of production relate to the wider economic structure, the legal, financial and political structures to upgrade and enhance quality and efficiency?

- How do traditional and modern and private and public, civil society and community organisations learn and innovate?

- How coherent or incoherent are the inter-linkages within national economies of such components as the financial, legal, policy, educational, scientific and social institutions?

- How effective are the institutional components in performance and how well do they co-ordinate and work in concert? – these are key issues that require systemic modelling, adjustment and resolution.

- How do processes between the macro-level co-ordination with the micro-level activities take place? How strong is the symbiosis between the institutions or networks of innovation and the micro-production processes that use and master knowledge and learning to create active comprehensive life-long learners?

- Who are the users/Who are the producers? And what is their potential in creating Pan-African user-producer interactions?

- How well do the degree of functioning and capacity of the modes and mechanisms that communicate knowledge accelerate learning between different parts of a Pan-African systems of innovation?

These are indicative questions. They, by no means exhaust the issues. They are offered to sensitise researchers to take them up in their research considerations.

Concluding Remark

We cannot transform Africa by denying its agency. The oft repeated assertion that Africa is a failure has to be countered by fostering an innovation and learning culture. It is not difficult to present Africa as a tragedy by itemising its numerous troubles like in a shopping list. We need a new economics of hope to attack Africa's problems with parallel solutions. Learning, innovation and imagination provide the heuristics and interpretative orientations to convert problems into solutions. Why lament? Anger at the unwholesome conditions yes; but courage to change that condition should supplant the easy option to give in to despair. There is thus a great need to search for innovative approaches to generate solutions that may create further problems, which, in turn lead to more and better solutions. Knowledge about what is, needs to be supplemented with what could be.

If Africans strive to possess innovation and learning as a problem-solving culture and build competence, the ability to form a self-reliant orientation to solve problems will grow. Received policies from development economics have the in-built external orientation making Africans look for ideas and finance to the outside world when they should be looking mainly to mobilise Africa's own initiatives, resources and possibilities.

The policy relevance of the systems of innovation concept is related to the provision of new knowledge, new attitude, new conceptual frameworks and new policy mechanisms in the search for uncovering hitherto untried learning paths or old ones with new and original approaches. Such learning paths emerge from a detailed empirical examination of Africa's own human base, nature, institutions, systems, organisations, markets, practices, successes and failures in the context of transforming Africa from poverty to prosperity. Rather than making policy selection amongst prescribed poverty reduction options, the policy direction for structural transformation is best realised through the application of innovative capabilities in guiding Africa's structural change through the deployment of national, systemic and innovation properties, functions and performance.

Africa's comparative advantage to move away from poverty to enter prosperity requires marrying knowledge, learning and innovation to its main resources: people, nature and in refining continuously its native institutions. The national system of innovation approach has been suggested as a heuristic guide on ways of stimulating possible learning paths for transforming the human, natural, attitudinal and institutional base of the African world. Future research should employ emancipatory methodology on how to embed the total learning and innovation culture to stimulate Africa's structural transformation on the basis of Pan-African integration and imagination.

Notes

[1] My thanks to Adey, Mikael, Negash, Meti, Kelvin for their personal support; to colleagues in Middlesex, Dennis Parker and Abby Ghobadian, Baskaran, for their professional support; and to co-editors Bengt-Åke and Peter for their resolved peer review of the chapter.

References

Adeboye, Titus (1997), The African Technology Policy Studies Network: ATPS, *Technology Analysis & Strategic Management*, Vol. 9, No. 2.

Bayart, Jean-Francois, Stephen Ellis, and Beatrice Hibou (1999), *The Criminalisation of the State in Africa*, Oxford: James Currey.

Belshaw, Deryke and Ian Livingstone (eds.) (2002), *Renewing Development in Sub-Saharan Africa: Policy, Performance and Prospects*, London: Routledge.

Chang, Ha-Joon, (ed.) (2003), *Rethinking Development Economics*, London: Anthem Press.

Scerri, Mario (1998), 'The Parameters of Science and Technology Policy Formulation in South Africa', *African Development*, Vol. 10, No. 1: 73-89.

Freeman, Christopher (1987), *Technology Policy and Economic Performance: Lessons from Japan*, London: Pinter Publishers.

Freeman, Christopher (1994a), *The Economics of Hope: Essays on Technical Change, Economic Growth, and the Environment*, London: Pinter Publishers.

Freeman, Christopher, (1994b), The Economics of Technical Change: *Cambridge Journal of Economics*, Vol. 18, No.5: 463-514.

Kaplan, David E. (1999), 'On The Literature of the Economics of Technological Change: Science and Technology Policy in South Africa', *South African Journal of Economics*, Vol. 67, No. 4: 473-490.

Katz, Jorge (1987), *Technology Generation in Latin American Manufacturing Industries*, London: Macmillan.

Kim, Linsu and Richard R. Nelson (eds.) (2000), *Technology Learning and Innovation*, Cambridge: Cambridge University Press.

Kuhn, Thomas (1962), *The Structure of Scientific Revolutions*, Chicago: University of Chicago Press.

Lal, Deepak (1983), The *Poverty* of *Development Economics*, London: Institute of Economic Affairs.

Lall, Sanjaya (2001), *Transfer of Technology: Kenya, Tanzania, Uganda and Ghana: Study for UNCTAD*, Unpublished Manuscript, Geneva.

Leys, Colin (1996), *The Rise and Fall of Development Theory*, Bloomington: Indiana University Press.

Lim, Youngil (1999), *Technology and Productivity: The Korean Way of Learning and Catching Up*, Cambridge: MIT Press.

Lundvall, Bengt-Åke (1992), *National Systems of Innovation: Towards a Theory of Innovation and Interactive Learning*, London: Pinter Publishers.

Lundvall, Bengt-Åke (2002), *Innovation, Growth and Social Cohesion: The Danish Model*, London: Edward Elgar.

Mandela, Nelson (1998), 'Address at Harvard University', September 1998, quoted in *East African*, September 1-7, 2003.

Muchie, Mammo (1999), *Final Report for DFID: Barriers to the Uptake and Effective Use of Environmentally Sensitive Technologies in Kenya's Leather Industry*, Autumn, Unpublished.

Muchie, Mammo (2001), *Towards a Theory for Re-Framing Pan-Africanism: an Idea Whose Time Has Come*, DIR Working Paper, No. 83, 2000, Aalborg, Also Translated in Chinese by the Chinese Academy of Social Science, *Journal of West Asia and Africa*, March-April, 2001.

Muchie, Mammo (2003), *The Making of an Africa-nation: Pan-Africanism and the African Renaissance*, London: Adonis-abbey Publishers.

National Academy of Sciences (NAS) (1996), *Lost Crops of Africa: Volume 1, Grains*, Washington, D.C.: National Academy Press.

Nelson, Richard R. (1993), *National Systems of Innovation*: Oxford: Oxford University Press.

Oyelaran-Oyeyinka, Banji, G. O. A. Laditan, and O. A. Esubiyi, (1996), 'Industrial Innovation in Sub-Saharan Africa: The Manufacturing Sector in Nigeria', *Research Policy*, Vol. 25, No. 7: 1081-1096.

Wamahiu, Shiela and Grace Bunyi (2001), *A Policy Forum on Innovations in Higher Education in Africa*, Ford Foundation, October 1-3.

Reinert, Erik (2003), 'Increasing Poverty in a Globalised World: Marshall Plans and Morgethau Plans as Mechanism of Polarization of World Incomes', in Ha-Joon Chang (ed.), *Rethinking Development Economics*, London: Anthem Press: 451-477.

Sertima, Ivan Van (1999), 'The Lost Sciences of Africa: an Overview', in Malegapuru W. Makgoba (ed.), *Africa Renaissance*, Cape Town: Mafube Publishers: 305-330.

Stein, Howard (2003), 'Rethinking African Development' in Ha-Joon Chang (ed.), *Rethinking Development Economics*, London: Anthem Press: 153-179.

Swatuk, Larry A. and David R. Black (1997), *Bridging the Rift: The New South Africa in Africa*, Boulder: Westview Press.

Toye, John (1987), *Dilemmas of Development*, Oxford: Basil Blackwell.

Wamahiu, Shiela and Grace Bunyi (2001), *Policy Forum Report: Innovations in Higher Education in Africa*, Nairobi: Ford Foundation African Higher Education Initiative.

World Bank (2001), *World Development Report 2000/2001: Attacking Poverty*, New York: Oxford University Press.

4

What is Innovation Policy All About?

Andrew Jamison

Introduction[1]

Over the past thirty years, there has developed a rather fundamental bifurcation, or contradiction, in the ways in which we think about science and technology, and a rather large gap has developed among those who formulate policies for research, development and innovation. In most national governments, as well as in many supranational organisations, such as the European Union, the United Nations, and the Organisation of African States, science and technology tends to get integrated into (at least) two very different types of political discourses, namely one that has to do with economic growth and an other that has to do with what might be termed social and environmental well-being.

At times, of course, participants in the different discourses meet and discuss together, but, for the most part, their deliberations, as well as the so-called expert advice that enters into their deliberations, are conducted separately from each other. In most decision-making systems, innovation policy – as science and technology policy has come to be called – is primarily a part of economic, or industrial policy, while all the other societal applications of science and technology are relegated to a residual welfare policy sphere. As the experiences of the past ten years in seeking to implement the call for sustainable development amply illustrate, it has proven difficult, if not impossible, to do anything particularly meaningful about this bifurcation, that is, to combine in a serious way the pursuit of economic growth with the fulfilment of other social and environmental objectives.

The doctrine of sustainable development was originally meant to refer to an integration of environmental concern into economic decision-making, but on its way into reality, it has confronted enormous institutional, practical, as well as ideological barriers (Jamison 2001). With the coming to power of neo-liberal parties in the United States and several European countries in 2001, all of which have come into office with a clear anti-environmental agenda, the possibilities for achieving more sustainable development policies have suffered a serious, if not fatal, setback. And yet, if innovation policies are to be formulated and implemented in ways that can be of benefit for society as a whole, and also serve to sustain the carrying capacity of the natural environment, we must find ways to recombine the now separated discourses and build some meaningful bridges across the very real bifurcation that exists.

In general terms, we can characterise the bifurcation, or contradiction, that has emerged as being between economists and sociologists, between those who think of scientific and technological change primarily in terms of economic activity, and those who think of innovation within science and technology as variegated processes of social construction (see Table 1).

Table 1 What is Innovation Policy?

	Economic Approaches	Sociological Approaches
What is it about?	Commercialisation	(Social) construction
How is it analysed?	Technological trajectories Systems of innovation	Actor-networks/hybrids Contextual tensions
What is studied?	Firm strategies Learning processes	'Laboratory life' Mediation/construction
What methods are used?	Surveys Economic modelling	Case studies Story-telling
What needs to be improved?	Competitiveness Policy instruments	Public participation Accountability procedures
What is it based on?	Instrumental rationality	Communicative rationality

In the following, I will try to unpack some of the assumptions and biases on both sides of the divide, in a modest attempt to open some space for mutual reflection and perhaps even some fruitful cross-fertilisation.

The Economic Approaches

In our contemporary world, technology is primarily seen as the source of marketable innovations and new products, which has brought into being new fields of expertise in such areas as technology management and industrial innovation, as well as new theories and concepts of evolutionary economics, innovation systems, technological dynamics, learning economy, etc. What is at issue here is not whether science and technology satisfy any particular social or human need or, for that matter, help solve any particularly pressing social, environmental or human problem; the overriding, and more or less exclusive, concern is rather whether a market can be found for new innovations, and, if so, how shares in that market can be increased for the purposes of corporate expansion and growth. In this perspective – what we might call the dominant technology discourse – scientific and technological change is seen as a key factor of economic competitiveness and successful business performance. The discourse is especially dominant in relation to firms that are actively promoting the so-called 'high' technologies, but its influence is much more general and all-encompassing. In some formulations, there is the idea that technological change is the core activity of business behaviour in general, and it is only by understanding the learning processes and selection mechanisms involved in technological innovation and in the marketing of innovations that companies will be able to survive in an increasingly globalising economy (see (Archibugi and Lundvall 2001).

In this sense, the meaning that is attributed to technological change is essentially commercial, and the processes of technological change are incorporated into the broader processes of economic development, or the accumulation of capital, or, more simply, activities of money-making. This meaning, or role, of technology in our societies has been around for a long time, but it is only recently – in the past fifty years or so – that it has taken on what we might term hegemonic proportions. It is as commerce, as 'exchange value', that technology and technological change are most understandable and meaningful in the contemporary world.

Over the past twenty years, it has been primarily under the political influence of neo-liberalism and globalisation that this commercial meaning of technology has taken on hegemonic status. But it is important to recognise that the dominant discourse also reflects important changes that have been taking place within the practices of science and technology.

While certainly not all technological change has become a matter of science-based innovation, there can be no denying that both information technology and genetic technology have become significant contributors to economic growth in many industrial countries. And as is readily apparent, these types of technology distinguish themselves from other types of technology in at least three major respects. Firstly, they are scientific or laboratory-based technologies, that means that they require major expenditures on scientific research for their eventual development. And unlike the science-based innovations of the early 20th century, which were, for the most part, applications of a scientific understanding of a particular aspect of nature (microbes, molecules, organisms, etc.), these new technologies are based on what Herbert Simon once called the sciences of the artificial. Information technology is based on scientific understanding of man-made computing machines, and biotechnology is based on scientific understanding of humanly modified organisms.

Secondly, we are dealing with technologies that are generic in scope, which means that they have a wide range of potential applications in a number of different fields, sectors and life-worlds. As opposed to earlier technologies, which were primarily based on finding solutions to identified or, at least, identifiable problems, these are solutions in search of problems. In this respect, both information technologies and biotechnologies are idea-driven, rather than need-driven, which means that, in relation to their social uses, they are supply-driven, rather then demand-driven. That is one of the reasons why they require such large amounts of marketing and market research for their effective commercialisation, and indeed for their development. Their generic nature means that the process of innovation is dependent on a specific 'trajectory' being defined, articulated, planned and implemented.

Finally, these advanced technologies are transdisciplinary in what might be called their underlying knowledge base; that is, their successful transformation into marketable commodities requires knowledge and skills from a variety of different specialist fields of science and engineering. In earlier periods of technological development, there were clearer lines of demarcation between the specific types of competence and knowledge that were relevant; indeed the classical categories of engineering are based on the particular types of scientific and technological theories that were utilised (chemical, mechanical, combustion, aerodynamic, etc.). Genetic engineering and information technology, however, require expertise and skills from a wide range of scientific fields, and, even more crucially, a competence in combining knowledge from different fields: hybridisation. The genetic engineer is neither (merely or exclusively) a scientist or a engineer, but rather a kind of hybrid combination of the two previously separated identities or roles. For this reason, the new technological fields have been characterised as being a part of a new 'mode' of knowledge production, which is sometimes referred to as technoscience (Gibbons *et al.* 1994).

Table 2 Changing Modes of Knowledge Production

	Little Science Before WWII	Big Science 1940s-1960s	Technoscience 1970s-
Type of knowledge	Disciplinary	Multidisciplinary	Transdisciplinary
Ideal of knowledge	Artisan craftsmanship	Industrial production	Commercial innovation
Organisational form	Research groups	R&D institutions	Ad-hoc networks
Dominant values	Academic	Bureaucratic	Entrepreneurial
Examples	Chemistry, biology	Atomic energy	Genetic engineering

Sociological Approaches

While economists have tended to dominate the field in recent years, there has nonetheless been a range of quite different activity within the social study of science and technology, or science and technology studies which is broadly sociological. During the 1960s and 1970s, several approaches developed both within sociology itself, as well as in neighbouring fields like history, psychology, anthropology and philosophy. Particularly influential within sociology was what might be called the rediscovery of the sociology of knowledge, especially in the book by Peter Berger and Thomas Luckmann, *The Social Construction of Reality*. Together with a number of other contributions, published in the tumultuous 1960s, such as Thomas Kuhn's *Structure of Scientific Revolutions* and Herbert Marcuse's *One-Dimensional Man*, Berger and Luckmann helped to open up the previously closed world of science and technology to sociological investigation. In the 1970s, it was primarily the natural scientists that were the objects of this attention, but by the late 1970s, technology also began to be seen as a legitimate topic for sociological scrutiny, and a range of sociological approaches to technology started to develop.

What all sociological perspectives on technology share is an explicit focus on actors, and on their actions, in relation to technological development (see (Bijker, Hughes, and Pinch 1987)). For some, actors are characterised as translators, and their actions are seen primarily in relation to particular projects of hybridisation, by which humans and non-humans construct reality. This sociology of translation puts emphasis on actions of enrolment and mobilisation, and has been developed by the Frenchmen Bruno Latour and Michel Callon to show why certain technological projects fail (the French electric car is one favourite example), while others transform society in fundamental ways (Latour's 'pasteurisation' of France). The point here is that technological change is a kind of lever, or vehicle, of broader social changes, and in order to be successful, technological 'actors' must build networks both with the non-human things they are interested in, as well as with other humans. Underlying it all is a kind of entrepreneurial model of human behaviour, and a rather instrumental view of social action.

For other sociologists of technology, the actors are seen as pursuing one or another kind of interest, be it personal, political, or religious, and the social construction of technology is viewed as a kind of negotiation process, by which interests are either in a state of conflict, or are combined in one or another compromise. The interest resolution is, for Wiebe Bijker, one of the most influential social constructivists, seen as a process of 'closure' by which a particular meaning or interpretation of technology is stabilised. His examples range from the safety bicycle to the electric light bulb and the industrial material bakelite (Bijker 1995). On a more systemic level, the historian Thomas Hughes has focused attention on the actors who construct large technical systems, like electricity distribution and production systems. Hughes and many other historians of technology, such as David Noble and Donald Hounshell, emphasise the actors who work, so to speak, at the interface between technology and society: the academic engineers and management scientists, the funding agencies of technological projects, the corporate executives who create links between various institutions, and so on.

In general terms, we can think of all social action in relation to technology as a kind of network-building, by which various brokers or mediators establish connections between different fields of knowledge and different types of people and organisations. As a general term, mediation includes both the translation and enrolment that is so important for Latour and Callon, as well as the flexible interpretation that is emphasised by Bijker. What is primarily involved in mediation is the construction of new kinds of 'hybrid' identities, literally new forms of action that cross over previously separated domains or areas of social activity. In this sense, technology as social construction focuses on practices, as well as role and identity formation.

A Cultural Approach

While most discussions of technological innovation have been framed within the language and terminology of economics or sociology, other meanings have recently been given new significance and actuality, particularly with the coming to market of products based on genetic engineering. The techniques of genetic manipulation have brought to the surface of public consciousness a number of critical ideas and perceptions, which we can characterise as a wide-ranging 'cultural critique of technology' (see (Baark and Jamison 1986)). The very real lack of interest and even distaste that many people feel towards genetically manipulated organisms is quite visible, and it is difficult to understand those processes – of rejection, resistance, dissatisfaction, and annoyance – within the vocabularies and theories of economics or even sociology. According to economics, those products should never have been developed if there had not been a recognisable 'demand' for them; and according to sociology, they should never have gotten as far as they have without important social groups being interested in them.

But in most of the world, and for a great many people, these technologies are seen primarily in negative terms, threatening traditional beliefs and ways of life, as well as forms of livelihood and employment, particularly in relation to agriculture, but also in relation to the integrity of the human body. What is so characteristic of the opposition to genetic technology – both in Denmark and the United States, as well as in many developing countries – is the feeling of powerlessness, the sense that decisions about technologies are made by far too few people. Also involved of course is the generalised notion of risk and the idea that we are living in a risk society, which means that these technologies are intrinsically not 'goods' that people really need, but they are more like 'bads' that simply produce all sorts of dangers and uncertainties (Beck 1992). That is why terms like trust, ethics and accountability are so much a part of the discourse about genetic technologies, and why so many different kinds of people are seeking to establish more direct forms of empowerment and public accountability. If money is to be spent on new technology, then it has to be made clear why; and even more importantly, it has to be shown that those technologies are useful. Technological change, in this context, is seen from the perspective of the user, rather than the producer. Of course, as we see in many of the contemporary debates about genetic technology, this can lead to strange sorts of alliances and campaigns, but what links the third world critics of genetic technology with the representatives of the small farmers and shopkeepers throughout Europe and North America is what might be called an interest in whether these technologies are 'appropriate' or not.

These debates about genetic technologies make abundantly clear that something has gone wrong in the processes by which technologies are integrated into society. On the one hand, there are problems at a discursive level; the idea of genetic manipulation runs counter to many important idea traditions in our societies, both in relation to the meaning of life, but, even more importantly, to the very idea of being human. If our beings are reducible to a genetic code, which can be manipulated and recombined and 'cloned', then many people react negatively.

On an institutional level, our societies have great difficulty in establishing appropriate organisational forms and, more generally, normative principles to deal with these technologies. There are of course a range of ethical councils and agencies of technology and risk assessment, but the problems with genetic technologies and their acceptance have not gone away for that. Even more significantly, genetic technologies have not entered into everyday life worlds, in terms of becoming integrated into customary behaviour patterns, and internalised in personal identities.

What have yet to develop are, we might say, adequate forms of appropriation for these new genetic technologies, and it makes it important to develop frameworks of understanding that can help us understand the relevant social processes. Much can be learned from previous technological transformations – mechanisation, electrification, auto-mobility, for example – when similar technologies, frightening at first, were made to fit into society through what might be termed a multilayered matrix of cultural appropriation processes. Understanding these processes of technological change requires insights primarily from the cultural sciences, rather than the economic or social sciences – from such fields as cultural and intellectual history, anthropology, linguistics, etc. It is the discourses and organisational cultures, the everyday life experiences and language games that are essential to grasp, as technologies are appropriated into societies.

Table 3 The Cultural Appropriation of Technology

Analytical Level	Phenomenal Level		
	Structures	Systems	Artefacts
Discursive	Language *Assimilation*	Grammar *Disciplining*	Semantics *Familiarisation*
Institutional	Rules, standards *Normalisation*	Corporations *Organisation*	Media *Dissemination*
Practical	Customs *Habituation*	Routines *Domestication*	Behaviour, identity *Internalisation*

What are involved, at different phenomenal levels, are different types of appropriation processes, and it is important to recognise that these processes tend to occur in a fragmented way. They do not occur all at once; they overlap and interact with one another in complicated ways. Mechanisation, electrification, computerisation, genetic engineering, to take some typical examples, affect both the ways we talk and think, as well as the ways in which we carry out our practical activities. Our language takes on new words and alters old ones, as technical artefacts are adapted to our discursive codes and frameworks. In our day, *information* and the *genetic code* have become central metaphors for all sorts of phenomena, and new words and concepts have entered our vocabularies while familiar ones have taken on new meanings. Our societies develop new forms of organisation and interaction, of regulation and governance, as technologies impose their systemic and infrastructural requirements on the social order. We now have genetic counsellors and the scientific field of genomics, biotechnology companies and genetic forensic experts. And in our everyday life-worlds, we take on new identities and must learn new skills, as our practices are altered by technological and scientific innovations. We have to learn what is in the food we eat and the seeds we plant, and we have to reflect on the choices we make in the supermarket. Cultural appropriation is thus a variegated and highly differentiated set of reformation processes, and they are seldom discussed in an integrated manner.

The sheer variety and range of these processes makes it difficult to generalise or identify typical patterns. Much depends on the specific process of technological change that is being discussed. Science-based processes, such as atomic energy and genetic engineering, where solutions are developed in search of problems, follow rather different patterns than needs-based processes, where problems, be they environmental, health, social, or technical, generate efforts to come up with meaningful solutions. Similarly, we can think of activity-based processes, driven, for the most part, by those responsible for particular functional areas in our societies – transportation, communication, sanitation, defence. Here, appropriation is a process of selection, among both ideas and artefacts, and primarily consists of social innovations by which improvements are made to various infrastructures.

Within each of these typical areas, there are characteristic patterns and different stages, or phases of appropriation, as acceptance and familiarisation accompany diffusion and increased use. There are also significant geographical differences. Technologies are appropriated not just on a global or general level, but rather they are filtered into national traditions and languages, as well as into regional and locally distinctive organisational and institutional cultures. What is considered appropriate behaviour in one neighbourhood or community can often be ruled out in another. Energy and transportation use, to take two obvious and current examples, while similar in many respects, nonetheless differ from place to place, due to particular local contingencies but also due to different ways of life, and different patterns of culture. The effort to change one's behaviour into more ecological directions is shaped by one's station in life, as well as by a range of practical demands of existence and circumstance. A focus on processes of appropriation is thus a way to bring out the multifarious and multicultural character of technological change.

Conclusion

From a theoretical perspective, it can perhaps be useful to consider these different meanings of technology and these different processes of technological change in relation to what Raymond Williams once termed cultural formations (Williams 1977). For Williams, social and cultural change involve, at their core, the emergence of new 'structures of feeling', new sensibilities, new mixtures of ideas and practice, or what Williams termed 'social experiences in solution'.

According to this terminology, capitalism, for instance, could be considered an emergent cultural formation that developed in struggle against the dominant, or hegemonic religious culture of the medieval church, on the one hand, and the pre-Christian pagan cultures, on the other. As an emergent cultural formation, capitalism, and somewhat later, industrial society, established what we might term a particular mode of technological appropriation, including a discourse of instrumental rationality and science-based progress, an institutional structure of industrial research and development, and an integration of these ideas and practices into everyday life. In the 19th century, socialism emerged as a cultural challenge to the dominant capitalist formation, on the one hand, and the residual formations of rural life and Christian religion, on the other. But in the course of the 20th century, the socialist challenge was largely incorporated into the dominant cultural formation, even though, in many countries, certain ideas and practices did exert an influence on the dominant culture. In our day, environmentalism has developed as an emergent cultural formation, and like socialism in the 19th and 20th centuries, environmentalism – or what I like to call an ecological culture – faces both the pressures of incorporation from a dominant commercial culture, on the one hand, as well as the resistance of the residual groups of populists and neo-populists, both in Denmark and elsewhere, on the other (see (Jamison 2001)).

In this sense, we can think of the transnational corporate culture, with its reduction of technology to economic innovation, as the contemporary hegemony, the dominant cultural formation, or technological regime, that seeks to incorporate all technical developments into its greedy, accumulative grasp; and we can think of the 'anti-modern' forces of resistance to globalisation in its many forms as residual cultural formations, a technological regime which is trying to adapt technological development to older ways of life and belief systems. Where the one tends to adopt an attitude of technological determinism, seeing a kind of fundamental driving force for social change in technological innovation, the other seeks to impose its own values on the pace and direction of technological change. An ecological – and perhaps also a Pan-African – sensibility can then be considered part of an emergent, or emerging cultural formation, a new sort of regime that, as in the past, must struggle both against the dominant and the residual cultural formations in its efforts to affect meaningful technological change, but which also is neither economic nor cultural in its underlying meaning, but more synthetic, contextual, and pragmatic in its relation to technological change.

Table 4 Contemporary Technological Regimes

	Residual	Dominant	Emerging
Social Process	Appropriation	Innovation	Construction
Type of Agency	Local	Transnational	Hybrid/synthetic
Form of social action	Resistance	Commerce	Mediation
Type of knowledge	Traditional/factual	Scientific/professional	Situated/contextual
Tacit forms	Personal	Disciplinary	Experiences

Notes

[1] This chapter has been published in a revised form, with Mikael Hård as co-author, in Technology Analysis and Strategic Management, No. 1, 2003.

References

Archibugi, Daniele and Bengt-Åke Lundvall (eds.) (2001), *The Globalizing Learning Economy*, Oxford: Oxford University Press.

Baark, Erik and Andrew Jamison (eds.) (1986), *Technological Development in China, India and Japan*, London: Macmillan.

Beck, Ulrich (1992), *Risk Society*, London: SAGE Publications.

Bijker, Wiebe (1995), *Of Bicycles, Bakelites, and Bulbs*, Cambridge: MIT Press.

Bijker, Wiebe, Thomas Hughes, and Trevor Pinch (eds.) (1987), *The Social Construction of Technological Systems*, Cambridge: MIT Press.

Gibbons, Michael, Helga Nowotny, Camille Limoges, Martin Trow, Simon Schwartzman, and Peter Scott (1994), *The New Production of Knowledge*, London: SAGE Publications.

Hård, Mikael and Andrew Jamison (eds.) (1998), *The Intellectual Appropriation of Technology*, Cambridge: MIT Press.

Jamison, Andrew (2001), *The Making of Green Knowledge*, New York: Cambridge University Press

Latour, Bruno (1993), *We Have Never Been Modern*, Cambridge: Harvard University Press.

Mackenzie, Donald and Judy Wajcman (eds.) (1999), *The Social Shaping of Technology*, Buckingham: Open University Press.

Williams, Raymond (1977), *Marxism and Literature*, Oxford: Oxford University Press.

PART II

Adapting the Innovation System Concept to African

Development

5

African Systems of Innovation

Towards an Interpretation of the Development Experience

Samuel M. Wangwe

Introduction[1]

Africa is a latecomer in development relative to other regions of the world economy. Studies on the development experiences in latecomer countries reveal characteristics and patterns that are quite different in many ways from those experienced by the forerunners. Latecomers are supposed to have an advantage in that they can initiate their development process through utilising technological and institutional backlogs created by the forerunners. Even close followers of forerunners such as Germany are shown to have benefited from the existence of the forerunners (Keck 1993). In the 19th century Germany turned to Britain and even Belgium for new machinery and for skilled workers to bring advanced technology to its industries. In general the latecomers have had some advantages in the catching up process resulting from the very fact of their relative backwardness (Shin 1996). Latecomers are able to import and exploit technologies already developed elsewhere and can derive extra scale economies from leapfrogging in plant size (Shin 1996). It is notable, however, that within the group of latecomers in development the paths taken and the degree of catching up[2] with the advanced countries reveal a mixed picture. While some countries have capitalised on the advantages of late-comers and made promising progress towards catching up with the forerunners others have been less successful in taking advantage of their position as late-comers. The latter group has continued to be laggards and in certain respects the gap between them and the developed countries has widened.

Most of sub-Saharan Africa falls in the category of countries, which have generally not succeeded, in narrowing the gap between them and the developed countries. SSA has received considerable amounts of technical assistance yet this resource does not seem to have been utilised to build the capacity of local human resources. Embodied technologies have been imported through capital goods and intermediate goods from the more developed countries yet they have not taken full advantage of this potential advantage as latecomers. Limits seem to have occurred in such aspects as the low

75

capacity to import (inhibited by the little progress in export development) and in the role of aid and the related procurement policies (influencing patterns of technology imports) and various imbalances, which have resulted. What explains this relative failure? In the context of policy reforms which started in the 1980s what prospects exist for initiating and sustaining the catching-up process in SSA? These are some of the questions that need to be put on the development agenda for Africa.

If the process of technological learning has been such a central driving force in development in recent years then there is a case for exploring these processes with a view to applying appropriate insights to SSA as a latecomer in development. The national systems of innovation (NSI) is one recent approach, which puts technological processes at the centre of analysis of development. It is in this context that this chapter argues that the NSI approach can be useful in interpreting the African development experience.

This chapter will examine the approach of NSI in section two and then proceed to discuss the application of this approach to explaining development among selected late-comers in development in section three. Section four will address more specifically developments in SSA with special reference to the case of Tanzania. Section five concludes by identifying areas for further research.

The National Systems of Innovation Approach as a Framework of Analysis

The neoclassical school treats technology as basically exogenous to the development process. The Marxists and Keynesians analyse technology in relation to capital accumulation. The neo-Schumpeterians investigate more directly technology and technological progress. Considering that the third industrial revolution and successful cases of catching up have been based on technological learning the framework of analysis that addresses technological progress more directly seems to be worth exploring with a view to applying it to development in SSA. It is in this context that the national systems of innovation is preferred as a framework of analysis for this study. According to the NSI approach, the most fundamental resource in the modern economy is knowledge and therefore it follows that the most important process is learning. It is understood that the NSI approach has been developed and applied to date in analysing development in the advanced countries. There has been little application to developing countries and hardly any to SSA. This study proceeds on the premise that the NSI approach permits diversity in addressing different contexts. The challenge is to find ways of adapting the NSI framework to suit the development conditions in SSA.

The development of the NSI approach is a recent phenomenon and the approach is still being developed and refined. It is only in the 1970s that the IKE group at Aalborg University began to integrate a French structuralist approach to national production with the Anglo-Saxon approach to innovation in order to explain international competitiveness. This new combination has come to be reflected in the concept of the national systems of innovation (Lundvall 1992).

The NSI approach exhibits variations with different authors in terms of approach and tools of analysis. Freeman (1987) adopts a historical approach based on modern innovation theory focusing on the interaction between the production system and the process of innovation to explain the development process in Japan. This method of analysis combines organisation and innovation theory. Nelson (1982) focuses on production of knowledge and innovation and upon the innovation system in a rather narrow sense of organisations and institutions involved in searching and exploring such as R&D institutions. Later, however, Nelson (1993) broadened the scope of NSI by including not only the description of the allocation of R&D activities and its sources of funding but also characteristics of firms and important industries, roles of universities and government policies aimed at spurring and moulding industrial innovation of each NSI. He invoked tools of law and economics to examine how well different institutional set-ups address the public-private dilemma of information and technical innovation. Porter (1990) adopts tools of business management to address four determinants (firm strategy, factor conditions, demand conditions and support industries) of competitiveness. NSI is viewed as the environment in which single firms and industries operate. Lundvall (1992) focuses on learning and innovation, recognising that learning is predominantly an interactive process.

In spite of the different approaches to NSI, the core of these approaches is distinguished by the way they look at innovation system as a crucial subsystem of an economy or society. NSI approach to development is essentially evolutionary associated with three main features: dynamic and process views, uncertainty and learning. The dynamics and process views are reflected in the characterisation of interactions (especially the centrality of firms), networks and linkages among agents of innovation and between them and the environmental factors (internal and external). Hence it puts emphasis on learning and innovation processes and their institutional dimensions. NSI is a set of interrelated institutions the core being those which generate, diffuse and adapt new technological knowledge. These institutions may be firms, R&D institutes, universities or government agencies. Institutions mark boundaries, which have an influence on uncertainty. In addition institutions influence the intensity and direction of learning. Learning is the key dynamic mechanism for knowledge accumulation, innovation and growth. Innovation is central to the learning process.

Innovation is a cumulative process gradually making use of pre-existing possibilities and components according to the principle of path dependence. However, innovations may result in radical break from the past rendering obsolete a substantial part of accumulated knowledge. Schumpeter referred to this process as creative destruction. The structure of production has influence on learning through routine activities generated in production, distribution and consumption. These activities produce important inputs to the process of innovation transmitted through at least three channels:

- Learning by doing (Arrow 1962) increasing the efficiency of production operations.
- Learning by using (Rosenberg 1982) increasing the efficiency of use of complex systems.
- Learning by interaction (Lundvall 1988) involving users and producers in an interaction resulting in product innovations.

From the point of view of the firm, innovation includes all those processes by which firms master and practise product designs and manufacturing processes that are new to them, if not to the nation or even to the universe (Nelson and Rosenberg 1993). The national systems of innovation (NSI) approach defines the nation as the appropriate level of analysis. The level of analysis is *national* in the sense that focus is on the behaviour of actors not necessarily at the forefront of world's technology but on the factors influencing national technological capabilities. This choice of nation as the level of analysis recognises the importance of central state authority and national and cultural idiosyncrasies. National systems are postulated to differ in respect of the structure of the production system and the institutional set-up hence the national idiosyncrasies. These may include internal organisation of firms, inter-firm relationships, the role of the public sector, institutional arrangements in specific sectors such as the financial sectors and R&D activity.

Late-comers in Development: Distinguishing Characteristics

Cases of forging ahead are fewer than cases of catching up. The countries that are known to have forged ahead are more exception than the rule. Britain forged ahead from late 18th century to mid-19th century and the US forged ahead in the first half of the 20th century. The cases of catching up, however, are more common. The study of all cases of catching up reveals that the context in which latecomers developed is often different from that experienced by the forerunners. It is against the background of these facts that development experiences may not be repeated without undergoing adaptations to specific conditions. Latecomers develop with constant reference to forerunners and can learn from the experience of the forerunners in both complementary and competitive aspects. The key here is the interaction between them. This interactive process is important in understanding the process of development of the latecomers.

Latecomers in development have potentially several advantages at their disposal. First, latecomers are more likely to start on a clean slate with no deep commitments to any particular technology or approach. This means they are able to purchase the latest technologies and benefit from improved efficiencies. Second, latecomers can be driven by a clear strategy of catching up which can be an effective guide of their efforts. Third, at least in the initial stages, lower wages can be a source of competitiveness as they prepare to move up the ladder in technological learning.

Fourth, latecomers have the possibility of leveraging knowledge such as through subcontracting.

Latecomers in development have generally evolved as learners. They have had to develop by borrowing and improving technology already developed. They had to grow without the competitive asset of new products or processes (Amsden 1989; Hikino and Amsden 1994).

The most important factor explaining development among the latecomers seems to be technological change flowing from the accumulation of technological capabilities over time. Technological learning starts with simple or duplicative imitation and develops into more advanced forms of creative imitation and innovation driven by increasing investments in domestic technological capability building.

The experiences of development among the latecomers also demonstrate that in order to operationalise the identified driving forces an effective NSI is needed. Such an NSI should be able to develop an array of well-balanced public programmes that create a conducive environment for foreign technology to flow in (reducing the cost of learning) and encourage domestic firms to learn. The NSI should also be able to bring about productive interactions between the government and the private sector and between suppliers and buyers.

Developments in Africa and the Nature of Learning Processes

Africa's Development in Comparative Perspective

Africa is the least developed among all regions of the world. In the dynamic context, Africa has performed below other regions in terms of socioeconomic transformation. In the 1950s and 1960s the level of development of some countries in Africa was comparable to that of South Korea and Taiwan but by the 1990s the gap between them had widened considerably. Median growth in per capita income for the 1960-99 period was 0.7 per cent per annum for SSA compared to 2.4 per cent for other developing countries. Slow capital accumulation and stagnant or deteriorating total factor productivity have contributed to slow growth in Africa. This relative lack of dynamism in Africa begs for an interpretation of the development process that has taken place in Africa guided by a comparative perspective.

The development process in Africa was expected to be facilitated by advantages of a latecomer. Embodied technologies have been imported through capital goods and intermediate goods imports yet the utilisation of these imports for technological development has been limited. But limits seem to have occurred in the capacity to import (export development) and in the capacity to harness technology imports for purposes of technological development through learning. The role of aid in influencing technology imports and various imbalances which have resulted seem to have hindered rather than facilitated technological learning. SSA has received considerable amounts of technical assistance but there are indications that this resource has not been utilised to build local human resources. There are even indications that technical assistance has played the role of capacity erosion than capacity building. Skilled labour has been imported but also brain drain has occurred.

Overall, it can be argued that SSA countries have not succeeded to catch up. It is suggested that the NSI methodology can be useful in explaining the development process in Africa. The development challenge in SSA countries entails first to initiate the development process by raising the levels of investment and returns to such investments, sustaining this process and upgrading it over time. Rapid learning is necessary for initiating and sustaining catching-up process. The NSI approach is relevant here to the extent it focuses on institutions, incentives and policies which influence technological learning and innovation.

Pre-independence Developments: The Case of Tanzania

The colonial period 1885-1961 not only interrupted the development of local initiatives and accumulation processes, but also several indigenous domestic activities were discontinued in favour of the production activities which were introduced during the colonial period.

These activities had the potential of putting in place a dynamic accumulation process. Attempts to upgrade production processes into processing or manufacturing were effectively discouraged by the colonial government on the grounds that such activities would compete with industry in the metropole (Barker et al. 1986). Some of the industries which had been set up to meet high demands during World War II (e.g. paint, chemicals, bricks, fibre board, tent dye and camphor) were closed down after the war. The point emerging here is that the cumulative learning process in colonial Africa, as shown by the case of Tanzania, was deliberately interrupted or curtailed by the colonial administration. This is contrary to the experience of South Korea and Taiwan where Japan, the colonial power, is reported to have played a more supportive role.

Post-independence Development Model: Special Reference to Tanzania

Policy choices in the post-independence period were influenced by factors such as the rejection of colonial policy tools, legacy of the Great Depression, legacy of the Russian Revolution and economic thought favouring planning, government ownership, intervention and control.

The Great Depression exhibited pervasive market failures, a situation which led to considerable mistrust of markets. That experience undermined the core of traditional economic theory based on efficient allocation of resources and comparative advantages (Krueger 1992). At the time when most SSA countries were getting their independence in the 1950s and 1960s, their development designs were guided by the conventional wisdom of the time.

These experiences and ideas led SSA countries to adopt planning using input-output based models and other planning models, addressing the problems of a dual economy, focusing on the elasticity pessimism and the need to industrialise through state-led import substitution. The policies which were adopted largely discounted the value of traditional commodities, gave a large responsibility to government and adopted import substitution industrialisation with protection as a response to weak domestic activities, lop-sided economic structures and inability to compete.

Tanzania had a small industrial base at independence in 1961. In the early years of independence industrial development followed private sector led import substitution. The slow rate at which local and foreign investment was responding led Tanzania to change strategy and engage the state more actively in undertaking industrial investments. Tanzania followed this route since 1967 when a state led industrialisation programme was adopted in the framework of import substitution industrialisation. Nationalisation was adopted, like in many countries in Africa. Nationalisation alienated the former owners and other actors who possessed detailed knowledge of the history and developments in the factories. Nationalisation resulted in yet another phase of interrupted development and accumulation of technological learning. A situation of discontinuity was forced on the operations and technological learning. In this connection, it is observed that the experience of state enterprises is contrasted with that of the smaller private sector enterprises. For instance, unlike the case of parastatals, in firms owned by domestic and East African private capital the owners kept a close watch on choice of technology by participating directly or engaging experienced expatriates for technical help including in setting up the factories (Barker et al. 1986).

From the mid-1980s Tanzania has undergone major shifts in policy representing yet further dislocation and discarding of some useful capabilities which had been built during the post-independent state led development model. The resulting policy reforms can be identified with three main characteristics which are important in influencing the national system of innovation. First, economic reforms have been associated with efforts to reduce budget deficits and therefore a tight expenditure regime has emerged. In other words, the trend has been to shift from a soft budget constraint for public enterprises towards a hard budget constraint. Second, economic reforms have led to a shift from administrative controls and central planning approach to economic management to market orientation. Trade liberalisation and import liberalisation in particular has led to intensified competitive pressures on domestic production activities. Third, the role of the state in direct economic operations has been reduced. There has been a shift from state led to private sector led development. This has meant that the private sector which was discriminated against for one and a half decades had now to be promoted.

Role of Aid

One aspect of the macroeconomic environment in the last two decades has been that of pressure on the limited investment and foreign exchange resources. These became major factors in determining the pace and pattern of project implementation during this period. At enterprise level this pressure was manifested in project implementation delays and cost overruns (Wangwe 1993). During this period, however, resource allocation continued to favour expansions rather than efficient utilisation of the capacities which had been created already (Wangwe 1979). The bias of resource allocation in favour of capacity expansion rather than capacity utilisation was driven by particular forms of aid. Aid came predominantly in the form of project finance.

The foreign finances, available for development, were channelled into industrial expansion in the form of ill-designed projects and choice of technology that implied a high level of foreign exchange to sustain it. The particular form that this foreign finance took in Tanzania had far reaching implications on the pace and pattern of project implementation. The macroeconomic environment put so much pressure on foreign exchange and investment resources that the implementation of new investment projects largely depended on mobilisation of foreign finance. This took precedence over considerations of local technological capability (TC) building. The pressure to implement as many projects as possible overshadowed concerns about technological learning. The main public enterprise development institutions, NDC, usually left the choice of technology and other technology decisions for its projects entirely to the foreign partner (Barker et al. 1986). The foreign partner had considerable flexibility in the way management agreements were framed giving the foreign partner much latitude in terms of choice of technology and making technological decisions.

The incentive structure in place encouraged the top bureaucracy in the parastatals to put high priority on setting up as many enterprises as possible in the shortest time with very little attention on technological learning. James (1998) has invoked the public choice approach to bureaucracy to make two main arguments related to this observation. First, that bureaucrats have preferences for projects rather than technologies. Second, those managers of parastatals sought to initiate as many projects as possible mainly on the basis of foreign aid.

Concern has been expressed in recent years about the ineffectiveness of aid and the need to revisit aid delivery mechanisms, management and coordination. During the period of economic reforms, many countries made improvements in the soundness of their macroeconomic policies but at the same time undercut ownership of the policy agenda. Loss of ownership is one problem that has been recognised in the World Bank's Comprehensive Development Framework and the rationale behind the recently introduced Poverty Reduction Strategy Papers. This situation of loss of ownership did not augur well for S & T policy and associated activities. Aid-funded technical assistance has failed to build local human and technological capacities. In fact in some cases technical assistance has even played the role of replacing indigenous expertise and further reducing the capacity of government agencies to manage their own affairs effectively (Wuyts 1995). Indeed, the amount of technical assistance today, measured in the number of expatriate advisors funded by aid agencies, is larger than during the early decades of independence.

82

Training and Organisation of the Labour Process

The interaction between skills and technology has been central to the process of learning and gaining competitiveness in the latecomers which are catching up. In the post-independence period, especially during the period of dominance of state enterprises, it seems that Tanzanian firms have not attached due importance to training, especially formal and long-term training for the low cadre employees. A study by Wangwe, Semboja, and Kweka (1997) in the metal subsector indicated the need for upgrading of production skills for all levels of employment if they are to install new production technology, thereby indicating a low level of humanware.

Industry specific training institutes were established but even these were geared to cater for basic and low level skills oriented towards operation of existing plant rather than for adaptations, innovations and improvements on imported technology.

Despite the diversity of final products generated in manufacturing it has been found that most of the jobs undertaken require skills of a very low order usually associated with attending machines and doing repetitive partwork (Barker et al. 1986). The low level of skill development in Tanzanian industry is partly a function of the technology employed and partly a function of management of the labour process.

Top management in parastatals (state enterprises) was largely expatriates who outnumbered local counterparts by a factor of 2 to 3. Most of the Tanzanian managers were trained in fields other than engineering or science and technology. Most of them were trained in finance and administration. Bureaucratisation of industrial management was institutionalised in the parastatal sector whereby management lacked initiative towards design of new products, processes and maintenance systems.

A recent UNCTAD mission to Tanzania concluded that the human capital base in Tanzania is small: almost all indices of skill formation place it at or near the bottom of the world ranking (Lall 2000). The importance of raising formal skill levels is recognised worldwide and in particular the criticality of the interaction between skills and technology. Often training has to go together with the provision of new equipment, better layout, improved process know-how and more modern product technology. All these may need specific policies addressing their informational, financial and other needs (Lall 2000).

Policy Reforms and the Learning Process

RATIONALE

One implicit assumption of economic reforms and industrial restructuring is that enterprise level inefficiencies are a reflection of distorted or inappropriate macroeconomic policies. It is suggested that if appropriate adjustments could be put in place at macro-level, enterprises would receive the right signals through the market. This approach has been associated with the World Bank especially in its earlier publications (World Bank 1981, 1987). According to this approach, reform or restructuring of industry is essentially a macroeconomic issue along with restructuring of the supply side by putting in place appropriate macroeconomic and sectoral policies.

MACROECONOMIC REFORMS NECESSARY BUT NOT SUFFICIENT

Various industrial studies have revealed that restructuring the industrial sector entails much more than macroeconomic management. For instance, findings of various enterprise-level and sectoral studies in Tanzania in the 1980s lend support to this observation. Findings made by Tanzania Industrial Studies and Consulting Organisation (TISCO) of the techno-economic review of eight Tanzania Investment Bank (TIB) client companies in the mid-1980s revealed many operational problems at the sectoral and plant level.

RESTRUCTURING AND TECHNOLOGICAL CAPABILITIES

One major determinant of international competitiveness is investments in technology. This implies that technological learning and acquisition of technological capabilities cannot be ignored in the process of industrial deepening and the implied qualitative changes in the structure of production. During the import liberalisation phase there has even been a tendency to establish more and more of the finishing types of industries. Industrial deepening and technological change have not received any notable priority (Wangwe 1993).

A study of the early phase of industrial restructuring showed that technological capabilities remained low even in the post-reform period. Four indicators of technological learning were used in that study (Wangwe 1992): the degree of local participation in the identification and implementation of the rehabilitation programmes; the balance of the output and the learning objectives; the extent of upgrading technical and managerial skills through training; and implications for the domestic capability to manufacture spare parts and components. On all four counts, the rehabilitation programmes were found to have paid little attention to the question of raising the level of local technological capabilities.

There are few cases that indicate that adjustment or liberalisation could lead to dynamic upgrading and competitiveness in industry. The skill base, small to start with, shows little improvement in recent years, making it difficult for industrial enterprises to realise the value of new technology and skills, or to try and achieve the upgrading necessary. A few are doing this, and benefit in market performance (Deraniyagala and Semboja 1999), but they are tiny islands in a vast sea of technological backwardness (Lall 2000).

RESEARCH AND DEVELOPMENT

Tanzania has a relatively elaborate S&T infrastructure. There are currently eight universities producing some 2000 scientists and engineers annually. There are also about 62 R&D institutes spread throughout the country, including the above universities.

Tanzania established a S&T policymaking body, Tanzania Commission for S&T (COSTECH). Similar developments were occurring in other countries in Africa during

84

the late 1970s. The role of COSTECH among others is to formulate S&T policy and recommend its implementation by the government and other stakeholders. Funding of S&T is largely done by donors and government. Under-funding of S&T has been a common problem and has worsened during the reform period.

The link between the S&T system and production has been weak. The sources of this weakness is two fold. First, R&D institutes often developed prototypes which were technically feasible but were not necessarily economically viable. Second, the environment governing investment decisions of manufacturers was not sufficiently competitive to induce search for new avenues of investments in the developed prototypes.

The development of prototypes from the R&D system has not shown growth over time. In many cases retrogression has occurred after the introduction of Structural Adjustment Programme (SAP) measures (Chambua 1996). These developments or retrogressions have been associated with overall tightening of budgets during the SAP period. In particular, the short-term preoccupation of SAPs has not favoured allocation of resources to activities which do not yield immediate returns. R&D activities fall in that category. Similar findings were arrived at by Enos (1995).

In 1998 the government categorically advised public R&D institutions to restructure in line with the market oriented private sector led economy. According to the government's directives, these are supposed to be self-reliant (Ministry of Industry 1998). As a result, prototype development activities have declined giving way to service/consultancy/jobs to industry as one approach to income generation for survival. This tendency has been observed by many scholars (see for example, (Hewitt and Wield 1997); (Mshana, Chungu, and Muller 1994); (Muller 1994)). During this time of restructuring of R&D institutions, there has been loss of key staff and some assets which are no longer needed (because functions for which they are used had been discontinued) have had to be sold off. R&D institutions therefore ran down in terms of budgets, staffing and physical assets.

Government policy has not been oriented towards technology capacity building. In the phase of trade liberalisation industry has tended to lose interest in local R&D preferring to import proven technologies. This shift of interest in industry is not supportive of the efforts R&D institutions are making to generate some income by selling their research output to industry.

RESOURCE ALLOCATION TO S & T

The major cry of developing countries, as far as S&T is concerned, is budgetary allocation. It has been noted that most of the least developed countries, Sub-Saharan Africa inclusive, allocate far less than 1 per cent of their GNP which is inadequate (Chambua 1996; Lalkaka 1995; RAWOO 1994; Vitta 1992). Worse still, low as they are, staff salaries alone take up to 90 per cent of the budget of the research institutions in SSA, leaving little for research and development activities (Enos 1995; Vitta 1992).

Financial resources in the form of credit or venture capital from the banking system are not forthcoming either. The financial sector reform has taken longer than had been envisaged. Lending to proactive sectors has remained sub-optimal. Failures in credit markets causes concern in the reform process but the situation is even more serious for technology finance. Capital markets in Tanzania, like in most SSA countries, are either missing or ill suited to meet demands of venture capital for the risky and uncertain learning associated with new technologies. Risk analysis for locally developed prototypes is difficult for firms. It is even more difficult for lenders who often know less about particular markets and technologies.

S & T AND THE MARKET

In the literature, nine categories of research laboratories have been indicated ranging from high government influences and low market influences to high market influences and low government influences (cited in Vitta 1992). During the pre-reform period (1970 and early 1980s) over 90 per cent of all research laboratories in SSA fell in the category of high government influences but low market influences implying the dominance of public science (Vitta 1992). However, during this era of policy reforms under SAPs, government influences have been reduced to a very low level. . However, alternative sources of funding have not emerged as reliable replacements. Therefore the R&D institutes must have slipped in the category of low market and low government influences reducing them to, as Vitta (1992) puts it, academic 'scribbling'.

Two major developments in relation to the market should be identified here. First, the R&D institutes have to face the market because of their decision to engage in manufacturing as an income generating activity. Second, the market environment has changed considerably with competitive pressures increasing following trade liberalisation and in particular import liberalisation. Competition from imports has become a major challenge even for enterprises whose capabilities were geared to manufacturing from the beginning. Trade liberalisation has brought in competition too strongly, too fast for the R&D institutions to manage the shift towards the market.

Linkages and Networking

Development of technological capabilities depends on a wide variety of linkages between users and producers and between knowledge-producing institutions such as universities and sectors which produce goods and services (e.g. industry). In general, these linkages are poorly developed in Africa (Hewitt and Wield 1997; Mytelka 1993). In the case of the innovation process, most important linkages are a network of relations between the industry and R&D's. The most obvious indicator of these is a vast commercialisation and widespread diffusion of products and processes either conceived or improved by R&D institutes and manufactured by the industry. This being a very important aspect of the NSI, it is conspicuously absent in SSA (Hewitt and Wield 1997; Mshana 1994; Vitta 1992). This may have accounted, to a considerable extent, for the low level of technological development in Africa.

During their interviews in Tanzania in 1996 Hewitt and Wield (1997) found a general lack of jointly conceived projects in which both industry and academia collaborate right from the early stage. Some sub-contacting occurs between firms but the resulting relationship does not include collaboration in technology. The practice of high technology foreign firms buying from local firms is insignificant (Wangwe et al. 1997). The possibility of technology transfer and technology upgrading through networking with FDI firms is remote.

Concluding Remarks

NSI provides a conceptual framework for describing, interpreting and acting on innovations. It establishes a positive feedback loop whereby to strengthen NSI is equivalent to improving the climate and feasibility of innovations. The fundamental policy question for SSA countries such as Tanzania is how to get the NSI started and in some respects 'restarted' following the erosion experienced during policy reforms and the associated budget cuts on S&T infrastructures.

The science and technology infrastructure is still in bad shape. The infrastructure is small and largely ineffective. It is now poorly funded and motivated, and has weak or no links with industry. Its ability to develop, adapt and disseminate industrial technologies is weak. It seems to have little awareness of the competitive and changing needs of Tanzanian industry, even less of how new technologies can be introduced to potential users. There are some attempts to reform and improve the main institutions, but in the absence of government support, the culture of dependence on external aid and the lack of involvement by the productive system (now led by the private sector) coupled with the incomplete financial sector reform (crippling credit for investment) are important handicaps to the revival of the national system of innovation.

While macroeconomic policy reforms have been adopted in many SSA countries, there is little evidence to indicate that policies at the meso and micro level have been put in place to mediate and facilitate the restructuring process in the national systems of innovation and promote new forms of technological cooperation and collaboration which would complement reforms and restructuring processes in SSA. Government policy has not yet been oriented to facilitate the process of restructuring of the NSI.

The use of the NSI framework must take into consideration the peculiarities of SSA and adapt it to suit the specific conditions of development in Africa in the 21st century. The specific conditions that need to be considered are:

- The colonial administration in most of Africa interrupted the cumulative learning process. Interruptions of various kinds (such as nationalisations and later privatisations) continued into the post-independence period.
- The process of development in SSA is taking place under conditions which are substantially different from those in the past. Internally, the new context is characterised by developments in economic and political liberalisation.
- There are challenges of globalisation driven by rapid technological advances of which the information revolution is probably the most pervasive.

Globalisation and the intensification of competition are bringing to the agenda the centrality of developing the capacity to be competitive to a greater extent than ever before. Globalisation has important implications on the national systems of innovation in Africa. The rising importance of trade and commercialisation and the intensification of competition, the emerging supranational policy accords (e.g. WTO), the rapid advances in technology led by information and communication technologies all present opportunities and challenges. These need to be understood better so that appropriate policy response can be designed.

- The role of aid in development and the donor-recipient relationships which developed out of aid processes has had substantial influence on the development of many countries in SSA. The role of aid and aid relationships need to be incorporated in the NSI analysis in consideration of their important position in SSA. The special position of Africa in relation to aid and aid relationships needs to be understood better in the context of the national systems of innovation. Africa has a special characteristic of having aid playing a large role in development and in the national systems of innovation. The place of aid needs to be understood better with a view to enhancing the understanding of the forms of partnerships in development that are likely to revive the NSI in Africa. In this context, the special position of development and aid relationships with development partners may need to be revisited with a view to promoting ownership and developing national systems of innovation.

- Development concerns on poverty, disparities in incomes, even if at a lower average level, and consumption patterns and the consequent diversity of technologies used in various sectors (e.g. peasant agriculture compared to large industry in urban centres) and by various actors (e.g. state and private actors) is likely to be more pronounced in SSA than the history of development in the more advanced countries has shown or would suggest. To what extent the proposed poverty reduction strategy papers are incorporating the revival of the damaged national systems of innovation is an open question. Preliminary indications show that little is being done in the context of PRSPs to promote innovations.

- The particular role of the state in development and its pervasive involvement in direct production and its relationship to other actors in the economy has led to developments which distinguish SSA bringing them closer to the experience of late-comers in Latin America than those in East Asia.

Notes

[1] The bulk of the work on this chapter was done during my visit to UNU-INTECH in Maastricht in April-July 1999. I therefore thank the UNU-INTECH for sponsoring my visit to the Institute and giving access to the necessary facilities making it possible to review literature on this subject.

[2] Catching up is a form of convergence narrowing the gap between groups of countries especially between the laggard and the leading economy. This is often distinguished from homogenisation as a form of convergence within some specific group of countries.

References

Amsden, Alice H. (1989), *Asia's Next Giant: South Korea and Late Industrialization*, New York: Oxford University Press.

Arrow, Kenneth J. (1962), 'The Economic Implications of Learning by Doing', *Review of Economic Studies*, June.

Barker, Carol E., M. R. Bhagavan, P. V. Mitschke-Collande, and D.V. Wield (1986), *African Industrialization: Technology and Change in Tanzania*, Vermont: Gower Publishing Company.

Chambua, Samuel E. (1996), *Endogenous Technology Capacity and Capabilities Under Conditions of Economic Policies of Stabilization and Structural Adjustment: The Case of Technology Generating Institutions in Tanzania*, ATPS Working Paper, No. 10, Nairobi.

Deraniyagala, Sonali and Semboja, Haji. H. (1999), 'Trade Liberalization, Firm Performance and Technology Upgrading in Tanzania', in Sanjaya Lall (ed.), *The Technological Response to Import Liberalization in Sub-Saharan Africa*: 112-147, New York: Macmillan.

Enos, John L., (ed.) (1995), *In Pursuit of Science and Technology in Sub-Saharan Africa: The Impact of Structural Adjustment Programmes*, London: Routledge/UNU Press.

Freeman, Christopher (1987), *Technology Policy and Economic Performance*, London: Pinter Publishers.

Hewitt, Tom and David Wield, (1997), 'Networks in Tanzania Industrialization', *Science and Public Policy*, Vol. 24, No. 6, December.

Hikino, Takashi and Alice H. Amsden (1994), 'Staying Behind, Stumbling Back, Sneaking up, Soaring Ahead: Late Industrialization in Historical Perspective', in William J. Baumol, Richard R. Nelson, Edward N. Wolf (eds.), *Convergence of Productivity: Cross-National Studies and Historical Evidence*, Oxford: Oxford University Press.

James, Jeffrey (1998), *Public Choice, Technology and Industrialization in Tanzania*, Paper presented at the conference on The Industrial Performance in Tanzania. Eindhoven, June 25-27.

Keck, Otto (1993), 'The National System for Technical Innovation in Germany', in Richard R. Nelson (ed.), *National Systems of Innovation: A Comparative Analysis*, New York: Oxford University Press.

Krueger, Anne O. (1992), *Economic Policy Reform in Developing Countries: The Kuznets Memorial Lectures at the Economic Growth Centre*, Yale University, Oxford: Blackwell Publishing.

Lalkaka, Rustam (1995), *Technology Entrepreneurship: The New Force for Economic Growth in South-South Cooperation*, Geneva: UNDP.

Lall, Sanjaya (1999), *Competing with Labour: Skills and Competitiveness in Developing Countries*, Discussion Paper, No. 31, Geneva: International Labour Office.

Lall, Sanjaya (2000), *Transfer and Development of Technology: Tanzania*, Study for UNCTAD under the Joint Integrated Programme of Technical Assistance for Selected Developed and other African Countries (JITAP).

Lundvall, Bengt-Åke (1988), 'Innovation as an Interactive Process: From User-Producer Interaction to the National System of Innovation', in Giovanni Dosi, Richard R. Nelson, and Christopher Freeman (eds.), *Technical Change and Economic Theory*, London: Pinter Publishers.

Lundvall, Bengt-Åke (ed.) (1992), *National Systems of Innovation: Towards a Theory of Innovation and Interactive Learning*, London: Pinter publishers.

Ministry of Industry (1998), *Budget Speech*, URT, Ministry of Industry and Trade.

Mshana, J. S., A. S. Chungu, and Maige Muller (1994), *Commercialization of Industrial Technology in Tanzania: Some Experiences from IPI*, Paper presented at a workshop on Roles, Relevance and Capabilities of Industrial R&D: Optimizing of the 21st Century. Dar es Salaam.

Muller, Maige (1994), *Reorientation of the Institute of Production Innovation*, IPI, Dar es Salaam: University of Dar es Salaam.

Mytelka, Lynn K. (1993), 'Rethinking Development: A Role for Innovation Networking in the 'Other Two Thirds'', *Futures*, Vol. 25, No. 6: 694-712.

Nelson, Richard R. (ed.) (1993), *National Innovation Systems: A Comparative Analysis*, New York: Oxford University Press.

Nelson, Richard R. and Nathan Rosenberg (1993), 'Technical Innovation and National Systems', in Richard R. Nelson (ed.), *National Innovation Systems: a Comparative Analysis*, New York: Oxford University Press.

Odagiri, H Hiroyuki and Akira Goto (1993), 'The Japanese System of Innovation' in Richard R. Nelson (ed.), *National Innovation Systems: a Comparative Analysis*, New York: Oxford University Press.

Porter, Michael E. (1990), *The Competitive Advantage of Nations*, London: Macmillan.

RAWOO (1994), *Development and Strengthening of Research Capacity in Developing Countries*, Hague: RAWOO.

Rosenberg, Nathan (1982), *Inside the Black Box: Technology and Economics*, Cambridge: Cambridge University Press.

Shin, Jang-Sup (1996), *The Economics of Late-comers: Catching-up, Technology Transfer and Institutions in Germany*, Japan and South Korea, London: Routledge.

Stewart, Frances, Sanjaya Lall and Samuel M. Wangwe (ed.) (1992), *Alternative Development Strategies in African Development*, London: Macmillan.

Vitta, P. B. (1992), 'Utility Research in Sub-Saharan Africa Beyond the Leap of Faith', *Science and Public Policy*, Vol. 19 No.4.

Wangwe, Samuel M. (1992), 'Building Indigenous Technological Capacity: A Study of Selected Industries in Tanzania', in Frances Stewart, Sanjaya Lall and Samuel M. Wangwe (eds.), *Alternative Development Strategies in African Development*, London: Macmillan.

Wangwe, Samuel M. (1993), 'Implications of Changing External and Internal Conditions for Industrial Restructuring in Tanzania', in Mboya S. D. Bagachwa, and Ammon V. Y. Mbelle (eds.), *Economic Policy Under a Multiparty System in Tanzania*, Dar es Salaam: Dar es Salaam University Press.

Wangwe, Samuel M. (1979), *Capacity Utilization and Capacity Creation in Manufacturing in Tanzania with Special Reference to the Engineering Sector*, Unpublished Ph.D Thesis, Dar es Salaam: University of Dar es Salaam.

Wangwe, Samuel M., H. Semboja, and J. P. Kweka. (1997), *Multi-Country Comparative Study of Private Enterprise Development in Africa: Report on Tanzania*, Dar es Salaam: ESRF.

World Bank (1981), *Accelerated Development in Sub-Saharan Africa: An Agenda for Action*, Washington, D.C.: World Bank.

World Bank (1987), *Industrial Development in Tanzania: An Agenda for Industrial Recovery*, Washington, DC: World Bank.

Wuyts, Marc (1995), *Foreign Aid, Structural Adjustment and Public Management: The Mozambiquan Experience*, Institute of Strategic Studies Working Paper, General Series, No. 206, November.

6

Human Capital and Systems of Innovation in African

Development

Banji Oyelaran-Oyeyinka and Lue Anne Barclay

Introduction

This chapter explores the human capital root of the slow pace of development in Africa within the system of innovation framework. We propose that historically generated institutions and persistent patterns of human capital formation condition emergent systems of innovation, and effectively determine the development trajectory of African countries. As Rodrik (1998: 5) observed, 'the way to reverse the trend (poor growth) is not to target the region's trade volume per se, but to raise overall growth rate'. For the relatively well performing African countries, Rodrik like others identified human resources and institutions as important predictors of growth.[1] We therefore argue that a fruitful way of understanding the African condition is an exploration of these growth predictors within the evolutionary technological change tradition pioneered and elaborated on by researchers including Freeman (1987), Nelson and Winter (1982), Rosenberg (1986) and Dosi et al. (1988).

Several explanations have been advanced to explain Africa's dismal economic performance. They range from policy-related issues (e.g., World Bank 1981); structural and institutional factors (e.g., Easterly and Levine 1997, Sachs and Warner 1997); the paucity of technological and managerial capabilities which result in the failure to effectively transfer technology and the under utilization of human and physical resources (Enos 1992; Lall 1992, 1993); and the long-term effects of historical factors (Engerman and Sokoloff 2000). While these factors explain parts of Africa's growth problems, a systemic explanation of the nature of institutions in long-run development is still lacking. Africa is far from having uniform initial conditions and varies widely in economic and political governance systems. This chapter therefore calls attention to the role of institutions broadly, and specifically, systems of innovation supporting technological advance in long-term industrialisation. In doing this, we combine the strand of literature on institutions and their persistence in shaping development with the literature on evolutionary theory and systems of innovation. The role of initial conditions such as levels of literacy and natural endowment, the structure of industry, as well as resource endowment have been emphasised (Abramovitz and David 1994; Sandberg 1982).

Considerable work has been done on the role of human capital in economic growth, from classical writers (Denison 1985; Schultz 1961); to others who link technological progress to human capital (Lucas 1990; Romer 1990). According to Lucas (1990), poor technology flow to poor countries is a result of poor human capital endowment. A number of scholars have also examined other dimensions of the human capital particularly the educational rates of return for a host of countries (Cohn and Addison 1998; Mincer 1998; Psacharopoulos 1994). The two broad conclusions from this wealth of empirical studies are that: the presence of large stocks of human skills tends to boost economic growth; and investment in schooling is an important prerequisite for effective human capital. This kind of emphasis on the explicit link between human capital and economic growth is lacking in the system of innovation framework.

The unique contribution of this chapter is its emphasis on human capital and institutions in shaping the evolution of the systems of innovation in Africa. This first tentative attempt to explore long-run development in Africa within the systems of innovation framework therefore follows the line of inquiry suggested by (Lundvall et al. 2002). The chapter is organized as follows: The next section reviews the role of education and human capital in development, and the institutional origins of Africa's present systems of innovation. The third analyzes the formation of human capital over time in Africa, while the fourth section presents empirical tests of the link between human capital and elements of systems of innovation followed by a concluding section.

Education, Human Capital and Economic Development

Easterlin (1981) and Sandberg (1982) presented systematic analyses of the linkage between basic education, economic growth and industrialisation. Schooling, according to human capital theory, is an investment that directly enhances the productivity of workers (Wolff 2001). However, an educated workforce without the necessary prerequisites of investment, training, research and development (R&D) and, 'a receptive political structure and low population growth may not lead to growth'.[2] Wolff (2001: 736) ascribes the relatively weak performance of less developed countries to their 'failure to keep up with, absorb and utilize new technological and product information, and to benefit from international dissemination of technology'. In other words, underdeveloped areas perform poorly as a result of underdeveloped systems of innovation, which fail to absorb, diffuse and adapt by imitation, available process and product innovation.

At the heart of Easterlin's analysis is the impact of education on technological change and the institutional structures and incentives – the national system – that facilitate or constrain progress. There are a number of important factors identified by these authors. The first is the role of formal schooling. There seems to be a direct correlation between schooling of the appropriate content and a country's ability to master new technologies (e.g., Easterlin 1981). Empirical justification for this consists of the high literacy rates in Western Europe and North America from 1850, and the virtual absence of mass literacy in countries outside of these regions.

Secondly, the combined rate of technology and human capital, the latter transmitted through educational attainment, are seen to be ultimately connected. Third, the pre-existing supply of human capital as different mix of skills at the onset of the industrialization process is an important prerequisite for rapid growth.

This factor is vitally important for latecomer countries starting from a very low or non-existing base of technical skills and managerial capabilities. The efforts to accumulate both physical technological capitals simultaneously with human capital starting from basic education to industrial skills could be enormous, and may well prove daunting for poor countries. Further, low levels of human capital will tend to slow down the rate of income growth. This is so because in addition to contributing directly to skill formation, high literacy rates tend to be correlated with the growth of financial services and formal banking systems, all of which have important implications for industrialization (Gerschenkron 1962).

The relevant questions relating to the non-dynamic Systems of Innovation and limited human capital development in Africa are these: what explains the relatively difficult process of implanting S&T institutions in Africa? How do initial conditions pattern the growth of technology in building up national systems in Africa? In what ways does the pattern of educational development influence the evolution of industrialization and in doing so, contribute to the observed structure of the national systems in Africa?

Institutional Origins of Systems of Innovation

Institutions are conceptualized narrowly or broadly[3] but in both contexts they take on the functions of the management of uncertainty, the provision of information, the management of conflicts, and the promotion of trust among groups (Edquist et al. 1997; North, Summerhill, and Weingast 1998).[4] For these reasons, institutions are necessary for innovation for two reasons. First, is the uncertainty that characterize innovative activities. Institutions act to provide stability and to regulate the actions of agents, and to enforce contractual obligations. Second, learning and knowledge creation, validation, and distribution are prerequisites of modern economic change mediated by institutions as organization (R&D laboratories, finance and investment institutions) and as rules, such as intellectual property rights, patent laws and so on. In this study, we employ the broader concept of institutions in addition to locating institutions within a historical context, which admits the evolution of institutions themselves (David 1994; Zysman 1994). Coriat and Dosi (1998) called attention to another set of issues in understanding institutional evolution. First is the origin of the institutions, and the need to explain institutions that preceded them and the mechanisms that led to the transition. Secondly, is what they refer to as the degrees of intentionality of institutional constructions. In other words, whether institution arose out of a *self-organizational* process or derived form a collective *constitutional* process. Third and last is the concern for institutional efficiency. The point is whether institutions are merely 'carriers of history' in the sense of David (1994) and simply 'path-dependently reproducing themselves well beyond the time of their usefulness (if they ever had one)' (Coriat and Dosi 1998: 7). Clark (2000) gave examples of Africa's higher education institutions established at a time for a purpose far different to what the current objectives of Africa's development presently demand. The founding initial objectives persist while the developmental requirements have radically changed and this constrains organizational effectiveness.

Path Dependence in Africa's Human Capital Formation

One key factor behind the phenomenal economic success of latecomers such as the South East Asian economies was their emphasis on human capital formation and a dynamic system of innovation. These countries, employing a mix of selective and functional policies, developed an education structure that effectively provided the requisite skills for their industrialization initiative. The governments, to varying degrees, intervened in curriculum development to ensure that it was compatible with the needs of their evolving industrial policy. To this end, they *inter alia* encouraged private sector involvement in universities. Additionally, some countries, notably Singapore, imported expatriate skills where domestic capabilities were limited (e.g. (Lall 1992: 1994)).

By contrast, the current educational structure in Africa has been described as being 'unsuitable for industrialization' (quoted in Lall 1992: 119). Several reasons have been advanced to explain this. First, some researchers argue that the present education system in Africa is a legacy of colonialism (e.g. (Blakemore and Cooksey 1982)). It seems that the metropolitan powers, implemented a highly academic, subject-centred curriculum in Africa. This curriculum, with its focus on producing an academic elite, was largely irrelevant to Africa's development needs.

Only a privileged minority benefited from this elite education. In 1960, the gross primary enrolment in all of Sub-Saharan Africa was a mere 36 per cent. This was roughly half the levels found in Latin America (73 per cent) and Asia (67 per cent) (World Bank 1988). In an attempt at social control, access to education was deliberately limited, particularly secondary education, among the Africans. Academic education was conceived not as a means of industrializing the countries but rather for creating an elite supply of white-collar African workers for the administration of the colonies.

African governments sought to remedy this situation in the post independence era. As Diagram 1 demonstrates, performance has been impressive. Within two decades, the gross primary enrolment ratio[5] tripled from 39 per cent in 1960 to an astounding 81 per cent in 1980. This ratio subsequently declined by 8 per cent in 1999. However, gross secondary and tertiary enrolment steadily increased during the period reviewed.

Secondary enrolment ratios rose six-fold during the years, 1960 to 1995, while tertiary enrolment ratios increased seven-fold during the years, 1960 to 1990.

Diagram 1 Gross Enrolment Ratios in Africa, 1960-1995

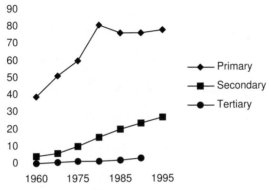

Source: World Development Indicators (CD-ROM).

Enrolment levels vary considerably at the three levels, with the largest variation found at the tertiary level and smallest at the primary level. Diagram 2, 3 and 4 show the mean, standard deviation and the coefficients of variation of the enrolment ratios at the primary, secondary and tertiary levels. The last variable, the coefficient of variation, is a relative measure of variation and can thus be expressed in percentages or ratios. The analysis of the data for primary education reveals that the standard deviation rose for the period 1960 to 1980, declined over the next ten years before rising again, while the coefficient of variation declined over the 1960 to 1990, showing a tendency of a rising mean enrolment. For both secondary and tertiary enrolment, standard deviation rose over the whole period, with the coefficient of variation for secondary enrolment first declining sharply in the first ten years, then rising and assuming a steady but slight decline from 1970 to 1990, again showing a steady rise in mean enrolment. Changes at the tertiary level were not as significant, however.

Diagram 2 Primary Education: Mean, Standard Deviation and Coefficient of Variation

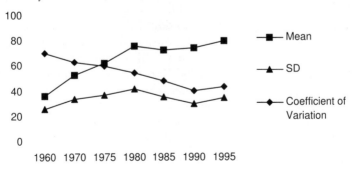

Diagram 3 Secondary Education: Mean, Standard Deviation and Coefficient of Variation

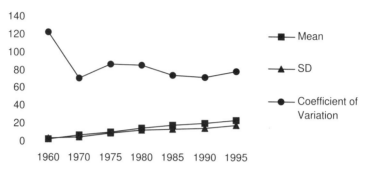

Diagram 4 Tertiary Education: Mean, Standard Deviation and Coefficient of Variation

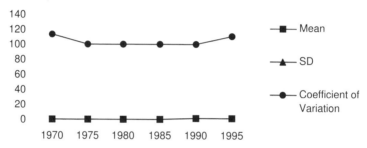

The average annual growth rates in school enrolment vary widely among different groups of African countries. The rate in primary education declined in the post 1990 period for only the middle-income, oil exporting countries such as Angola and Cameroon, and the middle-income, oil importers including Botswana and Senegal. However, all of these countries experienced sharp falls in the average annual growth rates of tertiary education. For example, in the low-income, semi arid countries such as Kenya and Burkina Faso, average annual growth rates in tertiary education were halved in the post 1990 period, from 6.5 per cent in 1985 to 1990 to a mere 3.5 per cent in 1990 to 1996.

This period also witnessed considerable fluctuations in public expenditure on education. This variable slightly declined during 1970 to 1975, recovered in 1980, but dropped dramatically to half of its 1970 value in 1996. Data on public education expenditure reveals wide variations among African countries. For example, low-income countries such as Benin, Ghana and Kenya, experienced the greatest decline in public expenditure on education: government expenditure on education as a proportion of total expenditure precipitously fell in two decades from 16 per cent in 1970 to 12 per cent in 1990. It was only in 1995 that government spending on education was restored to 1970 levels. Other groups of African countries also experienced fluctuations in public expenditure on education. Most experienced steep declines in spending in 1985 with some recovery in 1995.

Since education is generally publicly financed in Africa, it is subject to the availability of government revenues, many of which are heavily dependent on primary commodity exports. Nonetheless, the severe economic decline experienced in Africa in the 1980s adversely affected government spending on education.

However, while African governments recorded substantial progress in educational attainment, they made very little changes in the structure of the education system from what had existed during the colonial era. King (1991) suggests that African policy makers had very little influence over the development of their education systems. The education system, specifically at tertiary level, produces an inappropriate mix of skills. African institutions of higher education presently enrol 60 per cent of students in the arts and humanities, and 40 per cent in science and engineering. Enrolment in technical subjects presently lags behind that of other regions. While in 1995, only 0.04 per cent of persons as a percentage of the population were enrolled in technical subjects such as engineering and mathematics, the figure for four Asian Tigers was 1.34 per cent (Lall 2001). In a set of technical enrolment index constructed by Harbison Myers[6], while Norway ranked first with 73.52, South Africa, the most industrialized country in Sub-Saharan Africa (SSA) had a total of 23.61, Nigeria, 5.85 (less than 9 per cent of the Norwegian figure) with most SSA ranged from 1 to 5. However, this skill mix has remained unchanged for the past four decades despite the declining demand for arts and humanities graduates, and the rising and unfulfilled demand for science and engineering graduates (Dabalen et al. 2000; Fabayo 1996; World Bank 1988).

The situation is compounded by the quality of education offered in Africa, which is said to be well below world standards. Education standards are increasingly becoming poor with the gap in achievement between African students and those in industrialized countries 'widening to unbridgeable proportions' (Clarke 2000: 82). Over time, the student/teacher ratios at the primary and secondary schools have steadily increased in the post 1990 period especially for low-income, semi arid countries such as Gambia and Chad, as well as middle-income, oil importers such as Botswana and Zimbabwe. Expenditure on tertiary education fared no better: spending per student dropped from US$5,054 in 1980 to US$1,185 in 1990.

Empirical Analysis of Systems of Innovation and Human Capital

In this section, we carry out statistical tests of human capital and system of innovation using the variables identified in the study as being particularly strong predictors of national economic development. Figure 1 provides a simple diagrammatic representation for the test. While we employ enrolment rates, we are aware of their limitations in explaining growth of per capita income.[7] There is a generation lag between investment in education and economic growth, which we take account of in this exercise by using enrolment variables with a 20 and 30 year time lag.

Figure 1 Human Capital Variables in the System of Innovation

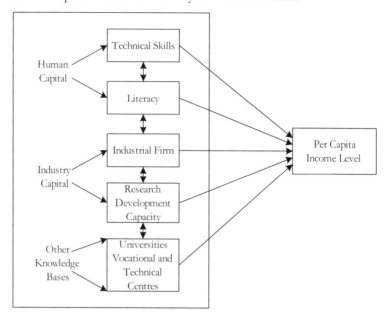

Statistical Analysis of the Variables

Table 1 presents the descriptive statistics. The equation represented by diagram 1 is hypothesized to be in the following form:

LOG of PER CAPITA INCOME (LOGPCAP) = f(PRI70+SEC70+TERT70 +AGL90+ILA90+TECHSUB+RD96).

For other years tested we simply substitute the dates.

To obtain an overall picture of industrial capability, we divided the SSA countries into two groups (low and medium) using a median value of per capita income. Significance levels show the considerable differences between the two groups except for tertiary level enrolment in 1970 which shows no difference between low and medium income countries. High levels of percentage of labour force working in agriculture tend to be associated with low per capita income, while MVA values show almost no association, implying little contribution to wealth contribution except for a few countries notably South Africa and Mauritius. Confirming our hypothesis, enrolment in primary education thirty years earlier is correlated with per capita income. Further tests are carried out employing Ordinary Least Squares (OLS).

We assume a log function of per capita income to eliminate non-linearity as a result of the wide variation in per capita income. All variables were included in the first model, and two additional models were tested. The correlations presented in Table 2 are the results of a pooled regression of the education enrolments. There is high correlation among the variables. PRI70, ILA90 display significant positive correlation with the dependent variable at 1 per cent and 5 per cent levels of significance. SEC70 and TERT70 display negative but non-significant correlation while AGL90 shows negative and significant correlation.

Table 1 Descriptive Statistics

	Mean	Standard Deviation
LOGPCAP	6.3970	.9851
PRI70	57.2818	32.0868
SEC70	7.5182	4.6089
TERT70	.4364	.2976
AGL90	65.4545	23.0927
ILA90	11.0000	8.6487
TECHSUB	5.373E-02	4.675E-02
RD96	.2622	.4406

Note: PRI70: Primary Education gross enrolment ratio in 1970. SEC70: Secondary Education gross enrolment ratio in 1970. TERT70: Tertiary Education gross enrolment ratio in 1970. AGL90: Percentage of labour force working in agriculture. ILA90: Percentage of labour force working in industry. TECHSUB: Technical subjects enrolments as a percentage of population in 1996. RD96: Total Research and Development (R&D) personnel per million in 1996.

Correlation of the Three Levels of Enrolment

Correlation coefficients indicate tendencies rather than causation. In this exercise we examine the relationship of the three levels of enrolments. Additional enrolment variables eliminated by the model were used in the correlation. Correlation coefficients are as high as 0.873 between SEC85 and TER90, and also between PRI60 and SEC75, SEC85, and also between the like variables. One way of interpreting this might be that the higher the primary enrolment, the greater the demand for and provision of secondary education and as the latter increases, the pressure on tertiary enrolment increases. The reverse tendency may be said to hold, that is, the lower the primary enrolment rates, the less the tendency for secondary education provision and less so the pressure for university education. This findings implies that: *a low initial schools enrolment, and for that reason, a low level of initial human capital in African countries will tend to perpetuate a condition of illiteracy and beyond that, a level of human capital that is unable to sustain rapid accumulation of technological capability.* The persistence of initial human capital conditions may well be partially proved by this correlation, conditions which we suggest are likely to have widespread impact throughout the system of innovation, thereby leading to the non-dynamic SI that is pervasive throughout the region.

OLS Regression of Human Capital and Per Capita Income

Due to high correlation among some of the variables and the incidence of collinearity, we tested several models. Further, we separated the variables and used pooled regression for the enrolment data since the data is both time series and cross sectional. Table 2 shows the regression outputs.

Table 2 Pooled Regression of the Three Levels of Enrolment

Source	SS	Df	MS			
I.1 Tertiary						
Model	8.797858	1	8.79785787	Number of obs. = 61		
Residual	48.031390	59	.81409135	F (1,59) = 10.81 Prob>F = 0.0017		
Total	56.829248	60	.94715413	R-squared = 0.1548		
				Adj R-squared = 0.1405		
				Root MSE = .90227		
Logcap	Coefficient	Std. Err.	t	P>/t/	95% confidence Interval	
Tertiary	.198026	.060238	3.287	0.002	0.774901	0.318562
Constant	5.512774	.174694	31.557	0.000	5.163212	5.862335
I.2 Primary						
Model	18.335997	1	18.3359971	Number of obs. = 72		
Residual	40.615244	70	.5802178	F (1,70) = 31.60 Prob>F = 0.0000		
Total	58.951241	71	.830299169	R-squared = 0.3110		
				Adj R-Squared = 0.3012		
				Root MSE = .76172		
Logcap	Coefficient	Std. Err.	t	P>/t/	95% Confidence interval	
Primary	.004507	.000802	5.622	0.000	.002908	.006105
constant	5.109552	.174570	29.269	0.000	4.761383	5.457721
I.3 Secondary						
Model	12.881252	1	12.8812523	Number of obs. = 72		
residual	46.069989	70	.6581427	F (1,70) = 19.57 Prob>F = 0.0000		
Total	58.951241	71	.8302992	R-squared = 0.2185		
				Adj R-squared = 0.2073		
				Root SME = .81126		
Logcap	Coefficient	Std. Err.	t	P>/t/	95% Confidence Interval	
Secondary	.15553	.003516	38.938	0.000	.008541	0.225645
Constant	5.49240	.141055	4.424	0.000	5.211075	5.773724

Parameters of the models assume that enrolments in primary, secondary and tertiary education in 1960, 1970, and 1975 will be positively associated with per capita income in 1996 and that the three enrolment levels in 1960, 1970, 1975, and 1980 will influence per capita income in 2000 The variables are highly correlated and cannot be used in a single model as explanatory variables.

This finding is consistent with (Mironov 1990) and (Sandberg 1982) that economic development is significantly correlated with society's human capital in a time lag of 20 to 30 years. Mironov suggests a periodicity of 20 years when there are no wars and a longer period when society endures a war, leading to delayed development of about a decade. We find that a time lag of 25-35 years exists between the initial investment in primary and secondary education. Tertiary level enrolment is also significant. They all appear in the pooled regression models.

OLS regression without the enrolment variable shows that TECHSUB is a significant predictor of development, but the total R&D personnel was not significant. This means that development is positively associated with the growth of technical personnel while personnel in R&D make no contribution to income growth. This is intuitively correct as very little local R&D is carried out in Africa. However, technical personnel may be fully engaged in production, and maintenance functions but not with R&D. Firms within the SI in Africa are engaged in imitative product innovation that requires marginal investment in formal research other than quality assurance. On industrial skills, we equally found statistical significance with labour force in industry but non-significance with agriculture labour force. This variable is indicative of a country's level of development. However, one may not read too much into this finding as the variable might have been subsumed by the other skills factors such as TECHSUB but it may also find consistency with the continuing poor contribution of agriculture to wealth creation in African economies. Read in conjunction with the contribution of technical personnel, our interpretation is that the economies of SSA do benefit from a significant local but relatively low-level technological regime. Conversely, it may be argued that the lack of significance found for research and development personnel might be reflective of the immaturity of the region's industrialization initiatives, which cannot yet fully utilize the skills of tertiary graduates in research. This is symptomatic of the dissonance between the education system and the stage of industrialization in the region.

Conclusion

We proposed that the slow development of African economies is explained in part by poor human capital endowment. First, we hypothesize that poor human capital formation could explain the lack of dynamism of the region's systems of innovation (SI), institutions that underlie the adoption, diffusion and adaptation of innovation. Secondly, institutions possess path-dependent characteristics influencing the growth rate of per capita incomes, our proxy for development. Third, path-dependent variables, codified loosely into the concept of systems of innovation, have institutional origins that have persistently impacted the evolution of African development. Fourth, we follow the notion that innovation is fundamentally shaped by social, historical, economic and political processes, outside the narrow domain of the firm, and the R&D system.

For these reasons, the nature of the state and its institutions determine whether dynamic or non-dynamic learning systems of innovation emerge. We suggest that the colonial origin and pattern of schools enrolment gave form to the current low technological base of African industry.

We did not attempt to test all variables pertinent to the SI due to poor systematic data over time. We have also not assumed undue causation as correlation simply implies tendencies. While we tested several models in arriving at these findings, we do not discount the possibility of finding other outcomes, particularly at lower levels of data disaggregation. We confine ourselves to one very important set of variables, the school enrolment at the three levels over the last 35 years (1960 to 2000), as well as technical enrolment, labour force in agriculture and industry, and R&D as per cent of population. We find statistical significance of the schooling variables, a very significant correlation of the three levels of schooling, suggesting a persistence of the initial enrolment and as such, its impacts on the national system thenceforth. We also confirm the correlation of enrolment with per capita income (1960, 1970 and 1995 respectively) with a periodicity of some 25 to 35 years, consistent with the findings in the literature with regions starting from low levels of technological development. This means that investment in education impact national wealth, after a generation of some two decades or more. Path-dependence and persistence of initial condition may well be implied, as failure to invest in basic primary and secondary education at the minimum would foreclose the development of modern industry exemplified in the system of innovation. The specific form for individual countries is not indicated and our data relates largely to the aggregate of SSA.

However, considerable work remains in understanding the systemic origin of Africa's non-dynamic innovation systems and research may take several forms. First, we need to understand the key elements of the system of innovation that are the most influential and how much they contribute to building the SI. Secondly, we should understand the nature of interactions, not only within the narrow domain specified for firms and industry, but at a wider socio-economic level. For instance, how do we intensify interactions of economic actors and make them more effective? Third, research should explore the specific ways in which the institutional origins of the SI influence development. The SI approach suggests that the skills and knowledge bases of seemingly unrelated components can be fruitfully brought together to promote development. Capacities outside the productive firm for instance may well be as crucial for firm growth as the capacities within. *As institutions and policies demonstrate persistent characteristics, African policy makers need to take a long-term view. Getting the institutions right is certainly more crucial than getting the prices right.*

Notes

[1] See Easterly and Levine (1995, 1997) and Sachs and Warner (1997) among others. However, none of these studies work within the system of innovation framework and none situate their analysis within the evolutionary technological change theory where emphasis is placed on institutions supporting innovative activities.

[2] Studies by Griliches (1970), cited in Wolff (2001) estimated that increased school attainment contributed one-third of the Solow residual. Denison (1979) also estimated that 20 per cent of the growth in US national income per person between 1948 and 1973 is attributable to the educational levels of the labour force.

3 In a narrow sense, institutions correspond to organisations such as universities, technological service organisations, while in broad terms, it includes political context governed by constitutions and the rules regulating innovation activities.

4 Coriat and Dosi (1998) refer to the broad meaning of institutions as having three components which are: (a) formal organizations (ranging from firms to technical societies, trade unions, universities, and state agencies); (b) patterns of behaviours that are collectively shared (from routines to social conventions to ethical codes); and (c) negative norms and constraints (from moral prescriptions to formal laws).

5 Enrolment ratio is defined as the ratio of the number of persons enrolled in school to the population of the corresponding age group by educational level.

6 See (Lall 2001). Technical enrolment index is tertiary enrolment (times 1000) plus tertiary enrolment in technical subjects (times 5000), both as percentage of population.

7 Formal schooling is a superior measure of human capital stock compared to literacy as Sandberg observes due to the additional advantage of numeracy training in formal schooling, which may be missing from education outside the school system. In addition to Sandberg's evidence, we have other studies that found a good correlation between schooling and per capita income levels, for example, see Mironov (1990) and Nunez (1990). While Sandberg (1982) conclusively establish strong correlation for literacy-on-income causation in a period of 120 years, the other studies show the strongest association in one generation, about 20-30 years.

References

Abramovitz, Moses A. and Paul A. David (1994), *Convergence and Deferred Catch-up: Productivity Leadership and the Warning of American Exceptionalism*, CEPR Publication, No. 41, Stanford: Stanford University.

Blakemore, Kenneth and Brian Cooksey (1982), *A Sociology of Education for Africa*, London: George Allen and Unwin.

Coriat, Benjamin and Giovanni Dosi (1998), 'The Institutional Embeddedness of Economic Change: an Appraisal of 'Evolutionary' and 'Regulationist' Research Programmes', in Klaus Nielsen and Björn Johnson (eds.), *Institutions and Economic Change: New Perspectives on Markets, Firms and Technology*: Cheltenham: Edward Elgar Publishing: 3-31

Clark, Norman (2000), 'Public Policy and Technological Change in Africa: Aspects of Institutions and Management Capacity', *Journal of Economic Studies*, Vol. 27, No. 12: 75-93.

Dabalen, Andrew, Bankole Oni and Olatunde A. Adekola (2000), *Labour Market Prospects of University Graduates in Nigeria*, unpublished World Bank Report, November.

David, Paul A. (1994), 'Why are Institutions The 'Carriers of History'?: Path Dependence and The Evolution of Conventions, Organisations and Institutions', *Structural Change and Economic Dynamics*, Vol. 5, No. 2: 205-220.

Denison, Edward F. (1979), *Accounting Slower Economic Growth: the United States in the 1970s*, Washington, D.C.: Brookings Institution Press.

Dosi, Giovanni, Christopher Freeman, Richard R. Nelson, Gerald Silverberg, and Luc Soete (1988), *Technical Change and Economic Theory*, London: Columbia University Press.

Easterlin, Richard A. (1981), 'Why isn't the Whole World Developed?', *The Journal of Economic History*, Vol. 41, No. 1: 1-19.

Easterly, William and Ross Levine (1995), *Africa's Growth Tragedy: A Retrospective, 1960-89*, World Bank Policy Research Working Paper, No. 1503.

Easterly, William and Ross Levine (1997), 'Africa's Growth Tragedy: Policies and Ethnic Divisions', *Quarterly Journal of Economies*, Vol. 112, No. 4: 1203-1205.

Edquist, Charles (ed.) (1997), *Systems of Innovation, Technology, Institutions and Organizations*, London: Pinter Publishers.

Engerman, Stanley L. and Kenneth L. Sokoloff (2000), 'Institutions, Factor Endowments, and Paths of Development in the New World', *Journal of Economic Perspectives*, Vol. 3: 217-232.

Enos, John (1992), *The Creation of Technological Capabilities in Developing Countries*, London: Pinter Publishers.

Fabayo, J. A. (1996), 'Technological Dependence in Africa: Its Nature, Causes, Consequences and Policy Derivatives', *Technovation*, Vol. 16, No. 7: 357-370.

Freeman, Christopher (1987), *Technology Policy and Economic Performance: Lessons from Japan*, London: Pinter Publishers.

Gerschenkron, Alexander (1962), *Economic Backwardness in Historical Perspective*, Cambridge: Harvard University Press.

King, Kenneth (1991), 'Education & Training in Africa: The Search to Control the Agenda for their Development', in Douglas Rimmer (ed.), *Africa 30 Years On*, London: The Royal African Society, London: James Currey and Portsmouth, New Hampshire: Heinemann: 73-90

Lall, Sanjaya (1992), 'Structural Problems of African Industry', in Stewart Frances, Sanjaya Lall, and Samuel Wangwe (eds.), *Alternative Development Strategies in Subsharan Africa*, London: Macmillan: 103-144.

Lall, Sanjaya (1993), 'Trade Policies for Development: A Policy Prescription for Africa', *Development Policy Review*, Vol. 11, No. 1: 7-65.

Lall, Sanjaya (2001), *Competitiveness, Technology and Skills*, Cheltenham, UK: Edward Elgar Publishing.

Lucas, Robert E. (1990), 'Why Doesn't Capital Flow From Rich to Poor Countries?', *American Economic Review*, Vol. 80: 92-96

Lundvall, Bengt-Åke, Björn Johnson, Esben S. Andersen, and Bent Dalum (2002), 'National Systems of Production, Innovation and Competence Building', *Research Policy*, Vol. 31: 213-231.

Mincer, Jacob (1988), *Job Training, Wage Growth and Labour Turnover*, NBER Working Paper, No. 2690.

Mironov, Boris N. (1990), 'The Effect of Education on Economic Growth: The Russian Variant, Nineteenth-Twentieth Centuries', in Gabriel Tortella (ed.), *Education and Economic Development Since the Industrial Revolution*, Valencia: Generalitat Valenciana: 113-122.

Nelson, Richard R. and Sidney G. Winter (1982), *An Evolutionary Theory of Economic Change*, Cambridge: Harvard University Press.

North, Douglas C., William Summerhill, and Barry R. Weingast (1998) *Order, Disorder and Economic Change: Latin America vs. North America*, Unpublished manuscript, Hoover Institution, Stanford University.

Nunez, Clara E. (1990), 'Literacy and Economic Growth in Spain, 1960-1977', in Gabriel Tortella (ed.), *Education and Economic Development since the Industrial Revolution*, Valencia: Generalitat Valenciana: 152-178.

Psacharopoulos, George (1994), 'Return to Investment: a Global Update', *World Development*, Vol. 22, 1325-1343.

Rodrik, Dani (1998), *Trade Policy and Economic Performance in Sub-Saharan Africa*, NBER Working Paper, No. 6562.

Romer, Paul M. (1990), Endogenous Technical Change, *Journal of Political Economy*, Vol. 98: 571-593.

Rosenberg, Nathan (1986), *Perspectives on Technology*, Cambridge: Cambridge University Press.

Sachs, Jeffrey and Andrew Warner (1997), 'Sources of Slow Growth in African Economies', *Journal of African Economies*, Vol. 6, No. 3: 333-376.

Sandberg, Lars G. (1982), 'Ignorance, Poverty and Economic Backwardness in the Early Stages of European Industrialization: Variations on Alexander Gerschenkron's Grand Theme', *Journal of European Economic History*, Vol. 11: 675-697.

Schultz, Theodore W. (1961), Investment in Human Capital, *American Economic Review*, Vol. 51, 1-17

Wolff, Edward N. (2001), 'The Role of Education in the Postwar Productivity Convergence Among OECD Countries', *Industrial and Corporate Change,* Vol. 10, No. 3: 735-759.

World Bank (1981), *Accelerated Development in Sub-Saharan Africa: An Agenda for Action*, Washington, D.C.: World Bank.

World Bank (1988), *Education in Sub-Saharan Africa: Policies for Adjustment, Revitalisation and Expansion,* Washington, D.C.: International Bank for Reconstruction and Development.

Zysman, John (1994), 'How Institutions Create Historically Rooted Trajectories of Growth', in Charles Edquist and Maureen McKelvey (eds.), *Systems of Innovation: Growth, Competitiveness and Employment, Volume II,* Aldershot, UK: Edward Elgar Publishing: 243-263.

7

Impact of Social Ties on Innovation and Learning in the

African Context

John Kuada

Introduction

During the last three decades, a number of researchers adopting National Systems of Innovation (NSI) as an analytical framework have presented persuasive policy guidelines for the development of enabling contexts for innovative actions of firms and institutions in different countries. Their aim has been to guide the development of national innovation policies that can help create, store, transfer and actively use knowledge, skills and artifacts within an economy. Most of them see innovation as a process of technological and organisational change. In their view, innovation processes are not triggered exclusively by endogenous initiatives. They depend, in part, on international knowledge inflows through inter-firm relations and development assistance as well as other forms of network effects. Despite an increasing volume of publications on the subject, there has been a dearth of studies focusing attention on innovative processes in the developing or emerging market economies. The few researchers with developing country focus have addressed the subject from the perspective of the need for developing countries to 'catch up' with the developed world.[1]

Parallel to the NSI studies, another stream of research by 'new institutional' economists and economic sociologists has shown that there exists a wide range of recipes of successful economic development. These successful national recipes are usually unique, emerging out of a combination of planned and fortuitous circumstances, and therefore non-transferable. Thus, successful innovation in the developing countries does not necessarily entail the adaptation of systems that have proved successful in the developed countries. A recent contribution to this perspective is Whitley's (1992, 1994) business systems model. The main thrust of Whitley's studies is to reveal the plurality of successful business systems in the world today and to gain insight into the foundations of the various distinctive systems. He argues that each system is a product of the societal norms, values and practices that have evolved throughout the history of a given society and are, as such, unique.

This chapter builds on the insights gained from these previous studies and intends to make two contributions to the literature on innovation systems in developing countries. First it argues that the contemporary literature has paid limited attention to the human side of innovation, discussing innovation mainly from a systems perspective. The chapter therefore discusses the resources that emerge through social ties and trust building mechanism and how they impact on knowledge creation, acquisition and usage (and thereby innovation) within work organisations. The chapter then proposes an integration of inter-personal relations into models for analysing innovation systems. Second, and building on the discussions proposed in the first point, the chapter argues that the cultural context within which innovation takes place is critical in understanding the nature and tempo of innovation and the extent of its diffusion within a national economic system. With these discussions as a background, the chapter explores the processes of learning and innovation within African societies and presents a set of propositions for future research.

The chapter is organised as follows: After this introduction the second section discusses the concepts of knowledge and learning as they pertain to the process of innovation. The third section discusses the concepts of social ties and social capital, again explaining their implications for national innovation processes. The discussions are then combined to form a foundation for the analytical framework proposed in the chapter. This framework is then used in section four to explore the processes of innovation in African countries and to suggest propositions that can guide future empirical investigations on the continent.

Knowledge and Learning

Knowledge and learning are foundational concepts in NSI research. The general understanding is that successful national innovations depend on the intensity of institutional learning. A quick review of the concept and approaches to learning is therefore undertaken in this section of the chapter to serve as a platform for exploring the impact of learning on innovation.

The learning literature informs that learning may occur when people give new meaning to (i.e. re-interpret) the information they receive. The more varied the interpretations given to the information the greater the probability that they will result in organisational innovation. Re-interpretation entails the process of unlearning. Hedberg (1981) defines unlearning as a process through which learners discard obsolete and misleading knowledge, replacing them with new knowledge. Some writers therefore see unlearning as the first step in a learning process. Unlearning opens the way for new learning to take place. But in terms of time, both unlearning and learning may take place concurrently. Unlearning is not an act of forgetting; it rather means acceptance of the failure of existing ways of doing things.

The concept of unlearning can be further explained with reference to Argyris and Schön's (1974) distinction between *single loop learning* and *double loop learning*. Single loop learning refers to the conscious, intentional and rational activity undertaken by people in organisations, institutions and societies to detect and correct discrepancies that appear between their expectations and the results produced by their actions. Correcting errors provides a learning experience that ensures that similar problems can be effectively addressed in the future. But if a person who corrects the error were to ask himself what in this organisation, institution or society has prevented the members from questioning practices that have resulted in the errors in the first place, he will be setting off a process leading to double loop learning (Argyris 1994). Double loop learning therefore allows an organisation to change its current mindset and the direction of its actions and fortunes. This forms the foundation of discontinuous innovation.

An interesting question discussed in organisational learning theory is whether organisations learn at all or whether organisational knowledge is merely a summation of the knowledge of the employees of the organisation. The literature generally endorses the view that an organisation learns just as much as its individual employees. Although an organisation does not have a brain, it has memory. Organisational memory may be defined as stored information from the organisation's history that can be brought to bear on present decisions. Organisational memories may not store exact replications of events but just strong impressions from the events. As such they are shared interpretations of the events and may therefore be regarded as socially constructed.

An organisational 'memory' allows for the creation of an organisational mind-set or a shared frame of reference built on past events, experiences and stories. This constitutes standard operating procedures that guide employees' course of action in the future. It therefore influences the capabilities that an organisation exhibits in the sense that it defines the nature of tasks that employees are capable of performing in a reasonably coherent fashion.

The concept of organisational memory can be conveniently applied in the study of learning processes in public institutions and societies as well. All human institutions develop mechanisms for storing information from events that its members have experienced. These experiences are usually stored in the form of norms and values, which are shared with new members of the institutions and societies through stories and sagas. The knowledge flow resulting from these sources is critical to the members' capacity to learn and innovate.

The organisational and institutional memories may also produce rigidities in the behaviour of its members. It has been argued that people often remain prisoners of their conceptual frameworks (Hedberg 1981). That is, they exhibit a general reluctance to leave old ways of thinking for new ones. In effect most people tend to believe that successful recipes of the past will continue to be successfully applied in new situations.

The rigidities in people's behaviour may also take the form of what Argyris (1994) called organisational defensive routines. The concept 'defensive routines' covers those types of behaviour that aim at ensuring that individuals minimise their vulnerability, risk of embarrassment and appearance of incompetence in the company of others. Defensive behaviour may also be motivated by the anxieties produced by prospects of change or by communication bottlenecks, rigid structures as well as persistence of some myths within the institutions and organisations. As we shall see subsequently, the cultural values and rules of accepted behaviour in African organisations tend to reinforce such defensive behaviours and thereby constrain innovation in these organisations.

Knowledge constitutes another important concept that has a significant bearing on innovation. Knowledge theorists usually draw a distinction between tacit and explicit knowledge. Tacit knowledge is that kind of skills and competencies that cannot be easily articulated. Explicit knowledge is the reverse. Winter (1996) suggests that there are differences in degrees of 'tacitness' of knowledge; some types of knowledge are less easy to codify or abstract.

The general understanding is that tacit knowledge has two sources. The first source is the skills and capabilities that individuals apply in their work. These are built up continuously through personal experiences. The second source is the routines that people develop over time on the basis of the shared meanings that have evolved through their interactions in specific contexts such as at the work place. In the case of manufacturing companies, for example, routines, may shape engineers' work process and assumptions that are encoded in a piece of machine they design. Some of the assumptions and routines can be decoded and made explicit in the form of manuals and handbooks. But these codified procedures still assume certain work habits found in the cultures within which the technology has been designed and used. Thus, from a learning perspective, effective transfer of technology requires a 'de-contextualisation' of the technology's inherent tacit knowledge and its subsequent 're-contextualisation' in the technology recipient organisation. Said differently, the knowledge contained in the technology must be transformed from its original context-specific form and fitted into the new context for it to be effectively applied. Herein lie the difficulties in attempts to trigger technical innovation process in developing country firms through the importation of technology from developed countries.

'Re-contextualisation' of imported technology can, however, occur if the employees of the recipient organisation are able to develop their own tacit knowledge in technology transfer process through reflexive experiences gained from the use of the piece of technology. The sharing of the reflexive knowledge among workers is facilitated by developing organisational structures and work routines that bring employees together to learn jointly through intensive collaboration and social construction of the realities confronting them (Brown and Duguid 1996).

Social Ties, Social Capital and Innovation Systems

In summary, the discussions above suggest that the ability to create new knowledge, share existing ones and apply organisational knowledge to new situations is critical to innovation. My second argument is that the motivation to create, acquire, share and use knowledge depends in part on the intensity of social ties within organisations and with individuals outside the focal organisations. For example, research has shown that 'sensitive information' and documents are shared among people who trust each other or have the moral duty of supporting the efforts made by their friends, relatives or colleagues (Portes 1998). These socially prescribed behaviours and benefits are termed in the academic literature as *social capital*. This concept therefore deserves an elaboration in the present chapter.

The Concept of Social Capital

The social capital concept draws on structural theory in sociology which is concerned with the role that relationships between individuals and groups in organisations. In most social structures 'certain people are connected to certain others, trusting certain others, obligated to support certain others, and dependent on exchange with certain others' (Burt 2000: 257). Obligations are based on feelings of gratitude, respect, and friendship. The connection, trust and obligations may be to the same group of people, thus resulting in cohesiveness or may be to different groups, thus resulting in loose-knit or fragmentary social structures. The central argument in the social capital theory is that networks of relationships constitute a valuable resource for the conduct of social affairs. Social capital is both the network and the resources derived through the network. Thus, in its initial usage in social science literature, the term social capital referred to the 'networks of strong, cross-cutting personal relationships developed over time that provide basis for trust, co-operation and collective action in communities' (Nahapist and Ghoshal 2000: 121).

Generally speaking, the concept of capital has played a prominent role in economic theories, policies and strategies. Distinction is usually drawn between pecuniary, physical and human forms of capital, with narrow variations in their degrees of tangibility. The introduction of social capital into economic thinking underlies the drift towards the acceptance of intangible factors in the understanding of the complexities and dynamics of economic systems. The prevailing understanding is that while physical capital can be created through material transformation and human capital through upgrading skill and knowledge, social capital derives from the underlying values within the society and can therefore not be created in the same way. Social capital is, therefore, the contextual supplement to human capital (Burt 2000).

As noted above, social capital derives from social ties. But the manner in which social ties develop and the nature of social capital they produce may be considered to be culture specific. Thus, there may exist differences in social capital endowments across nations. In the case of Ghana (and by extension, Africa as a whole) Kuada and Buame (2000) suggest that social ties can be classified into three categories in terms of social affinity. These are (1) *family-based* ties, (2) *ethnic-based* ties, and (3) *non-kin based* ties. The family and ethnic-based ties may be described as ascribed relationships, in the sense that people are born into them. The non-kin ties are self-initiated. People enter into such relationships voluntarily and are, for the most part, characterised by low exit barriers. Prominent examples in Africa include Old Boys/Girls Associations, and membership of religious and charity groups. People in these associations engage

113

in joint actions in pursuance of their interests. Their activities condition interactions that may lead to the development of interpersonal trust and loyalty towards each other. Friendship and close ties that people develop allow them to vouch for each other in business matters. Social ties in this regard create the social infrastructure that enhances people's ability to negotiate and/or create and share economic and non-economic resources (Coleman 1988; Hyden 1994; Sabel 1993; Unger 1998). For example, bank credits and acquisitions of pieces of equipment on credit become a lot easier with the intervention of trustworthy friends between the negotiating partners. Social ties therefore smooth the rough edges of the market mechanism and indirectly stimulate innovative processes in organisations.

Another important aspect in which social ties facilitate innovative economic behaviour is their promotion of interpersonal and inter-organisational collaboration. As noted above, social ties help build trust among people. Trust is a social attribute that generates willingness among people in dyadic relations to sacrifice their short run individual self-interests for the attainment of joint goals or longer-term objectives (Sabel 1993). Stated differently, it is an indication of one's belief that one's partner will behave in such a way that one gains rather than loses from the relationship.

Trust building involves learning and increases progressively. That is, collaborating parties will increase trust in each other over time when they have demonstrated to each other that they will not betray the confidence reposed in them by their partners. It is therefore analytically more purposeful to conceive trust in terms of degrees rather than in binary terms of either complete trust or zero trust. Different degrees of trust are required for different kinds of economic activities. To Humphrey and Schmitz (1998: 32) minimum trust is enough to ensure effective market transactions while extended trust is required for 'deeper kinds of inter-firm co-operation to work'.

In sum, the concepts presented above underpin the behaviour of actors in an innovation process. Figure 1 provides a schematic overview of the contribution of these factors to national innovation and the outcomes they generate. It shows that knowledge, learning, social ties and trust shape individuals' behaviours in firms, institutions and the civil society. They enable people to access information and leverage external resources, where necessary and therefore influence the content and tempo of technical, organisational and institutional innovation. Successful innovation produces outcomes in the form of improved firm-level performance and general economic growth.

Figure 1 A Schematic Presentation of the Antecedents and Consequences of Human Relations in an Innovation Process

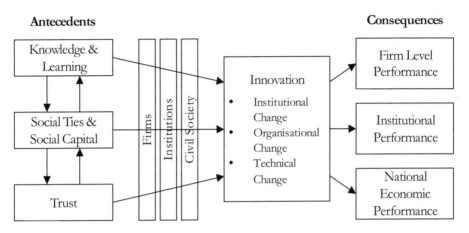

Implications for Innovation in Africa

Following the framework suggested above, we can now take a look at some of the preponderant characteristics of innovation processes in Africa. The focus of the discussions in this section will be on the cultural factors that influence individual and organisation learning and thereby shape the innovative processes in African organisations. Particular attention is drawn to the implications of the collectivist characteristic of African culture and the degree of trust inherent in the African business environment.

Family-Level Collectivism and Learning

African societies are generally described as having moderate collectivist characteristics (Hofstede 1980, 1991). Collectivist societies are characterised by the willingness of all individuals to contribute to collective gains. This motivation to contribute inheres in the cultural values and norms of society. The concept of social solidarity is employed by some scholars to describe this phenomenon, carrying the understanding that personal relationships and trust among in-group members underpin people's behaviour.

In Africa, the family is the primary social unit in relation to individuals. Values are determined by reference to the maintenance, continuity and functioning of the family, and the collective moral rules and obligations of the family bind individual family members to one another. Within such a social framework, all purposes, actions, gains and ideals of individual members are evaluated by comparison with the fortune of the family as a whole. The family, therefore, limits, influences and, in some situations, determines the individual's activities in organisations, institutions and the broad society. But the high degree of homogeneity of the in-groups implies solidarity against out-group members.

Within the business context, researchers have noted pockets of emergent industrial collectivism in some African countries (McCormick 1999). Collectivism of this kind has positive influences on learning. It promotes informal and non-planned changes in small entrepreneurial types of business. The mere fact that people live and work within close geographical proximity means that they can learn from each other through observations and imitations, a phenomenon referred to as 'technological spillovers'. This provides a cost-effective means of upgrading technology particularly within simple industrial clusters. As McCormick (1999: 1533) explains 'at its most basic level, clustering seems to encourage information sharing and opportunities for learning new techniques and designs'.

The rapid knowledge acquisition process in informal clustered businesses combine with strong social ties to promote business formation, particularly in sectors where entry barriers are low. It is not unusual for young people to set up their own businesses immediately after the end of their apprenticeship by relying on resources from family members and the social capital appropriated through the network of relationships they have developed within the community (Kuada and Buame 2000).

With limited financial resources at their disposal, knowledge becomes an important asset. People tend to have a pragmatic view of the knowledge they acquire and the mode of its acquisition. The significance of the knowledge acquired by African entrepreneurs depends on the extent to which it can translate directly into economic activities. That is, Africans who work in the informal sector will tend to learn as practitioners rather than learn about practice. The implication of the pragmatic orientation to knowledge acquisition is that knowledge transfer models that isolate learning from practice have limited chances of succeeding within these business environments.

Collectivism and its impact on the behaviour of business owners and employees appears to be eroding in some African societies. It has been noted, for example, that Ghanaians tend now to adopt individualistic and opportunistic attitude to life in general, exhibiting disregard for collective rules and codes of conduct. Business owners are increasingly reluctant to employ their family members, paying greater emphasis on individual professional competence rather than family attachment in their staff recruitment policies and strategies. This also implies that knowledge-based organisational development strategies are receiving priorities in African firms, especially in the larger organisations. Individualistic dispositions of employees in African organisations have not, however, been an unqualified blessing. In some organisations, it has taken the form of instrumental orientation to work and the decline in an entrepreneurial zeal to learn through observations, imitations and experimentation or to show commitment to their work.

Jones' (1986) analysis of management practices in Malawi provides some empirical support for this observation. According to him, Malawian workers basically have instrumental orientation towards work; they expect their jobs to bring substantial benefits to themselves but show very little (if any) loyalty and commitment to the organisation. Similarly, Montgomery (1987) observes in his analysis of the management practices of African executives in Southern African countries that they typically see their positions in their organisations as personal fiefdoms. They are more concerned about personal matters than about organisational goals. 'Even arguments and negotiations over public vehicles, housing and equipment centred about the convenience of the individual user more than about the mission of the organisation to which they were assigned' (Montgomery 1987: 917).

Added to this, research into African management practices has revealed the prevalence of paternalistic disposition of managers to their subordinates (Kuada 1994). Employees tend to act with undue caution while at work in order not to invite the anger of the superiors for any mistakes that they may make in the course of their work. The principal function of the loyal employee in Africa is to serve as a buffer for the immediate superior. If anything goes wrong, the loyal subordinate must do anything to blame all others, including himself, in order to protect his boss. Such behaviours imply that employees are highly reluctant to question existing practices in their organisations even if this would help rectify operational inefficiencies.

This kind of superior-subordinate relationship further implies that most employees are reluctant to embark upon experimentation as a method of learning. They may therefore stand passively by and watch inefficiencies persevere. Their general attitude is that acting and failing is more risky than failing to act. The fear of failure fosters immobilisation and the use of dysfunctional performance routines.

Promoting and Understanding Social Ties in Africa

The discussions above carry implications for both policy formulation and strategies aimed at promoting innovative behaviour in African organisations. This section of the chapter draws attention to some of the policy and academic implications.

Policy Implications

The policy implications discussed here are predicated on the awareness of the paucity of financial and technical resources in Africa. This underscores the need to enhance the use of non-financial resources, for example, human and social capital, in the economic development policies and strategies in these societies. Social ties must be encouraged through deliberate and purposive strategies by all institutions in Africa. The role of the government is highly important in this regard. First, there is the need for stimulating the development of mechanisms that can enhance social ties outside family and ethnic groups. As noted above kinship-based relationships tend to be ethnocentric and reduce the search and acceptance of novel ideas and knowledge. Access to new networks (and thereby social capital) increases the diversity and richness of the social relationships.

Second, African governments must also take deliberate actions to encourage cross-national ties rather than merely applaud successful isolated efforts by some individuals or groups. Cross-national interactions must be encouraged even where distrust is potent. Distrust is not always a problem. Where institutional arrangements are in place, even people who distrust each other may consider co-operation worth their while. The starting point is interaction. As indicated above, social capital is a social structural resource and therefore inheres in interpersonal relationships. Interactions between people improve accessibility to knowledge and information and encourage the formation of a collective attitude to learning.

Third, the promotion of civility in national and cross-national social systems must be a priority goal of African governments. The term civility is used in this context to mean a strong sense of civil obligation among Africans to fulfil their individual needs without jeopardising the chances of others to fulfil theirs or the entire community to survive and progress. Where such civility is lacking, jungle rule tends to prevail, creating a de-constructive social structure.

Research Implications

The concept of social resource carries useful academic implications as well. It lends intellectual credence to behaviours hitherto commonly accepted within business circles. That is, trust, empathy and affectivity underlie the success of all economic transactions and promote innovation. By recognising latent economic resources embedded in social ties, research investigations can direct entrepreneur's attention to the sources of the superior performance of their competitors and guide them in how to invest, harness and manage these resources.

In addition to these observations, the present study also highlights the following issues requiring further research attention.

1. The relationship between innovative behaviour and social ties needs to be empirically addressed. This will test the relevance of the model presented above, reject it or provide modifications where necessary. Future research must also investigate the relationship between firms' performance and their attitudes to various types of social resources.
2. The extent to which gender influences the accessibility of the various types of social resources also requires attention. Are women more able to build social ties than men are? If so what implication does this carry for staffing institutions and choosing institutional leaders that will promote intra and inter-African network formation?

Conclusion

This chapter is based on the premises that there exists a variety of successful recipes of national systems of innovation in the world. Each nation must therefore design its own system based on the awareness of how learning and knowledge flows can be effectively promoted within a given national context. It has been further argued that social ties provide rich sources of information, knowledge and access to economic resources. Societies facing problems of inadequate economic resources must therefore actively encourage the creation of mechanisms of social ties in order to compensate for and stimulate the development of economic resources. Social ties are sources of what is referred to as social capital, a concept that describes the 'actual and potential resources embedded within, available through and derived from the network of relationship possessed by an individual or social unit' (Nahapiet and Ghoshal 2000: 123).

Intensive social ties encourage interpersonal trust, which has an important psychological effect on interpersonal relations and the motivation to engage in joint action and learning. Conventional wisdom holds that trusting the intentions, words and decisions of people one deals with raises individuals' level of confidence to make decisions that affect other people. The lack of trust creates the twin problems of misunderstanding and suspicion, which together reduce the joy with which people relate to one another.

Building on these arguments, it has been suggested that the reliance of African societies on family and ethnic ties as the key social network reduces their members' ability to share new information and learn. Family and ethnic ties generally create closed networks with rigid sets of norms that define 'in-group' boundaries. This hampers the in-flow and dissemination of information that may conflict with prevailing thoughts and values within the family and ethnic communities.

There is, therefore, a need for the development of bridging ties in African communities to promote the formation of cross-ethnic and cross-national linkages and networks. Sociologists describe a 'bridge' as a person, group of people or institutions that link disconcerted parts of society and thereby order the entire social structure. The existence of 'bridge-men' in the African societies will bring new information and opportunities to the attention of different social units (families, ethnic groups, communities) and promote the development of trust building mechanisms between them.

Differences in goals, capabilities, norms and values combine to make cross-ethnic and cross-national linkages a highly challenging undertaking. But the economic, political and social benefits more than compensate for the efforts that may be put into facilitating the formation of strong social ties.

Notes

[1] I count myself among those who hold the view that development is not a race. The concern of African nations must not be to 'catch up' with the developed countries but to develop appropriate means of improving the quality of life of its citizens based on their unique historical circumstances.

References

Argyris, Chris (1994), 'Good Communication that Blocks Learning', *Harvard Business Review*, July-August: 77-85.

Argyris, Chris and Donald A. Schön (1974), *Theory in Practice: Increasing Professional Effectiveness*, San Francisco: Jossey-Bass.

Brown, John Seely and Paul Duguid (1996), 'Organisational Learning and Communities-of-Practice: Toward a Unified View of Working, Learning, and Innovation', in Michael D. Cohen and Lee S. Sproull (eds.), *Organisational Learning*, London: SAGE Publications: 58-82.

Burt, Ronald S. (2000), 'Contingent Value of Social Capital', in Eric L. Lesser (ed.), *Knowledge and Social Capital: Foundations and Applications*, Boston: Butterworth-Heinemann: 255-286.

Coleman, James S. (1988), 'Social Capital in the Creation of Human Capital', *American Journal of Sociology*, Vol. 94, Supplement: 95-120.

Hedberg, Bo (1981), 'How Organisations Learn and Unlearn', in Paul C. Nyström and William H. Starbuck (eds.), *Handbook of Organisational Design*, Vol. 1, New York: Oxford University Press: 3-27.

Hofstede, Geert (1980), 'Motivation, Leadership, and Organization: Do American Theories Apply Abroad', *Organizational Dynamics* Summer: 42-63.

Hofstede, Geert (1991), *Cultures and Organizations: Software of the Mind*, London: McGraw-Hill Book Company.

Humphrey, John and Hubert Schmitz (1998), 'Trust and Inter-Firm Relations in Developing and Transition Economies', *The Journal of Development Studies*, Vol. 34, No. 4: 32-61.

Hyden, Goran (1997), 'Civil Society, Social Capital and Development: Dissection of a Complex Discourse', *Studies in Comparative International Development*, Vol. 32, No. 1: 3-30.

Jones, Merrick (1986), 'Management Development: An African Focus', *Management Education and Development*, Vol. 17, Part 3.

Kuada, John E. (1994), *Managerial Behaviour in Ghana and Kenya: A Cultural Perspective*, Aalborg: Aalborg University Press.

Kuada, John and Samuel Buame (2000), 'Social Ties and Resource Leveraging Strategies of Small Enterprises in Ghana', in Olav Jull Sørensen and Eric Anourld (eds.), *Proceedings of the 7th. International Conference on Marketing and Development*, Accra: Ghana.

McCormick, Dorothy (1999), 'African Enterprise Clusters and Industrialization: Theory and Reality', *World Development*, Vol. 27, No. 9: 1531-1551.

Montgomery, John D. (1987), 'Probing Managerial Behaviour: Image and Reality in Southern Africa', *World Development*, Vol. 15, No. 7: 911-929.

Nahapiet, Janine and Sumantra Ghoshal (1998), 'Social Capital, Intellectual Capital, and the Organizational Advantage', *Academy of Management Review*, Vol. 23, No. 2: 242-266.

Portes, Alejandro (1998), 'Social Capital: Its Origins and Applications in Modern Sociology', *Annual Review of Sociology*, Vol. 24: 1-24

Sabel, Charles F. (1993), 'Studied Trust: Building New Forms of Co-operation in a Volatile Economy', *Human Relations*, Vol. 46, No. 9: 1133-1170.

Unger, Danny (1998), *Building Social Capital in Thailand*, Cambridge: Cambridge University Press.

Whitley, Richard. (1992), 'The Social Construction of Organizations and Markets: The Comparative Analysis of Business Recipes', in Mike Reed and Michael Hughes (eds.), *Rethinking Organization*, London: SAGE Publications.

Whitley, Richard (ed.) (1994), *European Business Systems: Firms and Markets in their National Contexts*, London: SAGE Publications.

Winter, Sidney G. (1996) 'Organizing for Continuous Improvement: Evolutionary Theory Meets the Quality Revolution', in Michael D. Cohen and Lee S. Sproull (eds.), *Organisational Learning*, London: SAGE Publications: 460-483.

8

Changing the Outlook

Explicating the Indigenous Systems of Innovation in Tanzania

Pernille Bertelsen and Jens Müller

Introduction

The majority of peasants and artisans in Tanzania are reproducing their livelihood through innovative diversification and indigenous technology systems. They are highly knowledgeable and skilled, certainly not ignorant as the public is commonly told. It is our contention that a significant social and productive potential, primarily in the informal sector, is being disregarded and not considered by policy makers.

We relate to a research agenda, which attempts to uncover the impact of the globalisation and structural adjustment processes presently at work. A particular focus is on the structure and institutional setting of the national technology systems in general and of the systems of innovation in the informal sector in particular.

Selected case studies of technologies in tool making by village blacksmiths and in boat building are presented. Historical evidence of the formation of the concomitant local systems of innovation is collected in order to bring forward an understanding of the dynamics or otherwise of these systems.

The guiding research question of the research project referred to is:[1] *What social learning processes are reproducing and transforming the indigenous systems of innovation?*

Alarming Contradiction

As 'globalisation' penetrates the outlook of politicians, civil servants, and academicians in Africa, their perception of indigenous knowledge, becomes increasingly deceiving, certainly in Tanzania: only scientific knowledge and exogenous technologies that can be communicated by hitchhiking on the information highway appear to be worth banking on. The question of 'catching up' technologically is prominent on their agenda.

President Benjamin Mkapa is quoted for declaring an all out fight against ignorance as an important part of the anti-poverty drive of his government (Daily News, Tanzania, 05.09.00). He was implying the conventional – almost tautological – wisdom that poverty is caused by ignorance (and disease).

Our contention here is that it is not, we repeat *not*, the indigenous craftsmen that are ignorant; it is these other peoples – including ourselves! – of modernisation and exogenous inclination that know far too little about the ins-and-outs *and* potentials of the indigenous knowledge systems. Thus the research project recorded here sets out to explicate the indigenous systems of innovation.

Fortunately, the common man and woman in Tanzania, and the South at large, do have another perception of the indigenous craftspeople: whether they like it/them or not, the majority of the population reproduce their livelihood on the basis of their skills and innovative performance (Bertelsen 1997; Mihanjo 2001; Müller 1980, 1984, 2001). And it should be said that an emerging recognition is underway of indigenous knowledge, primarily in the environmental debates. We want to broaden the debate to production technology in general.

Granted, the majority of the craftspeople we are referring to may be semi-illiterate, and are thus barred from the universally most accepted form of knowledge and information transfers. But to regard ignorant as synonymous with illiterate is a serious mistake. This chapter invites for discussions of these issues.

Research Design

In brief, the social problem addressed can be expressed as follows: the growing population of Tanzania increasingly make their livelihood on the basis of production and exchange of goods and services from the expanding informal sector, the technologies of which conventionally are perceived as simple/low in contrast to those of the receding formal sector which are regarded as advanced/high. This observation appears to be a contradiction in terms.

More explicitly we find two sets of frustrations:

• Frustrations in the informal sector due to barriers for increased mobilisation of *already existing* technological innovation and skill formation capacities, resulting in forgone socio-economic progress;

• Frustrations in public sector interventions designed to promote technological transformation and skill acquisition both in the formal sector, but in particular in the informal sector.

Selected technologies of indigenous origin were studied and related to selected local geographical areas, and historical evidence of the formation of the concomitant local systems of innovation was collected in order to bring forward an understanding of the dynamics or otherwise of these systems.

Conceptual Clarifications

Part of the explanation for the contradictions referred to is the inadequate, largely Euro-centric, conceptions and taxonomies at disposal for dealing with non-western knowledge systems and technological transformations. We therefore start out by clarifying our basic conceptions.

Indigenous People

We realise that by choosing the term indigenous at the turn of the millennium we are up against what the International Labour Organisation (ILO) defines as indigenous peoples.[2] The 1990's were the decade that internationally brought focus on the term indigenous in the meaning of people's rights. However, in search for improving the livelihood of these people, the UN as well as development organisations found themselves developing a popular definition of the term indigenous. The danger that rose following that definition was the creation of a short-circuit, i.e. a romantic Euro-centric view on who were the right people to receive development assistance.

Indigenous Knowledge Systems

Briefly speaking, our interpretation of indigenous is 'locally embedded' or 'of local origin', in contrast to what has *not* been 'indigenised', i.e. is exogenous or 'of foreign origin'.

Indigenous technology refers to a technology that is being produced, used and reproduced in a local production system. The knowledge is locally available and being disseminated in the society, and the organisation of the production is embedded in the local institutional setting. Therefore, the term indigenous technology covers many technologies ranging from potters in Africa to space industry in India (Baskaran 2001).

Figure 1 Salient Components of Indigenous Knowledge

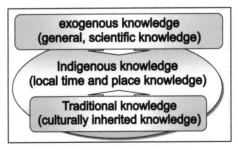

As shown in Figure 1 we conceive indigenous knowledge as including assimilated exogenous knowledge, as well as traditional knowledge.

At the same time as the term indigenous was used to cover the fight for marginalised people's rights, attention was created to rediscover how many of these indigenous ethnic groups practised a living different from what main stream politicians and planners have in mind.

Researchers and NGO's in local agricultural and forestry sectors began to study what is being labelled indigenous knowledge systems. Recognition of bio-diversity and local ways of managing 'traditional' agricultural production systems and local ways of solving conflict over natural resources is becoming an increasingly relevant and important driving force in agricultural research. However, the focus is still on the way things were done *before* 'modernisation' made its entry into the local and vulnerable cultural systems. (Labelle 1997)

In our definition and use of the term indigenous knowledge we are internationally and in particular among politicians and planners up against a perception of indigenous as synonymous with lack of progress and innovation. It is important for us to stress that our use of the term indigenous innovation systems should not be understood as a call for a return to, or glorification of static past. What we advocate is an active engagement with the present and future situation, and the formulation of critical alternatives to the hegemonising trends of the prevailing modernisation ideology and Euro-centric development discourse.

Technological Innovation

Technology and Innovation

We use a holistic definition of technology and innovation to stress that technology and innovation embraces much more than science based knowledge. It reads:

Technology is one of the means by which mankind reproduces and expands its living conditions. Technology embraces a combination of four constituents: Technique, Knowledge, Organisation and Product.

A qualitative change in any of the 4 constituents of technology that effectively leads to a transformative move and thus change in the other elements we denote a technological innovation. (Müller 2003)

National Systems of Innovation

Technological innovations have convincingly been argued to occur within various National Systems of Innovation (Lundvall 1992). The basic contention is that learning, searching, and exploring take place in practically all parts of the economy.

Lindegaard (1997) operationalises the systems of innovation conception. He starts out assuring that 'the notion of the knowledge-based and learning economy of innovation systems is applicable for the analysis of past as well as present economies, of less-developed as well as developed countries, and of traditional manufacturing as well as of high-tech industries'.

The model in Figure 2 depicts the important external actors around the enterprise organisation[3], where different forms of interaction transmit market and extra-market stimuli. At the same time, the external actors function both as valuable sources for knowledge and as selection environment where (often contradictory) preferences of performance are articulated.

The model is obviously set up primarily on the premises of formal sector operations in the North. But it is equally useful to apply it to the formal sector setting in the South.

However our intention is to adapt it to the *informal* sector setting in Tanzania. In other words, we claim that national systems of innovation are made up of two distinct and different systems. One set of systems of innovation in the formal sector, another set in the informal sector.

Figure 2 Model of an Innovation System

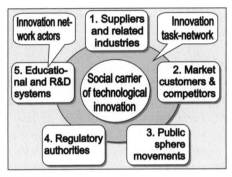

Source: Adapted from Lindegaard (1997: 18)

Diversification Processes at Work

Before proceeding with summaries of our empirical findings we need very briefly to refer to the changes in the social context that are taking place under the structural adjustment policies imposed. We restrict ourselves to the social scene apparently prevailing in the locations of our research, i.e. rural Tanzania in the 1990'ies. Here we primarily make use of the analysis and theoretical scope of Seppälä (1998) where he very elaborately highlights the intricate diversification processes at work.[4]

A salient feature of the diversification issue is that not only do we find numerous examples of horizontal diversification between the production units; diversification of productive activities also increases within the single household units. Even the individual craftspeople are taking up a number of different lines of production, as will be demonstrated below.

Indigenous Tool Making

This chapter gives a brief empirical record of our survey of village blacksmiths. For more details see Müller (2001) and Nsana et al. (2000 and 2001).

'If we go on strike, hunger will happen!'

This message was given to us in 1999 by one of the village blacksmiths in Ngindo Village, Mbinga district, at the end of a long interview. He had, till then, patiently told and demonstrated what tools his group are using, from where they obtain the raw materials, how they acquired their skills, how the group is organised, what products – including a lot of noteworthy repairs – they are making to what customers, and finally what network relations the group has to other agents in the area. But as we kept questioning him he got a bit frustrated; to him, the social significance of his trade is so obvious that *our* ignorance appeared overwhelming to him: why all these questions?

But where do these craftsmen come from? How can their present technology be characterised? How has their technology been transformed under the ever-changing contextual conditions?

Historical Recordings

Investigations and tests indicate that iron was excavated, smelted and forged in North Western Tanzania as far back as 500 B.C. (Schmidt and Avery 1978). However, whether 500 B.C. or A.C., blacksmiths have been at work before Arab and European intrusion (Schmidt 1996). Early European explorers of East Africa tell of powerful and prosperous kingdoms where much of the reason for this power and wealth was their mastery of iron works technology (Koponen 1988). A German lieutenant reported in 1892 that he estimated the number of hoes annually traded at Tabora market, the main centre for inland trading then, to be 150.000 (Kjekshus 1996).[5] In Mara region we were told that young men wanting to marry had to canoe all the way on Lake Victoria to Geita to bring back two special hoes as part of the bride price.

However, the German and later the British colonial authorities forbade the blacksmiths' trade in many districts. But these, largely administrative, measures to sabotage the blacksmiths were only partly successful. The smiths withdrew into hiding in the forests and swamps.

At independence the blacksmiths were ostensibly legalised. But as was the case with other rural non-agricultural activities, the blacksmiths were neither explicitly recognised nor registered. Another thing is that their very specialised skills did not fit into the code for skilled labour inherited from the British, e.g. a village master smith cannot pass a formal trade test even of the lowest grades. The smiths are thus still regarded as un-skilled by the system.

However, a survey of the blacksmiths in the 1970's gave reason to believe that the number of active village blacksmiths was declining, for the following main reasons (Müller 1980):

- The acquisition of raw materials in the form of scrap iron was frustrated by government restrictions.
- Restrictions on marketing of finished products were also introduced: (i) in some districts a ban on the traditional weekly rotating market system was

imposed. (ii) Regional trading corporations were assigned monopoly of marketing of a number of essential goods, including agricultural implements.

- A policy to promote small enterprises was carried out by the state agent Small Industries Development Organisation (SIDO) with a highly arrogant attitude towards the 'clients'. The indigenous technologists, if recognised at all, were referred to as being backward, lazy and crazy.

However, the study demonstrated that the village blacksmiths certainly were neither backward, nor lazy nor crazy. On the contrary, they were skilled, industrious and sane, *but their conditions of production were deteriorating to such an extent that we could only predict their early liquidation*:

> What the colonial state couldn't achieve by direct administrative means, viz. the liquidation of the smiths, the present independent state is on the point of accomplishing by indirect economic means. (Müller 1980:195)

It was therefore a big surprise to us when we happened to return to Bukoba town in 1994 and found that several market stands had a vide range of products made by the local village blacksmiths. In other words, the very pessimistic prediction from the 1970's appeared to be solidly contradicted.

A brief survey in Bukoba district was therefore done in 1995. A number of blacksmith's groups from the previous survey were revisited, in order to find an explanation for what had happened. In brief, the findings of this survey gave rise to the following main hypotheses (Müller 2001):

- The lift of the previous restrictions during the 1980's on commodity trade and performance of craft production has resulted in an increased room of manoeuvre for the unfolding of the technological capability of the village blacksmiths.
- The values and norms of the local society render the blacksmiths with a particular identity and give their technological capability a social 'meaning' which transcends the modernisation paradigm: the indigenous technology of the village blacksmiths is dynamically embedded in the culturally conditioned social division of labour.

Innovative Transformations

The research project recorded in the present chapter has verified these hypotheses. We further conclude as follows: although the core technique of the blacksmiths essentially has remained the same for thousand years, that of shaping red-hot iron by hammering, the tools and auxiliary instruments of labour have undergone recurrent innovations throughout. A dramatic change did occur when scrap metal from cars, the mining and other industries did replace the smiths' own iron excavation and smelting.[6]

Another example is that the heating of the charcoal-fired forging hearths is presently being changed from goatskin bellows to the use of labour saving bicycle wheel operated fans.

Evidences of product innovations are numerous, and the repair services of the blacksmiths are also undergoing ever-increasing diversification and sophistication. The main reason for the smiths to keep making hand-operated farm implements is that the demand is for such products (Poston 1994). In areas where animal-operated implements are in use, the smiths are making parts and vital repairs.

Our findings tend to concur with Boserup (1965/93) where she demonstrates the capability of local technological innovation. On the other hand, innovations are unlikely to take place in agriculture unless the community concerned is exposed to land shortage and pressure of population growth.

The blacksmiths are thus expanding their product diversification in close interaction with an increasing number of other craftspeople. They relate very closely not only to the farmers, but also to the local timber and log makers, canoe builders, carpenters, masons, wood grinder makers, cattle owners and bicycle repairers. In one case we came across a person that did not only master and practised the blacksmiths' craft, but also did canoe construction, carpentry works and fishing, apart from agriculture; making all the tools he is using himself.

Our cases also exhibit a wide variety of organisational set-ups of the enterprises. The point to make is that the micro-enterprises do change set-up from time to time, responding to changing circumstances. One year we may find an advanced manufacturing system, next year the same enterprise may have changed to a simple putting-out system of organisation.

A final observation is that hardly any of the blacksmiths had been trained at formal vocational training schools. A few had received extra and useful training through a donor-funded project connected to a formal vocational training provider, i.e. Folk Development Colleges. Some had had valuable contact to SIDO; but these were exceptions. On the other hand these cases are encouraging signs of a changed ideological attitude on the part of SIDO.

Finally we can confidently state that the number of active and visible smiths has increased since the 1970's. Although we found no way of making any exact count, we confidently claim that the number has tripled whilst the population has doubled.

Indigenous Boat Building

For centuries indigenous craftsmen along the shores of Lake Victoria in Tanzania have build different types of boats, i.e. first only using tools made by local blacksmiths and later supplemented by tools bought in hardware shops in the towns. They have thus enabled the fishermen to supply the surrounding communities with protein rich food. They have also provided the traders with means to transport of goods on the lake. Our field studies reveal that there has been, and still is a close relationship between the availability of fish species, the fishing methods applied, the design of boats, the technology used for their construction, and the fish market.

Historical Perspective

During the last century the indigenous boat builders have innovated and changed the boat designs and construction technologies from canoes made out of one big trunk, to up to 50 feet long container boats with two outboard engines. For all types of boats no drawings or other documentation in writing are being used neither for the construction nor for the training of apprentices.

The recent liberalisation of the economy has created a change in demand for fish, i.e. export market openings. The demand for boats has increased and new designs have come up. The local boat builders have responded innovatively to this situation. There has been no import of fishing or transport boats, and neither has any boat building training schemes been started by fish factory owners, government institutions or any other actors.

Without any external support at all, the indigenous boat-building craftsmen have been able to supply the new industrial sector with enough means of production (boats) to actually run an extremely profitable international export industry.

Table 1 shows the change in the types of boats being constructed that has been stimulated by either change in the market structure, the catch technology or the fish species available.

Seine Net Fishing Technology and Construction of Mitumbwi Boats

The first major change in the region's fishing technology appeared in the beginning of last century after the Europeans and Indians settled in the Kisumu area of Kenya and started to introduce seine net fishing in order to export fresh fish by rail to Nairobi.[7] The first few locally made nets are recorded to have been made by a Norwegian Mr. Aarup and his workers. Shortly thereafter seine nets were imported from Ireland by the local shops (Graham 1929: 16). Later cotton line was sold for the locals to make their own nets.

> In Tanganyika, generally speaking, imported nets are not used, their place having been taken by the fishermen's own make. (Graham 1929: 16)

The spreading of a new fishing technology from Kenya to Tanzania in 1920's made it possible to catch large amounts of fish, preserve these by drying them and export them nationally and regionally. The Dhow and the railways were used for the transportation.

Table 1 From Canoe to Engine Boat: Relations Between Fish Species, Fishing Technology and Type of Boats Constructed 1900-2000

Period	Type of boat	Dominant fishing technology	Fish conservation technology	Dominant fish market
1900-	Freight boat (Dhows) two types	Transport of preserved fish and agr. crops	Dried and smoked	Regional, National, East Africa
-1930	Canoe	Baited traps Barriers	Fresh, dried and smoked fish	Local petty commodity trade
1990-		Hand nets Tangle nets Seine nets Beach seines		
1920-	Mtumbwi Plank boats for max 7 person	Seine nets Beach seines	Fresh, dried and smoked fish	Regional, National, East Africa
1970-	Dagaar fishing boats	Dagaa nets	Dried fish	Regional, National, East Africa
1980-	Engine boat with ice containers to transport of fish	Seine nets	Frozen fish, factory fillet	International: EU Japan

The design and construction of the Mtumbwi boat with a capacity to carry up to 7 people, that could stay safely far from the shore for longer time than the canoe allowed, is closely linked to the change in fishing technology and the opening of up-country export of dried fish in the beginning of the 1920's. The construction of these boats was the work of the indigenous craftsmen and their apprentices, done by the use of local materials – wood and blacksmith-made tools, as well as few other tools bought in hardware shops in the towns.

Effects of Market Liberalisation

The recent liberalisation of the economy has changed the demand for fresh fish for export to EU and Japan. Simultaneously, the original way of organising the building of boats, i.e. a master boat builder working in his own workshop in the rural area assisted by apprentices, has changed to include at least the following types of works organisation:

1. Skilled boat builder working for a master boat builder in his workshop in the rural area or minor towns when orders are plenty.
2. Skilled boat builder working for a timber trader or a fish agent in his workshop in the town together with 8-10 other craftsmen and apprentices.[8]
3. Skilled boat builder working on the Indian fish factories to construct engine boats to collect fish on the lake.

Innovative Transformations

There is no doubt that the work of the indigenous boat builders plays a very important role in making it possible to supply the local population with valuable proteins for their diet as well as an income through the sale of fish to the export industry.

An important finding from our study is the observation of a high degree of diversification of the economic activities among the boat builders in the rural area. The boat builders are also farmers, fishermen, hunters, and carpenters. They practise their boat construction whenever an order is given, and in periods with few orders (as was the case when EU banned the export of frozen fish after incidents of use of poison as fishing method) they have other income activities to rely on. The artisan skills they practice are primarily socially and culturally inherited. However, changes are observed. Young boys from other families now get a chance to become apprentices.

The boat builders that we found in the regional towns of Mwanza and Musoma are less diversified in their income activities. Their income depends almost entirely on building boats. They have learned their skill from family relatives either in the rural areas or relatives that has already moved to town.

The boat builders are innovative and possess a high degree of change readiness. They have shown an ability to respond to the environmental and economic changes in society by changing their technology and develop new products. They adjust their production capacity to the market demand either when fish species availability change or when liberalisation of the economy makes fish export a lucrative income possibility.

Local blacksmiths still make a few of the tools and nails that the boat builders are using, but the majority is bought from hardware stores in the towns. The wood they use comes from timber dealers in the towns; shortage of hardwood in recent years has changed the price and/or the quality of the boats.

Synthesis of Findings

Since we explicitly aim to contribute to a changed theoretical paradigm in technology studies in the South, we now provide some of the theoretical deductions we find reason to make based on our empirical observations. For further explications see Bertelsen and Müller (2001) and Müller (2002).

Institutional and Technological Dichotomies

It is difficult to explicate the indigenous systems of innovation because these exist under informal institutional settings: *The rules of the game differ from those of the formal sector, and also vary from region to region within the same country.* See e.g. Raikes (2000: 65) explaining that it is a fatal mistake to think of one and only one market setting at work: several markets are instituted with very different 'rules of the game', and for analytical purposes, the national systems of innovation should not only be divided into formal and informal segments. They must also be split up between indigenous and exogenous segments with reference to the qualitative different technology systems at hand; we thus identify four segments of the national systems of innovation.

The distinction formal-informal sector is well described (Hope 2001; King 1996), although certainly not always emphasising the same features. Our contention is that the distinction primarily should be in regard of *institutional* disparities.

The distinction exogenous-indigenous needs some clarification. Exogenous technology is largely of foreign origin. It depends on all kinds of imported inputs in terms of technique and knowledge, and its organisation is also thereby to a large extent influenced by foreign management structures.

Indigenous technology is largely of local origin. Of course traditional artisan operations belong to it, but it also includes technologies that originally came from abroad. The distinction is whether or not the technology has been what we may term innovatively assimilated.

Figure 3 Matrix of the Four Segments of the National Systems of Innovation

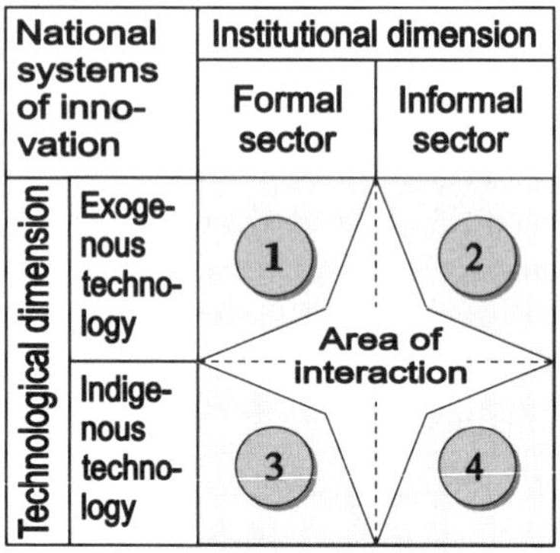

Put together, as we do in Figure 3, we are able to go into some more details of the dynamics or otherwise, both of what we call the institutional dimension and the technology dimension.

There has been a tendency to conceive the structure of production in the South as being situated either within segment [1] and [4]. Either we saw large-scale industries of foreign origin in the formal (modern) sector, or small-scale industries of local origin in the informal sector. What happened in segment [2] and [3] was blurred, and in any case the segments were looked upon as one rack bag of all kinds of activities.[9]

It is of course the area of interaction that is the most interesting to penetrate empirically when the focus is on innovation. What kind of interactive learning, user-producer communication and other exchange processes take place between the four segments?

The Intriguing 'By-Pass'

The conventional idea that exogenous technology inputs in the formal sector would be assimilated and eventually become part of the indigenous technology system, i.e. be directly transferred from segment [1] to segment [3] has largely been frustrated. Numerous examples of failed attempts to indigenise exogenous technologies have been seen in most development assistance regimes.

However, a noteworthy amount of hardware and software is transferred to and adapted in segment [2]. The knowledge and bits and pieces from this segment are then *assimilated* in the indigenous technology systems in segment [4]. Finally, some technologies in part or whole are gradually *domesticated* and embraced by the formal institutional setting in segment [3]. These movements from [1] via [2] and [4] to [3] we see as a 'by-pass' to the conventionally projected path from [1] to [3].

Figure 4 Dynamics of the National Systems of Innovation in the South

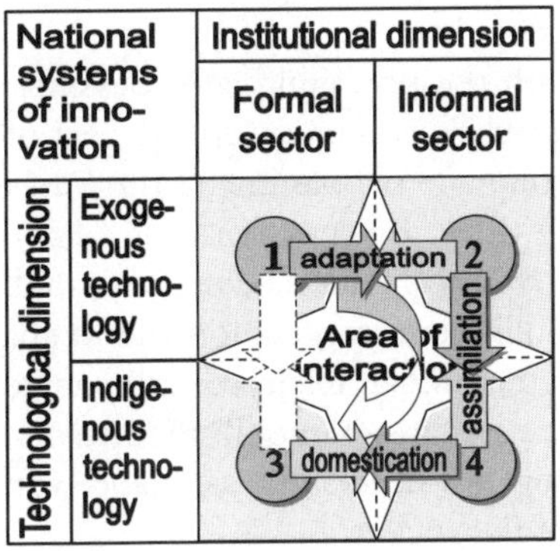

National systems of inno-vation	Institutional dimension	
	Formal sector	Informal sector
Exogenous technology (Technological dimension)	1 adaptation 2	
	Area of interaction	assimilation
Indigenous technology	3 domestication 4	

The indigenous and exogenous knowledge systems are apparently merging together at some spaces. And we would fulfil the proactive objective of our research project when and if we can point to policies that may one day reconcile the indigenous and exogenous technology systems.

Making Ends Meet

Having highlighted some of the details of the alarming contradiction presented in the introduction between the formal/exogenous and the informal/indigenous segments of the national systems of innovation, we may now be equipped to answer the question of how to further an effective reconciliation. Referring to Figure 3 the question can also be posed as a question of how to expand the area of interaction, i.e. *connecting* the formal with the informal systems of innovation, and making these ends meet.

We are very much aware that making policy proposals are highly tricky in the present era of globalisation, where economics are replacing politics. And most often, proposals made by academics from the North are fruitless anyway. We therefore content with recommending that dialogue be established between agents of all the four segments of the national systems of innovation with the purpose to facilitate an enhanced mobilisation of the technological knowledge and organisational capabilities of indigenous artisans.

In other words, we suggest that politicians and academics in the South do an effort to change their outlook. *A first step is recognition of the actual existence and change readiness of the indigenous technologists.*

Towards a Post-Pessimist Outlook

There is no reason whatsoever to indulge in craftsmanship nostalgia. The technology of e.g. the village blacksmiths is extremely labour demanding, cumbersome and low-level productive. But the specialisation potential of the village blacksmiths will be the 'just-in-time' backbone of the transformation processes needed. The parallel case of the boat builders is also illustrating this.

We need to do away with the prevailing 'afro-pessimism' (Bourenane 1992). Not because there is any reason to be optimistic on the part of the village blacksmiths or boat builders and other groups of people in the rural areas of Africa. We know too much of what is happening there. But pessimism is fatally hindering clarity and is an expression of lack of creative imagination. Our field studies fortunately gave us the necessary inspiration to mobilise our imagination and to abandon our pessimism.

What we can see is that the human resources in its widest sense are there, ready to be mobilised. The craftsmen are already mobilising these resources to some extent on their own initiatives, but they may need some assistance for doing so more expediently for the benefit of the rest of the society.

Notes

[1] The project is titled Indigenous Systems of Innovation in East Africa and done in collaboration with Department of History, University of Dar es Salaam, and Institute of Development Management, Mzumbe, Tanzania.

[2] ILO 'Convention Concerning Indigenous and Tribal Peoples in Independent Countries', Convention 169, Article 1, June 1989.

[3] The enterprise in the middle of the model Lindegaard calls the 'organisation'. We have chosen to call it the 'social carrier of technological innovations' (Müller 2003).

[4] Our analysis is moreover in line with Maliyamkono and Bagashwa (1990). We also refer to more recent theoretical discourses, e.g. Bryceson, Key and Mooij (2000) and Munck and O'Hearn (1999).

[5] This number corresponds to the production of hoes from the state own farm implement factory in Dar es Salaam, UFI, in the mid 1970'ies.

[6] The change from iron smelting to the use of scrap metal appears to have happened during the 1930's and 1940's. The latest changeover we have recorded is about 1949 in Ngindo village, Mbinga District (Nsana et al. 2001).

[7] Seine net fishing uses a large net with sinkers on one edge and floats on the other that hangs vertically in the water and is used to enclose fish when its ends are pulled together or are drawn ashore.

[8] A fish agent buys fish from the small fishermen at the lake and delivers the fish to a fish factory by agreement/contract. A fish agent may have several small fishing boats, which he leases out to fishermen against a share in the catch and an agreement that they only deliver fish to his collection boats.
[9] Transistor radios, video cameras, computers etc. are being repaired in numerous urban and semi-urban workshops belonging to segment [2]; repairs that are rejected in European shops ('why don't you buy a new one?') are nowadays effectively and very cheaply done in Dar es Salaam.

References

Baskaran, Angathevar (2001), 'Competence Building in Complex Systems in the Developing Countries: the Case of Satellite Building in India', *Technovation*, Vol. 21, No. 2.

Bertelsen, Pernille (1997), *Traditional Irrigation and Technological Change: Operation and Maintenance of Farmer-Managed Irrigation in Northern Tanzania*, Ph.D. Thesis, Department of Development and Planning, Aalborg University.

Bertelsen, Pernille and Jens Müller (2001), *Who are the Ignorant? Current Transformations in Tanzanian Indigenous Technology Systems*, Paper for Nordic Africa Days, Uppsala: Nordic Institute of African Studies.

Boserup, Ester (1993), *The Conditions of Agricultural Growth: The Economics of Agrarian Change under Population Pressure*, London: Earthscan Publications, (Reprint of original from 1965).

Bourenane, Naceur (1992), 'Prospects for Africa for an Alternative Approach to the Dominant Afro-Pessimism', in Peter A. Nyong'o: *30 Years of Independence in Africa: The Lost Decades?*, African Association of Political Science, Nairobi: Academy Science Publishers.

Bryceson, Deborah, Cristobal Kay, and Jos Mooij (eds.) (2000), *Disappearing Peasantries? Rural Labour in Africa, Asia and Latin America*, London: Intermediate Technology Publications.

Graham, M. (1929), The Victoria Nyanza and its Fisheries: a Report on the Fish Survey of Lake Victoria 1927-1928 and Appendices, London: Crown Agents for the Colonies.

Hope Sr., Kempe R. (2001), 'Indigenous Small Enterprise Development in Africa: Growth and Impact of the Subterranean Economy', *The European Journal of Development Research*, Vol. 13, No. 1.

King, Kenneth (1996), *Jua Kali Kenya: Change & Development in an Informal Economy*, Eastern African Studies, Nairobi: East Africa Educational Publishers.

Kjekshus, Helge (1996), *Ecology control and Economic Development in East African History, (Second Edition)*, Eastern African Studies, Nairobi: East African Educational Publishers.

Koponen, Juhani (1988), *People and Production in Late Precolonial Tanzania: History and Structures*, Finnish Society for Development Studies and Scandinavian Institute of African Studies.

Labelle, Huguette (1997), 'Global Knowledge and Local Culture', Notes for Remarks by Huguette Labelle, President of CIDA, at the Global Knowledge 97 Conference.

Lindegaard, Klaus (1997), *State of the Art of Innovation Systems Analysis*, SUDESCA research Papers, No.7, Centre for Environment and Development, Department of Development and Planning, Aalborg University.

Lundvall, Bengt-Åke (ed.) (1992), *National Systems of Innovation: Towards a Theory of Innovation and Interactive Learning*, London: Pinter Publishers.

Maliyamkono, T. L. and Mboya S. Bagachwa (1990), *The Second Economy in Tanzania*, Eastern African Studies, London: James Currey.

Mihanjo, Eginald P. A. N. (2001), 'Production and Exchange Transformations of the Kisi Pottery Enterprise in South West Tanzania', in Patrick O. Alila and Poul O. Pedersen, *Negotiating Social Space: East African Micro Enterprises*, Trenton: Africa World Press Inc.

Munck, Ronaldo and Denis O'Hearn (eds.) (1999), *Critical Development Theory: Contributions to a New Paradigm*, London: Zed Books.

Müller, Jens (1980), *Liquidation or Consolidation of Indigenous Technology: a Study of the Changing Conditions of Production of Village Blacksmiths in Tanzania*, Development Research Series, No. 1, Aalborg: Aalborg University Press.

Müller, Jens (1984), 'Facilitating an Indigenous Social Organization of Production', in Martin Fransman and Kenneth King (eds.), *Technological Capability in the Third World*, London: Macmillan.

Müller, Jens (2001), 'Consolidation and Transformation of Indigenous Technology: Prospects for a Revival of the Village Blacksmiths in Tanzania', in Patrick O. Alila and Poul O. Pedersen, *Negotiating Social Space: East African Micro Enterprises*, Trenton: Africa World Press Inc.

Müller, Jens (2002), *Making Ends Meet: Diversification, Infrastructure and Endogenous Systems of Innovation in Tanzania*, Paper for International Workshop on Globalisation, New Technologies, Inequalities and Social Well-being, Aalborg University.

Müller, Jens (2003), 'Perspectives on Technological Transformation', in John Kuada (ed.), *Culture and Technology Transformation in the South: Transfer or Local Innovation*, Copenhagen: Samfundslitteratur.

Nsana, Bernard, Eginald Mihanjo, Pernille Bertelsen, and Jens Müller (2000), *Indigenous Systems of Innovation in East Africa: Field Study Report, June-August 1999*, Department of Development and Planning, Aalborg University.

Nsana, Bernard, Eginald Mihanjo, Pernille Bertelsen and Jens Müller (2001), *Indigenous Systems of Innovation in East Africa: Field Study Report II, July-November 2000*, Department of Development and Planning, Aalborg University.

Poston, David (1994), *The blacksmith and the Farmer: Rural Manufacturing in Sub-Saharan Africa*, London: Intermediate Technology Publications.

Raikes, Philip (2000), 'Modernization and Adjustment in African Peasant Agriculture', in Deborah Bryceson et al., *Disappearing Peasantries? Rural Labour in Africa, Asia and Latin America*, London: Intermediate Technology Publications.

Schmidt, Peter R. and Donald H. Avery (1978), 'Complex Iron Smelting and Prehistoric Culture in Tanzania', *Science*, Vol. 201: 1085-89.

Schmidt, Peter R. (ed.) (1996), *The Culture & Technology of African Iron Production*, Gainesville: University Press of Florida.

Seppälä, Pekka (1998), Diversification and Accumulation in Rural Tanzania: Anthropological Perspectives on Village Economics, Uppsala: Nordiska Afrikainstitutet.

PART III

Regional Innovation Systems and Cross-Regional

Experiences

9

Differences in National R&D Systems between Early and

Late Industrialisers

Alice H. Amsden and Hyun-Dae Cho

Introduction

National R&D systems everywhere share much in common, but notable differences distinguish advanced economies and 'latecomers' (countries that industrialised initially by borrowing foreign technologies), and latecomers of two distinct types. One type, 'integrationists', rely for their advanced skills on the spillovers created by foreign direct investors (as in Argentina and Mexico). Another type, 'independents', invest heavily in their own technologies and national enterprises to generate proprietary cutting-edge skills (the principal cases are China, India, Korea and Taiwan) (Amsden forthcoming). With respect to 'integrationists', whose industries generally exhibit a high incidence of foreign direct investment, foreign firms have done very little R&D in such countries to date. In the future, foreign firms may be expected to subcontract only those R&D tasks that are 'non-strategic', as suggested by product cycle theory. Therefore, the R&D systems of 'integrationists' have tended to stagnate. In the case of 'independents', their R&D systems exhibit a high degree of government intervention relative to that of the prototypical innovator, the United States. Mechanisms to minimise 'government failure' thus differ between the two sets of countries as well. We first discuss the 'integrationist' case to perceive the role of foreign direct investors. Then we analyse the unique control mechanism governing R&D activity in Korea, Taiwan and China.

The Basis for Divergence

Governments the world over have played a larger role in R&D than in most other economic activities due to well-known market failures associated with knowledge and the technology that it underpins: knowledge is the quintessential 'public' good whose use by one individual or firm does not diminish its availability to others; knowledge exhibits large externalities; it requires complementary public investments in transportation and communications; its returns are only partially appropriable, and so forth (see for example (Branscomb 1993)). Governments in latecomer countries,

141

however, have intervened more in national R&D activity than such market failures have typically induced in advanced economies because of another imperfection associated with knowledge: the knowledge-based assets of leading firms in advanced economies constitute entry barriers. To penetrate such entry barriers and to create a space for national enterprises in those 'newly'-established industries (for latecomers) subject to scale economies and 'first mover' advantage, the activist role of latecomer governments in R&D has chartered a new course.

Knowledge is a special input because it is difficult to access in an advanced state, whether by 'making' or 'buying', and initially buying is what all latecomers must do. Unlike information, which is factual, knowledge is conceptual; it involves combinations of facts that interact in intangible ways. Perfect information is conceivable – with enough time and money, a firm may learn all the extant facts pertaining to competition. Perfect knowledge is inconceivable because knowledge is firm specific and proprietary (Amsden forthcoming).

Given 'entrepreneurial' or 'technological' rents from proprietary knowledge, there is a great reluctance on the part of a firm to sell or lease knowledge-based assets. Instead, their value may be maximised if kept proprietary and exploited inside the firm (Hymer 1976). The secrecy of these assets is typically protected by law. Even if such assets are offered for sale, as they are in technology transfers, diffusion from one production unit to another production unit is likely to be imperfect, and dependent on a high level of skills on a *buyer's* part. Whatever is sold may comprise merely the codified part of a technology. The knowledge about how a production process works, and how to improve that process, may never be divulged (Nelson 1987; Rosenberg 1976). Furthermore, in the event that unit production costs are sensitive to market size and subject to scale economies, incumbents enjoy 'first-mover' advantage over newcomers (Chandler Jr. 1990). Thus, the knowledge-based assets of incumbents create oligopolistic market structures and barriers to entry.

In a subset of industries, therefore, that are both (a) R&D intensive (competitiveness depends on innovation and innovation depends on R&D, as defined shortly), and (b) subject to 'first mover' advantage, latecomers face a choice. To develop such industries they may either rely on foreign direct investment, or try themselves to capture 'first mover' advantage using infant industry and *infant firm* 'protection', simultaneously investing heavily in proprietary, firm-specific skills, including those associated with R&D. The R&D system of early industrialisers (Case A) may consequently differ from the R&D system of latecomers in terms of either the degree of *foreign direct investment* (Case B) or *government activism* (Case C).

Foreign Direct Investment (Case B)

In Case B, latecomers compete in R&D-intensive industries by means of spillovers from foreign firms, an outcome of the 'crowding out' of national enterprise and successful capture by foreign firms of first-mover advantage. If knowledge was perfect, and if technological (or entrepreneurial) rents did not exist owing to competition and the diffusion of the proprietary skills on which above-normal profits depend, then no importance would attach to ownership, foreign or national. But ownership matters in a latecomer country if knowledge is imperfect and foreign firms crowd out national enterprise but conduct no local R&D. Because the nature of R&D in the two countries differs, the question in a latecomer country arises if income in an R&D-intensive industry (also subject to first-mover advantage) will ever equal that of the home country of a foreign investor. Other things equal, the answer about parity depends on whether the R&D that generates technological rents will be developed by the multinational firm outside its home base, in the latecomer country in which it has 'crowded out' national competitors.

By the year 2000, it had become a stylised fact that multinational firms invested *very little* in R&D outside their home base – on the order of 15 per cent.[1] Most R&D activity occurred at home: 'far from being irrelevant, what happens in home countries is still very important in the creation of global technological advantage for even (the) most internationalised firms' (Patel and Vega 1999: 154-55). The figure on R&D out-sourcing varied by region – some European multinationals, especially from Sweden, conducted as much as 40 per cent of their R&D in other European countries. But in latecomer countries, the R&D undertaken by multinational firms remained minuscule even though some multinationals had been operating in these countries since the 1920s. Even if multinationals invested in local learning in order to adapt the products they sold domestically to suit consumer tastes (as in Proctor and Gambols' customisation of Pampers for hot, low-income climates), and even if they transferred advanced production skills, as in the Mexican automobile industry, research for entirely new products or processes at the world frontier was rare.

Given the absence of national R&D in favour of a reliance on foreign spillovers, but given the absence of spillovers in the form of local R&D by foreign firms, then, other things equal, a latecomer could despair of ever attaining the income level of the home country of a foreign investor in R&D-intensive industries. Such industries, moreover, comprise a large – and rapidly growing – subset of service and manufacturing activity in high-income economies.

In Case B where a latecomer's R&D depends on investment by foreign firms that have 'crowded out' national enterprise, we encounter yet another paradox. Divergence internationally in R&D systems may be expected to stem from differences in both environment and actors. If the actor – a multinational firm – is the same in two countries, then divergence may be expected to be *smaller* than if the actor was also different. Yet R&D activity of a multinational firm tends to be different at home and overseas. Only two types of R&D activity are likely to be transferred (or replicated) abroad: generic basic research, assuming the availability of secure intellectual property rights and low cost scientists in latecomer countries (and the powerlessness of corporate-based scientists to prevent outsourcing); and production-related product and process adaptation and incremental innovation. Thus, incomes in R&D-intensive industries, as a consequence of the R&D systems of Case A (advanced countries) and Case B (latecomers dependent on foreign investment), are both likely to diverge.

In fact, major representatives of the B case (Argentina, Mexico and to a lesser extent Brazil) lend support to this divergence. In terms of *level* of R&D activity, it is far lower in these countries than in both 'independent' latecomers (China, India, Korea and Taiwan) and in the advanced countries in which the foreign investors that dominate production in Case B are head-quartered (see Table 1 for a comparison of integrationists and independents).

Table 1 Research and Development (R&D) Expenditure as a Percentage of GNP, 1985† and 1995*

	1985	1995
Korea	1.8	2.8
Taiwan	1.2	1.8
India	0.9	0.8
Chile	0.5	0.7
Brazil	0.7	0.6
Turkey	0.6	0.6
China	n/a	0.5
Argentina	0.4	0.4
Malaysia	n/a	0.4
Indonesia	0.3	0.1
Thailand	0.3	0.1
Mexico	0.2	0.0

* India, Korea 1994; Malaysia, 1992; Mexico, 1993.
† Brazil, Korea, Mexico, Turkey 1987; Chile, Taiwan, 1988; India, Indonesia 1986.
Source: All countries except Taiwan: (UNESCO (United Nations Economic and Social Council) various); Taiwan: (Taiwan (Republic of China National Science Council) 1996).

The integrationists also tend to have more foreign direct investment in their manufacturing sectors than the independents, as measured by the share of foreign direct investment in gross fixed capital formation and the share of foreign control of mergers and acquisitions. For all practical purposes the R&D expenditures of Argentina, Brazil and Mexico were nil in 2000, with the preponderance of R&D activity that did occur undertaken by the government, for non-commercial purposes (medical research, for instance), in sectors other than manufacturing (Alcorta and Peres 1998). There are, of course, other plausible reasons besides high shares of foreign direct investment why industries with first-mover advantage and R&D-intensity may exhibit low levels of R&D activity in integrationist latecomers. The 'crowding-out' effect, however, may be paramount but tends to be overlooked. Instead of local R&D, integrationists have relied strategically for their growth in productivity and their improvement in product quality on 'spillovers', or sales of technology and (unpaid) externalities from foreign firms, as noted later. Thus, the spillover approach inherent in Case B represents an alternative model.

Government Activism (Case C)

In Case C, which assumes zero foreign direct investment but positive flows of other forms of foreign technology transfer, latecomers compete in industries that are R&D-intensive and subject to 'first-mover' advantage by means of: (1) subsidising infant industries *and firms* to prevent 'crowding out' by foreign competitors; and (2) heavily investing in national R&D (and in national science and technology generally). The 'independents' belonging to Case C may be expected to pioneer an activist role for the state in R&D activity that is present in neither Case A nor Case B. The differences in the state's R&D role in cases A and C are now explored theoretically and empirically.

Nature of State R&D Activity

To differentiate sharply the state's R&D role in Case A, an innovative advanced economy, and Case C, a latecomer without foreign direct investment, it is helpful to make another assumption: that the 'first mover' to develop and dominate new industries in A-type countries is the private rather than public sector. This assumption is tantamount to assuming that in the advanced country model, private R&D activity precedes public R&D activity in chronological time. This precedence tends to truncate the state's organisational and entrepreneurial R&D role. The A-case is most closely approximated by the United States.[2] In advanced countries where this assumption does not necessarily hold (depending on the industry), such as many European countries and Japan, the state's R&D role will more closely approximate that of Case C latecomers.

The state's role in R&D in the pure A-case may be said to exhibit the following characteristics:[3]

- Government R&D activity is confined to the provision of public goods – defence and 'welfare' (basically health and the environment), and gains from such activity by private business largely take the form of 'crowding in' by virtue of spillovers.
- The government 'targets' (allocates resources selectively) only to those industries (firms) related directly to the provision of public goods.
- The government's relations with private firms are, in general, instrumental rather than strategic: they exist to the extent necessary to execute government policies related to defence and welfare rather than to enhance the market competitiveness of the firms in question.
- Government R&D activity includes no national master plan.

Restraint on the part of government in the A-case to use R&D to bolster national commercial (as opposed to military) competitiveness is illustrated by the US government's response to Japan's competitive market challenge in the 1980s. The US government promoted some R&D efforts to foil the Japanese threat, most notably

145

SEMATECH, a cooperative venture with the private sector to promote the American semiconductor machinery industry (Flaherty 1986). But the fundamental US response to Japan's competitive challenge at the micro level was political – to force open markets in Japan and to protect them in the U.S.[4]

State R&D activity in the C-case, by contrast, is the opposite of what it is in the pure A-case:

- Government R&D precedes the establishment of private R&D. Initially government R&D is oriented towards the provision of public goods, mainly defence, as it is in the A-case. Then, as manufacturing industry develops in tandem with government industrial policies (including protection of industries with 'first mover' advantage and controls in such industries on foreign direct investment), government R&D becomes more developmental in scale and scope, with 'catch-up' as its primary goal.
- Government science and technology policy is designed to subsidise private R&D in the form of preferential long-term credit, tax breaks, the provision of infrastructure and collaborative research ventures. Promotion of private R&D includes the targeting of strategic industries and firms.
- Government policy towards the private sector is designed explicitly and directly to advance the competitiveness of national 'champions' (or national 'leaders') by various means, such as the creation of science parks and the inauguration of national public/private collaborative R&D projects.
- Government R&D policy is guided by a master plan that is part of a more general plan to promote economic development.

The deliberate way in which governments in 'independent' latecomer countries build R&D capabilities is illustrated by the example of telecommunications, an industry whose technology is user-driven; it is heavily influenced by the provider of telephone service. In the U.S., the provider was private and government's role was largely regulatory (Vietor 1994). The symbol of R&D in this industry was Bell Labs, the private facility of the largest player, AT&T (American Telephone and Telegraph). In China, India, Korea and Taiwan, by contrast, the state was initially responsible for providing telephone service, and R&D was an outgrowth of this initial state role (similar to the state's role in another key sector, petrochemicals). Telecommunications exhibited both state direction and state entrepreneurship.

Strengths and Weaknesses of Different Cases

The three R&D models we have considered, A (Laissez-Faire Innovator), B ('Integrationist' Latecomer) and C ('Independent' Latecomer), each has its peculiar weaknesses and strengths.

Comparing B and C, or catching-up with and without foreign direct investment, the short-run weaknesses of independence and rejection of foreign direct investment may be hypothesised to be production inefficiency and product inferiority. These disadvantages will persist in the long run unless the independent learner achieves parity in costs and quality with the multinational firm, or at least achieves parity to the point that the learner's local advantages offset the multinational's disadvantages of operating at a distance. If independents do achieve parity in terms of skills – production, project execution and innovation – with firms at the world technological frontier, and if foreign firms continue not to undertake R&D outside their own national boundaries (the locale of their corporate headquarters), then technological skills and, hence, technological rents and national income may be expected to be higher in the independent compared with the integrationist country (*ceteris paribus*)

Comparing A and C, or catching up with an R&D system that is 'laissez-faire' (to the extent that the American R&D model is laissez-faire) compared with a state-directed R&D system (wherein the government initiates R&D activity according to a master plan and targets specific industries, firms and technologies for special support, as discussed above), then the short-term and long-term advantages of state intervention may be hypothesised to be earlier R&D, greater R&D and more rational (say, less duplicative) R&D activity than otherwise. The disadvantages, as discussed next, are likely to be: entombment of new technologies in government laboratories that are inept (by assumption) at commercialisation, with private R&D effort stagnating; and/or mistargeting of the wrong industries, firms and technologies for special support, along with general bungling and corruption.

We now examine how independent latecomers have acted to minimise these two drawbacks. Basically, efforts by governments to create actors at the firm level have helped overcome the entombment problem, while a control mechanism unique to latecomers, that imposes and monitors performance standards as a condition for R&D support, has professionalised government intervention, thereby reducing the likelihood of picking losers, inefficiency and venality.

Firm-Level R&D Activity

To push R&D activity out of public laboratories and into private firms, governments in Korea and Taiwan and to a lesser extent China and India have involved themselves in deliberate firm-formation. In the case of the Korean telecommunications industry, the Ministry of Communications established a state enterprise, the Korea Telecommunication Company, to produce M10CN switches with foreign technical assistance. The government then sold KTC to the private sector. After the government had established a public institute, Electronics and Telecommunications Research Institute (ETRI), to undertake the R&D necessary to develop an indigenous digital switching system (given the reluctance of private companies to assume the risk), ETRI premised its activities on collaboration with these companies. It transferred its technologies to them, provided them with technical assistance and training, and generally interacted with them organisationally so that they could eventually initiate new R&D projects independently. This hand-me-down approach, ending in 'collaboration' between public and private partners with roughly equal skills, has characterised virtually all Korea's Highly Advanced National (HAN) R&D projects (Cho and Amsden 1999). These are government-initiated ventures, but private firms share total costs, responsibilities and outcomes with government agencies.

In Taiwan, firm-formation under government auspices has been no less deliberate than in Korea and has been even more intense given Taiwan's relative scarcity of large-scale private firms with big R&D budgets. To compensate for this scarcity, the government has assumed the role of (state) venture capitalist (Amsden forthcoming). The public Industrial Technology Research Institute (ITRI), for example, formulated objectives for upgrading the electronics sector and agreed to assume the financial risk involved. In the early 1970s it founded an Electronics Research and Service Organisation (ERSO) with a strategy to develop Taiwan's integrated circuit (IC) industry by borrowing foreign technology (from RCA), establishing a demonstration factory, and then diffusing know-how to the private sector. To become technologically self-reliant, and to get 'first-mover' advantage over foreign firms in scale and personnel recruitment, ERSO founded the United Microelectronics Corporation to manufacture ICs. Then, with a view towards strengthening Taiwan's IC design houses, ERSO invested US$400 million jointly with Phillips of Holland to build a high-precision IC manufacturing facility, Taiwan Semiconductor Manufacturing Corporation (Mathews 1997). In personal computers (PCs), 'ERSO usually takes the lead in developing the crucial technologies, and then transfers these to the PC enterprises' (Chang 1992: 208). These PC enterprises are incubated in science parks that are government-created, as in the case of Acer Computers, a national champion. In 1995, firms in the Hsinchu Science Park accounted for only 4.2 per cent of manufacturing output but 17.5 per cent of total R&D.[5] Taiwan's second major science park, in Tainan, was founded in 1996 to build technological capabilities in the microelectronics, precision machinery, semiconductor and agricultural bio-technology industries. The park is designed to provide high-quality residential and recreational facilities for as many as 110,000 people. By 2005, employment is anticipated to reach 21,000 and sales of US$16 billion are expected (Tainan Science-Based Industrial Park 1996).

Control Mechanisms

The government-led R&D systems of Korea, Taiwan and China were disciplined by monitorable performance standards that were imposed on national and foreign firms[6] in exchange for government subsidies or contracts. These standards exceeded merely cost sharing. The build-up of technological capabilities was thus subject to reciprocal rules similar to those that were intended to minimise 'government failure' in the build-up of manufacturing capacity in general (Amsden forthcoming). In addition, government agencies involved in R&D promotion were subject to various forms of institutional monitoring.

In Korea's telecommunications sector, private firms that benefited from technology transfer from ETRI were obliged to send a stipulated number of their personnel to ETRI for training. ETRI also required firm participants to formalise and implement in-house project management systems as a condition for support. Korea Telecom in turn scrutinised the quality of ETRI's outputs.

To improve performance in terms of delivery time and quality, private companies in Korea's telecommunications consortium were also required to compete against one another in developing the same technology. Korea Telecom and ETRI evaluated the performance and technological capabilities of the manufacturers, and 'the proportion of purchase from each firm was set as the reward' (Lee 1993: 66). In Korea's semiconductor national project as well, the share of subsidised funding going to a private firm in a later stage of research has depended on its performance in an earlier stage (Cho and Amsden 1999).[7]

Three mechanisms have evolved in Korea to monitor the performance of government organisations, as distinct from private organisations, involved in R&D. First, promotion is firmly grounded in law. Laws prescribe in detail the procedures and criteria for planning, implementing and evaluating R&D programs, as well as the procedures for requesting, granting, using and evaluating R&D subsidies, and a transparent legal framework has helped to prevent arbitrariness and abuse. Second, the finance of R&D programs has some built-in incentives for good performance. For example, government agencies are generally empowered to use royalties earned from previous research, and cash from selling public enterprise stocks, the latter acting as an incentive to privatise, which presumably strengthens firm-formation. Third, inter-ministerial organisations, advisory ad-hoc committees and competition-cum-cooperation among different government agencies involved in the same research project all help to check imprudence and to increase technical assistance and technology transfer, at least in theory.

In Taiwan, performance standards applied to both public and private R&D players are easily as extensive as in Korea. To qualify for the benefits of a science park, a firm has to meet pre-screening criteria. Admission into Taiwan's Hsinchu Science Park has depended on the evaluation of a committee that consists of representatives from government, industry and academia. The major criterion for admission is the nature of the technology a firm is developing. According to the Hsinchu Park Administration 'an existing company would be asked to leave if it changed to labour-intensive operations and no longer met the evaluation criteria (which the Park Administration specifies)' (Xue 1997: 750-51). Benefits for companies in Tainan Science Industrial Park (TSIP) include grants of up to 50 per cent of necessary funds from government programs, tax exemptions, low interest loans, as well as special educational facilities. In exchange, companies seeking admission into TSIP have to meet criteria related to operating objectives, product technology, marketing strategy, pollution prevention and management (Tainan Science-Based Industrial Park 1996).

Thus, not only does the degree of government intervention appear to differ in the A and C models of R&D activity. The rules governing that intervention appear to differ as well. In the A case, the 'invisible hand' is still mostly the control mechanism. In the C case, reciprocal rules and conditionalities have superseded that means of discipline.

Despite the large amount of capital involved in some subsidised R&D projects of latecomers, there have been few cries of corruption, in contradistinction to the accusations of corruption that have characterised the subsidisation of large-scale investment projects in new physical capacity. Assuming the latter accusations are not exaggerated, and the former cries are not understated, then the differences between the two would seem to lie in: (1) more experienced firms and government administrators in the R&D case, which, among other virtues, enables more cost-sharing; (2) more competition among private firms and ministries (several private firms typically are involved in one large R&D project but only one firm at a time is involved in a large physical capital investment); and possibly (3) greater legal grounding and transparency, a function of (1).

Conclusion

We have suggested that three models of R&D now coexist in the world economy: A, found in innovators, quintessentially the United States; B, found in latecomers that have relied on foreign direct investment for their advanced technological capabilities; and C, found in latecomers that have relied on heavy government intervention and national firms for proprietary, cutting-edge skills. These models differ according to the role of major institutions (in terms of ownership, national or foreign, and government intervention) and the rules governing the behaviour of major organisations in R&D (contained in 'control mechanisms'). There is considerable overlap among the three models depending on the industry and country. The telecommunications industry in Brazil, for example, has exhibited many characteristics similar to those of the telecommunications industry in Korea and India, although other Brazilian industries conform more with model B. Still, we would argue that there is value in distinguishing the three models both analytically and policy-wise. More latecomers, such as Malaysia, Indonesia, Thailand and Egypt, are in the process of establishing their own R&D systems, and the ABC distinction, as opposed simply to A or B, the traditional choices, offers them a richer inductive menu from which to develop their own national R&D systems. Turkey is in the process of joining the European Union, and exhibits tendencies to follow the B model, of relying on foreign investment for advanced skills. Yet the experience of Argentina and Mexico is that such reliance will lead to negligible domestic R&D activity, given a stylised fact about multinational firms to date: they conduct little R&D outside their own home base, especially in latecomer countries. Two small foreign bases that have received abundant foreign R&D investment, Singapore and Ireland, warrant further study insofar as they deviate from this stylised fact.

All three R&D models are legitimate under WTO law. Whatever else the WTO stands for, it may be regarded as a champion of R&D. Subsidies to exporting are in most instances illegal, but subsidies to R&D are not. Hence, the C model, which deviates the most from free market principles, conforms with new trade rules. Nor, in future, would imposing conditionalities on foreign investors, in the form of R&D expenditure requirements, infringe on current rules. There is no international convention that prohibits imposing conditionality on foreign investors, and the foregoing considerations suggest that such a convention may not be in latecomers' best interests.

All three models hold potential gains as well as risks for countries still at a distance from the world technological frontier. Model A has never been observed other than in a country that industrialised on the basis of national proprietary innovations. Model B may be said to characterise Canada, for example, a high-income North Atlantic economy with a heavy dependence on foreign investment for both its productive capacity and skills. But B has never been tried in a low-income latecomer over a long time period. In any event, the evidence on spillovers, whether from Canada or other countries, strongly suggests that they materialise only in the presence of substantial national investments in capability building.[8] Model C has been observed in Japan, but may be the riskiest of all. Its success depends on nurturing national champions with sufficient knowledge-based assets to compete against the world's premier multinational firms. The jury is still out on the success of model C in Taiwan, Korea, and especially China and India. If, however, investments in national skills are regarded as the best predictor of the wealth of nations, then these countries are the most likely to succeed.

Notes

[1] See (Prasada Reddy 1993), (Patel and Pavitt 1995), (Archibugi and Michie 1997), (Doremius *et al.* 1998) and (Organisation of Economic Co-operation and Development (OECD) 1998).

[2] Empirically, there is some question about whether the government or the private firm (individual) in the U.S. was the mover and shaker behind early nineteenth century innovation. Government arsenals were crucibles of innovation that created large externalities for private entrepreneurs (Smith 1977, 1994). But there is little doubt that by World War II, after which time government R&D exploded, private R&D laboratories in the U.S. were already well established in most industries.

[3] The relevant supporting literature may be found in Center for Science and International Affairs (n.d.).

[4] The U.S. became 'the biggest user, among industrial nations, of new non-tariff measures during the 1980s' (Hufbauer 1991: 93). By 1988, 'special protection' covered some 23 per cent of imports. 'The cutting edge of US policy towards declining industries is trade policy' (ibid.: 99).

[5] For a discussion of the (positive) effects of R&D incentives on R&D spending in Taiwan, see (Gee 1995); (Wang and Tsai 1995).

[6] The main obligation imposed on foreign firms was the transfer of technology in exchange for government contracts. Regulation under way in the World Trade Organisation will require members to give equal opportunity to nationals and non-nationals in bidding for government contracts, but nothing precludes governments from imposing technology transfer requirements on *all* bidders.

[7] The advantages of competition, however, were reduced by the disadvantages of duplicative effort and a firm's refusal to share its research results and commit its best engineers to joint projects (Lee 1993: 66).

[8] For the importance of national capacity to absorb foreign spillovers in latecomer countries, see Amsden (forthcoming) for the nineteenth century and Blomstrom and Kokko (1998) for a review of the current literature.

References

Alcorta, Ludovico and Wilson Peres (1998), 'Innovation Systems and Technological Specialization in Latin America and the Caribbean', *Research Policy*, Vol. 26: 857-881.

Amsden, Alice H. (forthcoming), *The Rise of the Rest: Non-Western Economies' Ascent in World Industries*, New York: Oxford University Press.

Archibugi, Daniele and Jonathan Michie (eds.) (1997), *Technology, Globalisation and Economic Performance*, Cambridge: Cambridge University Press.

Blomstrom, Magnus and Ari Kokko (1998), 'Foreign Investment as a Vehicle for International Technology Transfer', in Giorgio Barba Navaretti, Partha Dasgupta, Karl-Göran Mäler, and Domenico Siniscalco, *Creation and Transfer of Knowledge: Institutions and Incentives*, Berlin: Springer: 279-311.

Branscomb, Lewis M. (ed.) (1993), *Empowering Technology: Implementing a US Strategy*, Cambridge: MIT Press.

Center for Science and International Affairs (n.d.), *Core Policy Documents: Bibliography*, Cambridge: John F. Kennedy School of Government, Harvard University.

Chandler Jr., Alfred D. (1990), *Scale and Scope: The Dynamics of Industrial Capitalism*, Cambridge: Harvard University Press.

Chandler Jr., Alfred D. (2000), *Paths of Learning: The Evolution of High-Technology Industries*, New York: Free Press.

Chang, Chung-Chau (1992), 'The Development of Taiwan's Personal Computer Industry', in N. T. Wang (ed.) *Taiwan's Enterprises in Global Perspective*, New York: M.E. Sharpe Armonk.

Cho, Hyun-Dae and Alice H. Amsden (1999), *Government Husbandry and Control Mechanism for the Promotion of High-Tech Development,* Cambridge: MIT, Materials Science Laboratory.

Doremius, Paul N., William W. Keller, Louis W. Pauly and Simon Reich (1998), *The Myth of the Global Corporation*, Princeton: Princeton University Press.

Flaherty, M. Therese (1986), 'Coordinating International Manufacturing and Technology', in Michael. E. Porter (ed.), *Competition in Global Industries*, Boston: Harvard Business School Press.

Gee, San (1995), 'An Overview of Policy: Priorities for Industrial Development in Taiwan', *Journal of Industry Studies*, Vol. 2, No.1: 27-55.

Hufbauer, Gary C. (1991), 'United States: Ajustment Through Import Restriction', in Hugh Patrick and Larry Meissner, *Pacific Basin Industries in Distress: Structural Adjustment and Trade Policy in the Nine Industrialized Economies*, New York: Columbia University Press: 88-128.

Hymer, Stephen H. (1976), *The International Operations of National Firms: A Study of Direct Foreign Investment*, Cambridge: MIT Press.

Lee, J. (1993), *Strategic Management of Large-Scale R&D Projects: Case Study* (in Korean). Taejon: Korea Advanced Institute of Science and Technology.

Mathews, John A. (1997), 'A Silicon Valley of the East: Creating Taiwan's Semiconductor Industry', *California Management* Review, Vol. 39, No.4: 26-53.

Nelson, Richard R. (1987), 'Innovation and Economic Development: Theoretical Retrospect and Prospect', in Jorge M. Katz (ed.), *Technology Generation in Latin American Manufacturing Industries*, New York: St. Martin's: 78-93.

Organisation of Economic Co-operation and Development (OECD) (1998), *The Internationalization of Industrial R&D: Patterns and Trends*, Paris: OECD.

Patel, Pari. and Keith Pavitt (1995), 'The Localized Creation of Global Technological Advantage', in Jose Molero, *Technological Innovation, Multinational corporations and New International Competitiveness: The Case of Intermediate Countries*, Germany: Harwood Academic Publishers: 59-74.

Patel, Pari. and Modesto Vega (1999), 'Patterns of Internationalisation of Corporate Technology: Location vs. Home Country Advantages', *Research Policy*, Vol. 28: 145-155.

Prasada Reddy, A. S. (1993), 'Emerging Patterns of Internationalization of Corporate R&D: Opportunities for Developing Countries?', in Claes Brundenius and Bo Göransson, *New Technologies and Global Restructuring: The Third World at a Crossroads*, London: Taylor Graham: 78-101.

Rosenberg, Nathan (1976), *Perspectives on Technology*, Cambridge: Cambridge University Press.

Smith, Merritt R. (1977), *Harpers Ferry Armory and the New Technology: The Challenge of Change*, Ithaca: Cornell University Press.

Smith, Merritt R. (1994), *The Military Roots of Mass Production: Firearms and American Industrialization, 1815-1913*, Cambridge: Mimeo.

Tainan Science-Based Industrial Park (1996), *Prospectus*, Tainan: Tainan Science-Based Industrial Park.

Taiwan (Republic of China National Science Council) (1996), *Indicators of Science and Technology*, Taipei: National Science Council.

UNESCO (United Nations Economic and Social Council) (various), *Statistical Yearbook*, Geneva and New York: United Nations.

Vietor, Richard H. K. (1994), *Contrived Competition: Regulation and Deregulation in America*, Cambridge: Harvard University Press.

Wang, J.-C. and K.-H. Tsai (1995), 'Taiwan's Industrial Technology: Policy Measures and an Evaluation of R&D Promotion Tools', *Journal of Industry Studies*, Vol. 2, No.1: 69-82.

Xue, Lan (1997), 'Promoting Industrial R&D and High-tech Development Through Science Parks: The Taiwan Experience and its Implications for Developing Countries', *International Journal of Technology Management, Special Issue of R&D Management*, Vol. 13, No. 7/8: 744-761.

10

Innovation Systems in Capacity Building in Maghreb

Countries

Abdelkader Djeflat

Introduction

Like many other developing countries, Maghreb Countries (MCs) face enhanced competition, vanishing trade barriers, more stringent intellectual property regimes and deeper concern for the environment. Trends in all these areas are expected to pose serious challenges for fragile components in the socio-economic systems of the region. In the face of these new challenges, S&T policies and strategies remain inadequate, at best immature, in MCs. Yet, these are essential prerequisites for attaining viable national innovation systems and competence building. In the Maghreb, as indeed elsewhere, the process of transition these countries are undergoing involves the active role of the state aimed at promoting scientific and technological innovation as a basis for the development of a market-driven competitive economy. The efficiency of the market in allocating resources is enhanced with the supply of more and better information about 'new combinations'. These latter presume vision of the entrepreneur as well as the flair for innovation. The emergence of the enterprise culture cannot, however, be expected to occur in a policy vacuum. To date, though, the state in the Maghreb has been far from effective in promoting technological progress, so that a major agenda for transition would now be the institution of regional initiatives to explore new policy trajectories based on a sound synthesis of theory and experience. All these issues have been explored within the MAGHTECH network.[1]

MCs have also their particularities both as a sub-region and as individual countries when it comes to the approach to S&T policy as a result of recent history, political systems and economic policy options. This chapter will examine in the first section, the national system of innovation (NSI) as an analytical tool when applied to LDCs in general and to MCs in particular in order to highlight some of the underlying conceptual and methodological issues. The second section will concentrate on the analysis the implementation of S&T policy in MCs. The objective is to examine what might have been the ingredients of a proper NSI. Finally, the third section will look at new initiatives taken to build S&T capacity in the future. By way of conclusion, we shall outline some of the standing heart issues.

The National System of Innovation: Prospects and Limitations as an Analytical Tool for MCs

The national innovation system (NIS) has benefited from a vast and varied literature in recent years: (Freeman 1982; Gilles 1978; Mowery and Rosenberg 1998; Nelson and Winter 1993; Niosi and Faucher 1991; Von Hippel 1976). Various aspects were examined: the role of R&D departments in firms, the process of technological innovation, the important historical works of science, the concept of the 'technical system', the role of science, the role of the state in promoting technological innovation, the importance of technical alliances and co-operation agreements among countries, etc. The integrated approach to the NIS was however put forward by Lundvall (1985) and revised in the nineties for LDCs. Figure 1 indicates the different interdependent components which make up the NIS:

Figure 1 National Innovation System

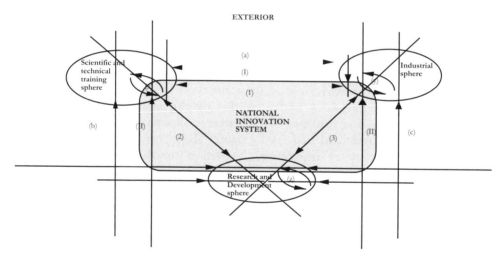

The three spheres identified are: the productive sphere, the training and education sphere, and the research sphere. Cooperation is essential between these spheres if the system is to work. In these relationships, the international aspect cannot be neglected particularly when the role of the most industrialised countries play in influencing R&D trends is significant even if the national aspect is central insofar as technological development is concerned. Social and political institutions and economic policies are also factors behind the homogeneity of the system.

From the empirical work, we have done on innovation, technology transfer and technical change in the Maghreb Countries, it appears that the NSI in its purest form, with the three poles approach cannot be a useful tool to analyse the weakness of innovation in LDCs and particularly in Maghreb countries. It needs indeed to take several factors into account and a broader methodological approach. Figure 2 shows that it is an open system whereby several sub-systems are in continuous interaction: the education and training sub system, the research subsystem and the industrial sub-system, each one of these sub-systems being itself open on other spheres both nationally and internationally (Bes 1995).

Secondly, it is far from being a set of simple straightforward relationships. It appears that a complex set of relationship between three poles exists: the education, the research and innovation, and the industrial one. The limited performances achieved in spite of the existence of these three poles in MCs are an indication of this complexity and the existence of other parameters and other actors. The linearity of the system is questioned not only by theoretical contributions such as the Kline and Rosenberg (1986) model but also by the multiple institutions and layers of decision-making involved in minor projects of introducing new products, services or new working methods namely in the public sector.

Figure 2 The National System of Innovation as an Open System

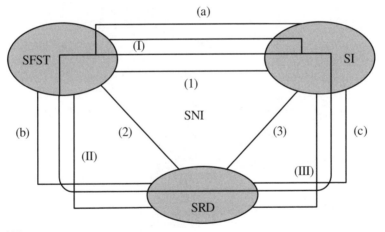

Source: Bes (1995).

This can be seen through a first attempt made to characterise the national innovation system in Morocco (see Figure 3). The variety of links and relationships, which are built and are necessary for innovative activities to take place involve necessarily several elements. As pointed out by Lundvall (1985), elements of trust, power, and loyalty characterise these relationships. These elements of cooperation and coordination are the only possible way to transfer the qualitative aspects much needed by innovation, as indeed the amount of tacit knowledge imbedded in each innovative activity cannot be catered for by the conventional contractual relationships. We have in a previous work highlighted the importance of similar dimensions in the interaction process:

notably commitment, distance, adaptation, conflict and cooperation in the process of technology transfer between high technology suppliers and low technology buyers (Djeflat 1992, 1998; Ford and Djeflat 1983). In the interaction model for technology transfer developed the interactive dyadic model. Thirdly, the treatment of the power ingredient is of great significance in the interaction between the various actors involved. Previous studies have revealed that technology transfer was greatly hampered by the unbalanced bargaining power, which exists between technology suppliers who hold dominant position in the negotiation process leading consequently to a variety of restrictive practices and restrictive clauses in the technology acquisition contracts. We have identified no less than forty-eight of these clauses in contracts between Algerian public companies and foreign firms and also between Egyptian firms and transnational corporations in the oil, gas and petrochemical sectors (see (Djeflat 1987, 1988)). Often these restrictive practices hold embedded the seeds of disintegration of the potential NSI as natural or organised forward and backward linkages cannot take place.

Figure 3 The NIS as a Complex System

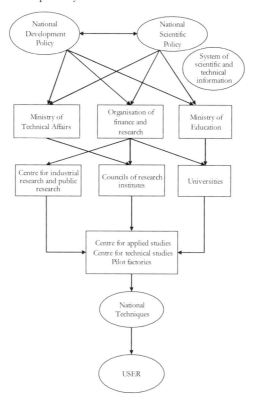

Source: Zekri (1990).

This is our fourth point: the importance of the international sphere as a highly relevant component of what might be an NSI adapted for LDCs, a 'fourth pole'. Our analysis of the Algerian industrial model a decade and a half after its implementation back in the eighties has shown that the so-called industrialising-industries model proposed by De Bernis and effectively implemented has led to a highly disintegrated model, with 'solid brick walls' being erected between the various sectors (Djeflat 1985). The explicit objectives of building an integrated economy with progressive inter-sectoral linkages could not take place. Steel produced could not be used by the mechanical industry with the exception of its usage for the production of pipes for the oil and gas sector and often at very high and non-competitive costs. This is also the case of zinc and sulphuric acid produced in the plant of Ghazaouet in the West of the country which could not be used locally but had to be exported to European markets at very low and unprofitable prices. At the same time Zinc and Sulphuric acid had to be imported for the needs of the local industry in a different shape. Many examples and case studies exist in this respect: they all convey the same results which are the difficult home-based linkages and the strong linkages both forward and backward with foreign firms namely suppliers of technological plants, equipment and services. Figure 4 shows how a need-driven system could be worked out.

Figure 4 University-Industry Link Based on Need-Driven System of Technology Development

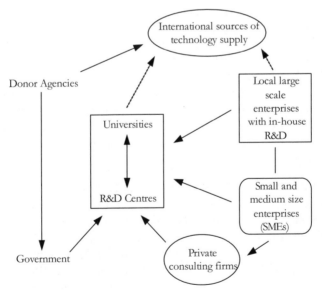

Components of the NSI in Maghreb Countries

In this section, we shall try to examine the components of the conventional NSI approach in the case of MCs while recognising that it is grossly non-adapted to their situation and specificities.

R&D Policy and Innovation

Most MCs have managed to establish R&D in various fields during the last two decades as mentioned earlier: agriculture, health, manufacturing and sometimes engineering. However, in comparison to other countries and in particular to the East Asian ones, there is a relative scarcity of innovative activities in the classical sense of the word. However, it is more and more admitted that considering the current situation of most LDC's they need to focus more on incremental, often within their reach, than on radical innovation. When looking at the MCs, a large proportion of innovations, particularly those within the productive sector, appear to be of the incremental type. Innovation in Maghreb enterprises can also thrive on informal R&D, although the extent of informal R&D is not fully documented. Some empirical studies of the iron and steel sector have for instance shown the existence of non-negligible innovation dynamism of this type (Djeflat forthcoming). Paradoxically, another type of research, more of the basic and fundamental type, is conducted in universities, in various areas of knowledge: physics, chemistry, biology, computer science, geology and biochemistry. The issue of fundamental versus applied is a highly debatable one. The Asian experience does not give evidence of the absolute necessity of fundamental research at an early stage such as the one the Maghreb is in. However, current circumstances are different from those of the early stages of Asian development. Other kinds of research are mostly research in agriculture where a limited number of centres are looking for certain varieties of crops to adapt them to the local environment. Although the issue of appropriate indicators of innovation in LDCs is an ongoing debate (Perrin and Abdelmalki 1998), we will use the classical approach to assess the R&D and innovation dynamism looking successively at the output and the input of R&D.

THE OUTPUT OF SCIENTIFIC RESEARCH AND R&D ACTIVITIES

Publications: In terms of publications, out of 1229 articles published by Maghrebians during 1981-86, only 5.1 per cent of the articles were concerned with engineering and technology. This is a relatively poor performance compared to NICs' rates. Over the last 28 years, the rate of growth of publications of MCs is higher than the Arab average. However, publications in international journals on S&T remain relatively limited: 0.32 per cent for Maghreb as a whole compared to 0.96 per cent for Latin America and 2.86 per cent for Asia; publications are dominant in chemistry and physics. Most of these were published in collaboration with foreign counterparts, mainly from France.

A great deal of the Maghrebian authors who manage to publish either live outside the sub-region or publish as co-authors with researchers from the North as shown in recent research. This pattern finds itself repeated in the field of industrial research. An empirical study based on a survey of electronic firms in Algeria showed that in the majority of cases, the research themes dealt mostly with the large multinational electronics companies rather than with the problems of infant national electronics firms (Dahmane 1994). Maghrebian researchers seem to perform better when the discipline are more abstract (physics) than when it is applied sciences (biology, medicine).

Patents: The number of patents registered by MCs at the local patent office in the last two decades and a half of the industrialisation process remains relatively low compared to Asian countries and to Israel in the Mediterranean sea. In Algeria 7930 patents were registered in the twenty-year period (1966-1986) at an average rate of 397 per year. The share of nationals did not exceed an average of 3 per cent, while 97 per cent were held by foreign firms (INAPI 1979/1988). This is the case of all three MCs, even though the figures have to be taken with some care considering the discrepancy, which sometimes exists between various sources. Except for Algeria where there was a major decrease since 1992 resulting no doubt from the crisis and the major fall in foreign investment, in the three others, the figures show a stagnation of the number of patents registered.

Table 1 Number of Registered Patents in the MCs

	1990	1991	1992	1993	1994	1995	1996
Algeria	592 (145)	617	83	-	- (110)	118	200*
Morocco	311 (300)	303	372	352	329 (400)	354	325*
Tunisia *	160 (220)	130	120	142	144 (120)	141	
Libya **	57	47	-	36	31	43	
Turkey	486	694	674	804	1102	722	
Switzerland	16152	16808	17967	20637	22306	20345	

Source: (*) INORPI Tunisia (**) WIPO Geneva, in Lahzami (1998); figures between the brackets are given by other sources.

Compared to a small country such as Switzerland or even a developing country such as Turkey, the gap is relatively high (Table 2). A substantial share of these patents is held by non-resident companies. The low share of patents from residents varies between 2 per cent to 30 per cent showing the crisis, which the R&D sphere knows, in spite of its gradual increase in both Algeria and Morocco. This is even more so, when we know that some of the foreign institutions have the status of resident in the country where they operate. The steady decrease in the Tunisian case in the period should be a point of concern if it goes on; it could indicate that the Tunisian manufacturing sector is more and more relying on external innovation and less on its own capacity.

From a sectoral point of view, it appears that the situation differs from one country to the other, however the most active and efficient sectors in Morocco are medicine and hygiene (16 per cent) and chemistry (43 per cent of all registered patents) probably due to its phosphate industry.[2] In Algeria, it is the hydrocarbon sector that has this leading position. While the issue of patent protection has been the centre of heated debate in the past, it does not raise as many controversies nowadays, as protection is seen as an incentive for local innovators and a way to protect them against big multinational firms.

THE INPUTS OF S&T RESEARCH

Three major inputs will be examined: financial and human resources and the institutional and organisational framework.

Funding of R&D activities: R&D expenditures as a percentage of GNP did not exceed 0.36 per cent for Maghreb, while reaching 1.94 per cent for America, 1.08 per cent for Asia, 1.79 per cent for Europe and 1.11 per cent for Oceania. It is recognised that some non-negligible efforts were made in the late seventies and early eighties, and funds allocated to research have effectively increased particularly in countries like Algeria, Tunisia and Morocco. The figures for the nineties seem to indicate however that the gap has widened: 0.25 per cent for Maghreb compared to 2.05 per cent for Asia (including Japan), Oceania 1.38 per cent, America (2.87 per cent) and Europe (2.21 per cent) (World Bank 1993).

Table 2 The Proportion of Research Funding

Country	R&D expenses	Share public (%)	Share enterprises (%)
Japan	3.05	18.2	72.7
Germany	2.66	36.5	60.5
Portugal	0.61	61.8	27.0
Turkey	0.47	71.3	27.6
Tunisia	0.31	90.0	6.0

Source: SERST (1996) cited by Attia and Houari (2000).

Taken individually, MCs compare unfavourably with newly industrialising countries: expenditure on R&D in Morocco amounts to 2 dollars per inhabitant per year (0.2 per cent of GDP in 1990). This proportion falls short of the ones achieved by India (0.9 per cent in 1987), Korea (2 per cent in 1987), Brazil (0.6 per cent in 1986) and Mexico (0.6 per cent in 1986). At enterprise level, the effort made in the area of R&D is weaker: only 10 per cent of total R&D funds are from enterprise budget while 90 per cent are funded by state. In Tunisia, internally funded R&D by enterprises does not exceed 6 per cent (Table 2). However, the trend shows that private funding has been growing at a higher rate (Lahzami 1998).

Moreover, few countries in the sub-region set aside a chapter of the national budget for a clear statement of expenditure on R&D and scientific and technological services due to lack of financial resources and lack of commitment on the part of policy makers in the face of competing and more urgent demands (food, health, education) and the huge military expenses, hence the relatively high involvement of external funding to support research (60 per cent of total budget). The problems related to external funding are numerous ranging from minor technical ones to more fundamental ones. The structural dependency of the national research systems, which are not given the chance to take off by themselves, may lead to a 'permanent infant research sector'. There are fears that this dependency on external funding increases over time now that there is a major squeeze on public funds as a result of the implementation of SAPs in nearly all countries. We shall come back to this later on.

Human resources for R&D: The personnel involved in R&D in the MCs is on average 10 to 20 times less than in Europe. For all countries of the Maghreb, the number of scientists and engineers involved in R&D is less than 400 per million inhabitants. In the same period Europe had 1735. In universities, personnel involved in academic research does not devote more than 10 per cent of its time to effective research: the increase in the demographic pressure and the increasing teaching loads at higher education levels had the effect of reducing their research effort. Beside the weak numbers, published figures show that the bulk are involved in science research while those involved in real technological research represent between 10 to 20 per cent in the countries for which statistical figures are available. In comparative terms, Korea had 54 per cent in 1983. Moreover, most of these scientists and engineers are either in universities or in ministries and public bodies and not in industry or specialised research centres. The relatively low status of researchers in MCs in particular has often been put forward as one of the most important problems facing S&T endogenising. This situation does not contribute to their stabilisation nor to their motivation. Other factors include the relatively bad research conditions, the reluctance to recognise the professional credibility of Maghrebian researchers forcing them to emigrate, the criteria used for career promotion which favour academic research and international recognition, the indigent state of universities, their relatively low material status compared to other professions. Marginalisation is made worse by the fact underlined earlier that in many MCs, most industrial undertakings are owned by TNCs which carry out their research in their parent companies rather than locally. This often constitutes an inhibition to the development of local research. Whenever research is conducted locally, foreign personnel is preferred to the local one. Moreover, a substantial share of the means used by researchers in the academic world comes from Europe and particularly from France with regard to MCs. The quantitative and qualitative shortfall of technical support personnel for scientific and technological activities in R&D constitute another obstacle: statistics for the beginning and mid-eighties concerning MCs indicate that only two countries (Algeria and Morocco) had two technicians for every researcher engaged in R&D. Moreover, whenever budgets are allocated to R&D activities, a great proportion of local funds are allocated to wages and salaries of the research institution's personnel. In Morocco, 95 per cent of the budget goes towards wages. However these competencies are often in a relatively short supply as seen earlier and also as a result of high pressures on production, which takes the priority over research and innovation as we have shown elsewhere. In the mid-eighties, only six countries in the whole Africa had more than 1000 scientists and engineers per million inhabitants and only two Maghreb countries, Algeria and Libya, were listed among the group according to Unesco Statistics.

The lack of technology policy has lead to the fact that innovation has never benefited from an explicit innovation policy or innovation strategy as it has in the case of a latecomer such as Korea. Many governments took part in the initiation of R&D activity; however, their action remains still below requirements. The R&D institutional framework (laboratories, sub-contractors, etc.) is below requirements. These institutions, whenever they exist, are not operated to optimum capacity for lack of adequate staff, equipment, operational funds, and scientific and technological support services. The equipment working life is shortened because of inadequate repair and maintenance facilities and the technical staff required to ensure that their scientific equipment will work well is limited. Waiting time to get the equipment repaired by foreign technicians ranges from five to 10 months. There is some evidence that the allowances for recurrent items are not even sufficient to keep some institutes ticking over.

Moreover, little seems to have been done with regards to support services. R&D needs a series of scientific and technological services ranging from documentation centres to institutions to valorise research results and which are an absolute must between research and the practical application of its findings in industry.

Scientific and technological research brings various agencies into play (political decision-making bodies, universities, research institutes and centres, etc.). In the Maghreb, each one appears to follow its own logic and its own path, which are not necessarily compatible with those of other agencies. One of the main problems of university research is its isolation from industry as it has a very limited and non-institutionalised relationship with the local industry. Finally, unlike Japan and Korea, the private sector is totally absent from R&D activities at least in the formal sector. Not much work has been done yet exploring the degree of interaction between the activities of the public research centres and that of R&D of private firms and on the mileage to Maghreb economies arising from these. The competitiveness of the technologies which Moroccan firms adopt and their commitment to R&D leave a lot to be desired (Lahlou 1999b). In brief, the policy for S&T has been too ambitious with little or no appeal to the direct and immediate needs of the economy. R&D made no impact on productivity. Instead it alienated the accumulated experience of the local industrial labour force. The R&D network also involved poorly integrated components that account for the dysfunctioning of the system.

Education and Training

The ultimate success of the technology transfer process remains however the development of endogenous technological capabilities. Growth is more and more dependant on the capacity of the economy to develop human capital and R&D and innovation capabilities as shown by recent endogenous growth theories (Romer 1990). Human capital rests on education and training and their capacity in responding to the needs of the productive sector. However, more relevant to S&T capacity building is the effort made in science and engineering subjects. Competitiveness rests on those with availability of the most qualified human resources because they possess the most scientific and technological mastery, skills and know-how. We will examine in the first paragraph the education and training system as a whole and in the second paragraph the scientific and technical training system in particular.

It is widely recognised that MCs have made a significant effort in the last two decades in educating and training their people: a large slice of government budget has been allocated to education and training period reaching as high as 36 per cent of total government expenditure. Regarding basic education, a large proportion of the work force in MCs is composed of young people and adults who have completed primary education only. However the situation differs from one country to the other. The average duration of full-time schooling is 5.1 years for all levels of education in Morocco, while it is 8.3 years for all other Arab countries and 7.4 per cent on average for countries elsewhere in the world with similar GNP per capita to that of Morocco. The percentage of Moroccan school age children in full time education is known to be the lowest in the region.

With regards to higher education, the total number of students has steadily increased over the last two decades with a significant increase in women participation. Local universities offer training for degrees up to masters level in all three countries (Alcouffe 1994). The number of students has been multiplied by 2.5 since with an acceleration for certain countries such as Tunisia. The output of higher education ranges between 10,000 and 30,000 each year. The rate of graduates on average reaches 120 to 220 per thousand compared to France, which scored 246 in 1994.

Abroad Maghrebian students follow specialised training or undertake Ph.D. studies. They were present in 45 countries with a big concentration in France: 67 per cent of the total. The strong dependence on France raises several specific problems. French higher education neatly separates the university, which is oriented towards the teaching of liberal professions and research, and the Grandes Ecoles, which are oriented towards production, and management of enterprises. Many of these graduates, particularly those from Algeria, tend to stay in France or to 'try their luck' in Canada, thus stocking up the 'brain drain'. However their proportion started stagnating after the economic crisis of the eighties.

Weaknesses of scientific and technical training: Graduates in social sciences and humanities represent between 50 per cent and 60 per cent of the total number of graduates of higher education. Graduates in natural sciences and technology are in a minority situation except in Algeria: their proportion is around 50 per cent are in science and engineering. In the last five years these proportions have not improved much showing that no one country has been able to increase its proportion of scientists, engineers and technicians. This indicates that crisis in policy and means has seriously occurred leading to an important deficit in terms of science and engineering personnel required by the opening up of the economies and the innovation-based competition they are likely to face fairly soon. In terms of effectiveness in the field of S&T, investment in higher education has not produced research and development capability in MCs or where it has, R&D initiatives have for the most part remained remote from industrial practice.

Difficulties in linking education and training to industry needs: in most MCs, education policy is not matched with economic policy; nor economic policy with resource endowment. Consequently, the influence of the education function on the production function of the economy, and hence on its absorption capacity, has been marginal; and this has left the economy fragile and potentially vulnerable to demographic pressures and other external shocks. Hence possibilities for 'growth reversals'. Growth reversal is associated with chronic balance of payments difficulties, the persistence of government deficits, indebtedness, massive unemployment and political instability. Education, and particularly higher education, has been a major facilitating factor in this exercise. But, as mentioned above, the effectiveness of education as a basis for technological progress would very much depend on the nature of economic policy and prevailing socio-economic circumstances. Like the technology function, the education function in MCs in general, has been supply-dominated. Both functions have been complementary, one providing the basis for the other; and in their interactive relationship, the two together have created and consolidated a monopolistic structure of production relations, thus effectively narrowing down the scope for competition, for innovative capacity building and, hence, for sustainable expansion of the economy. Consequently, while policy favoured 'supply push' technology and the education function, targeted at the modern sector of the economy, remained subservient to the prevailing technology function, the scope for indigenous innovation has been limited to learning by tinkering in the so-called 'informal sector'. But to the extent that the informal sector is disorganised and irregular, it cannot be expected to provide a robust basis for exploiting to the full the indigenous innovative potential. The fact that MCs spend a lot for a relatively modest result reflects the economic inefficiency of the manner in which the education system is organised and managed.

Higher education aimed at the accumulation of 'knowledge capital' can remain of marginal significance until the economy develops a sufficiently diversified structure. The transition is from supply push technology transfer and diffusion to need-oriented strategy of technology development.

In MCs, there is even a preponderance of employers, in particular those in the informal sector who prefer labour force without qualification in order to minimise cost preferring recruits trained in-house and which accept tough working conditions. A strong tendency is also observed for industrial firms in Morocco to recruit those with little or no education and to spend little, if at all, on research and technology

(Lahlou 1994a). There is not any significant evidence that higher education helped to meet industry needs through the production of relevant skills and the provision of R&D support (Zawdie 1994).

The Industry Sphere

The productive system in the Maghreb is generally characterised by a diffuse structure. For instance, the industrial system in Algeria consists of less than 200 public firms covering the essentials of manufacturing activities without which the industry/research liaisons appear more developed, save for energy and petrochemical sectors. In this case, it is the differentiation and subsequent isolation of production activities with respect to suppliers and clients in a very bureaucratic system which has blocked possibilities for R&D spin-offs for a variety of reasons: (1) The 'passive technology consumption' which can be assessed in several ways: the optimal use of installed equipment and their maintenance and repair. (2) Optimal use of technology, which is the first step towards the development of local know-how and technological expertise, has not always reached the required level. (3) The rate of utilisation of industrial capacities in industry has been remaining at below acceptable levels: between 30 per cent and 50 per cent on average. The causes mentioned however are not all of the pure technological type relating to the surrounding environment. (4) The limited capacity in local repair and maintenance of the equipment being used at enterprise level even though governments have encouraged the establishment of centres to repair scientific instruments and produce spare parts (Zawdie 1994). We have seen earlier how mismanagement of technology transfer as a whole contributed to this.

It is thus apparent from the Maghreb experience that technology transfer hardly helped directly or indirectly to enhance learning, and hence, the accumulation of knowledge and technological progress.

Concluding Remarks: Standing Issues in MCS

The important standing issues remain in Maghreb countries: institutional weakness and instability; the de-industrialisation process; prospects for an increase of unemployment of graduates; ambiguous role of foreign direct investment; general obsolescence of existing technological capacity; crisis of policy reforms; language and technological culture issue; many of the institutions set up to plan and implement S&T were incomplete and fragmented whenever they existed; institutional instability results also from the political instability of the sub-region. The flight of capital from productive to tertiary activities, seeking less risky, short term and relatively high return investments. This has of course contributed to worsen the situation of unemployed, notably of the youngsters and university graduates. The increase in qualifications does not necessarily improve the possibilities for the development of local capabilities. Several studies have shown that under an import substitution regime, the role of foreign direct investment in building local technological capabilities was limited. The new orientation towards the opening up of Maghreb economies in front of big international firms raises the issues of their effective participation in endogenous S&T capacity building. In this respect, the role of foreign direct investment, which is highly sought by all the countries, still remains relatively ambiguous when it comes to building S&T and innovation policy. One of the most relevant ingredients for an adapted NSI in MCs appears to be the establishment of adequate institutional instruments (laws, contracts, and governing bodies). The recently promulgated laws appear not to be sufficient to cater for the specific needs of innovation in these countries. Policies undertaken to launch a new growth dynamics need to be revised to integrate the issue of science end technology in broad terms and more precisely innovation. Added to that the existing components of traditional NSI need a real uplifting and to be put in a relatively adequate working order: this applies to the education and training institutions, the limited but existing research institutions and of course industry, namely the SMEs. Industry, and particularly the public sector, suffers a general obsolescence of existing technological capacity. While considerable learning has taken place in the last few decades since intensive industrialisation started back in the seventies, subsequent financial and economic crises have reduced the chances of renewal of both equipment and know-how. All this is happening in the context of the crises of policy reforms, which result partly from a deeply rent-seeking system not very keen on seeing its various advantages being annihilated or not even reduced.

On a more optimistic note, external stimuli to innovate do exist, not only because competitive pressures build up and pressure both SMEs and big state companies to innovate, but also because the World Bank increasingly integrates science and technology capacity building in their programmes.[3] The new system in the Maghreb should be capable of harnessing the enormous potential of creativity at a more decentralised and population level, suddenly revealed paradoxically by the various crises resulting from the implementation of SAPs.

Notes

[1] The MAGHTECH (Maghreb Technology) network we have initiated in 1994 is the first network of researchers and policy-makers in the field of S&T for the Maghreb. Initiated by the author in 1994, it has become the most important network on S&T policy in the sub-region. Strong of 350 researchers from the sub-region and Europe working in more than 12 disciplines, it has held 5 international conferences, and made several publications to contribute to renovating the thinking on the issues involved.
[2] Morocco is the first producer and exporter of phosphates in the world.
[3] The author is involved personally in building such programmes for the Maghreb sub-region.

References

Alcouffe, Alain (1994), 'National Innovation Systems: The Case of the Arab Maghreb Union', in Girma Zawdie and Abdelkader Djeflat, *Technology and Transition: The Maghreb at the Crossroads*, London: Frank Cass: 61-68.

Attia, Fethi and Mohamed Houari (2000), 'L'Innovation technologique en Tunisie', in Abdelkader Djeflat, Riadh Zghal and Mohamed Abbou (eds.), L'Innovation au Maghreb: Enjeux et Perspectives pour le 21ème Siècle, Editions Ibn Khaldoun, Oran.

Bes Marie-Pierre (1995), 'Les systèmes nationaux d'innovation des pays en développement dans la globalisation technologique', in Lahsen Abdelmalki and Claude Courlet, *Les nouvelles logiques du développement*, Paris: L'Harmattan: 73-86.

Dahmane, Madjid (1994), *The Relationship Between University-Research and Industry: Approach Through Communication*, The First International Conference MAGHTECH, Sfax: Tunisia, April.

Djeflat, Abdelkader (1985), 'Les difficultés de l'intégration inter-industrielle en Algérie et la Dépendance Technologique', *Africa Development*, Vol. 10, No. 3: 137-185.

Djeflat, Abdelkader (1987), *The Egyptian Experience of the Regulation of Technology Imports*, Joint Unit ESCWA/UNCTC (Unit on Transnational Corporations), E/ESCWA/UNCTC/87/4, 16 December, New York.

Djeflat, Abdelkader (1988), *Firmes Transnationales et Transfert de Technologie dans les industries Pétrochimiques en Afrique du Nord*, UN Economic Commission for Africa, United Nations, L/ECA/UNCTC/61, 31 August, Addis Ababa: Ethiopia.

Djeflat, Abdelkader (1992), *Technologie et Système Educatif en Algérie*, Unesco/CREAD/Medina.

Djeflat, Abdelkader (1998), 'High Technology Buying in Low Technology Environment: Issues in New Market Economies', *Industrial Marketing Management*, Vol. 27, No. 6: 483-496.

Djeflat, Abdelkader (forthcoming), 'L'innovation informelle dans le secteur formel: analyse empirique de quelques expériences dans le secteur public Algérien', in Abdelkader Djeflat and Lahsen Abdelmalki (eds.), *Les Conditions de l'Accumulation Technologique Endogène dans la Perspective du Développement*, Algiers: OPU.

Freeman, Chrisopher (1982), *The Economics of Industrial Innovation*, London: Penguin Books.

Ford, David and Abdelkader Djeflat (1983), 'Export Marketing of Industrial Products: Buyer-Seller Relationship Between Developed and Developing Countries', in Michael R. Czinkota (ed.), *Export Promotion*, New York: Praeger Publishers.

Gilles, Bertrand (1978), 'Histoire des techniques', *La Pléiade*.

INAPI (Institut National de la Propriété Industrielle, Algiers) (1979/1988), Reports 1979 and 1988.

Kline, Stephen and Nathan Rosenberg (1986), 'An Overview of Innovation', in Ralph Landau and Nathan Rosenberg (eds.), *The Positive Sum*, Washington: National Academy Press.

Mowery, David and Nathan Rosenberg (1998), *Paths of Innovation: Technological Change in 20th-Century America*, Cambridge: Cambridge University Press.

Lahlou, Mehdi (1994a), 'Performance of the Education System and Profile of Industry Demand for Skills in Morocco', in Girma Zawdie and Abdelkader Djeflat, *Technology and Transition: The Maghreb at the Cross-roads*, London: Frank Cass: 81-88.

Lahlou, Mehdi (1994b), 'Science, Technology and Society: What makes the Culture of Innovation?', in Girma Zawdie and Abdelkader Djeflat, *Technology and Transition: The Maghreb at the Cross-roads*, London: Frank Cass: 105-113.

Lahzami, Choujaa (1998), *Place et conditions de l'innovation technologique dans les pays du Maghreb à l'horizon du XXIe siècle*, The 3rd International Conference MAGHTECH, Sfax: Tunisia, April.

Lundvall, Bengt-Åke (1985), *Product Innovation and User-Producer Interaction*, Aalborg: Aalborg University Press.

Nelson, Richard R. and Sidney G. Winter (1993), 'An Evolutionary Theory of Economic Change', in Richard R. Nelson (ed.), *National Innovation Systems: a Comparative Analysis*, Oxford: Oxford University Press.

Niosi, Jorge and Philippe Faucher (1991), 'The State and International Trade: Technology and Competitiveness', in Jorge Niosi (ed.), *Technology and National Competitiveness*, Montreal: McGill-Queen's University Press: 119-141.

Perrin, Jacques and Lahsen Abdelmalki (1998), *Concevoir l'innovation pour un développement soutenable*, The 3rd International Conference MAGHTECH, Sfax: Tunisia, April.

Romer, Paul (1990), 'Endogenous Technological Change', *Journal of Political Economy*, Vol. 98, No. 5: 71-102.

Von Hippel, Eric (1976), 'The Dominant Role of Users in the Scientific Instrument Innovation Process', *Research Policy*, No. 3: 212-239.

World Bank (1993), *World Development Report: 1994*, Washington: Oxford University Press.

Zawdie, Girma (1994), 'Tertiary Education and Technological Progress in Transnational Economies: Whither Demand Pull', in Girma Zawdie and Abdelkader Djeflat, *Technology and Transition: the Maghreb et the Cross-Road*, Frank Cass, London: 151-162

Zekri, Ahmed (1990), Problématique de la Recherche Développement dans les PVD: le Cas du Maroc, Ph.D. thesis, Lyon: University of Lyon 2.

11

The Prospects for Regional Innovation System(s) Within

Sub-Saharan Africa

_____

Mario Scerri

The Argument

In this chapter I propose that history and the current reshaping of the global economy render the formation of viable national innovation systems in most of Sub-Saharan Africa virtually impossible and that consequently the only feasible analytical and planning context is a regional one. I will first explore the theoretical legitimacy of extending the national innovation system concept to that of a regional innovation system. In the process the conditions for the integrity of the regional innovation system concept will be examined. The chapter will then proceed to analyse the possibility for and the implications of a transition from a national to a regional innovation system in countries within Sub-Saharan Africa.

The Conditions of Viability of National Innovation Systems

The viability of innovation systems can be defined in terms of their potential for their own reproduction, growth and evolution.[1] Viable innovation systems often have lacunae, whether spatial or sectoral, that may place pressure on their positioning relative to other innovation systems.[2] They may also periodically fall back in the international ranking order of successful systems but their long-term survival is rarely in doubt.[3] The viability of successful systems derives from established levels of complexity, diversity and complementarities which yield a high degree of flexibility within established and emerging techno-economic paradigms. Overall, successful systems do not tend to exhibit long-term supply side bottlenecks. There are also numerous innovation systems, most of them in Africa, which are not viable and which have not, for various reasons, mostly historical, attained the threshold levels of the complex sets of preconditions for self perpetuation.

In order to approach, however tentatively, a specification of the preconditions for the viability of innovation systems, we have to examine the foundations of the national innovation systems concept. Proceeding from a specific examination of the objective of developing systems, i.e. the enhancement of the capability to innovate, to the contextual conditions that determine the limits of this potential, we can explore the contingency of the innovation systems concept. In the process we can identify those cases where the prospects for the attainment of the feasibility of specific national innovation systems are poor.

The definition of innovation has been extended and enlarged to incorporate any introduction of a product or production technique, which is new within a specific context. Again the context matters. Current usage limits the definition to technical or technological innovations. However, following Schumpeter (1954), the concept of innovation can be extended to institutional change. This extension is especially relevant to developing economies where the fundamental assumption is that development requires structural transformation. So basic is this tenet that development is occasionally equated to structural transformation. In the case of first world economies, which can, from an evolutionary perspective, be defined as viable innovation systems, there is little scope to incorporate structural transformation in the analysis of innovation. In fact the Eurostat Community Innovation Survey explicitly excludes organisational change, excepting those cases where it is the handmaiden of technical change, from the definition of innovation. Innovation therefore can represent radically different processes depending on the context in which the phenomenon is being studied.

Innovation, technological capabilities and core competences emerge from an institutional basis and it is the complex institutional web, which forms a specific system.[4] Institutions can again be defined in a variety of ways. At the more specific level they can be defined as formal organisations charged with the implementation of policies which have a bearing on the development of some aspect of the specific innovation system. On a broader level, institutions can be defined so as to incorporate both explicit and implicit sets of relationships among economic agents.[5] Again the relevance of alternative definitions differs according to the context. In the case of viable innovation systems the focus of national policy is usually on formal institutions which implement science and technology policy. The implicit institutions which govern underlying economic relationships are rarely the concern of policy since they are well established and are seen to provide the underpinnings of the specific innovation system. The accord between state and market, labour relations, managerial cultures and specific work ethics, whatever form they take, are assumed as given in viable innovation systems and explicit public policy takes these institutions as given. In those cases where these particular institutions are brought to the fore in the public planners' concern (the one notable example is the shift to Thatcherism and then to the New Labour economics under Blair) the implication is that of a deep-rooted *malaise* in the specific innovation system. These occurrences are however rare in industrialised economies. On the other hand, in the case of innovation systems which are yet to attain viability it is specifically the deficiency of these necessary underlying economic relationships which is one of the main constraints on development. It is in this area that the issue of the appropriateness of institutions is most problematic for public policy.

172

Discourses on institutions and their development are indissolubly bound with discourses on power.[6] In the absence of some egalitarian utopia institutions are the embodiment of asymmetrical distributions of power are designed to reinforce or to challenge existing power asymmetries and thus to act as the vehicle for the expansion of the power base that they represent, often at the expense of other existing or potential ones. An inter-institutional framework is therefore marked by a tension between the imperative of stability and the drive of specific institutions to augment the power base on which their members can draw. Stability is assured by a multiplicity of means ranging from a widespread acceptance of a social contract, due to effective socialisation processes and a real or perceived ability to exercise an effective voting right, to an overt and effective coercive dominance by one power base, or a conglomeration of such bases, over others. Of course the means that are used to attain stability affect the nature of the stability that is attained. It is often the case that the more overt the exercise of power, the less stable is the resulting power base. The production of knowledge and the choice of technology paths, both of which we now locate within an analysis of institutional development, are consequently transformed into a discourse on power; they become a narrative of the manifestation of power, a tale of reinforcement and conflict, of victories and defeats, where defeat often results in the eradication of the loser from the accepted strictures of the prevailing accepted rationality and ultimately from history. This discourse is obviously applicable to inter-system relationships and provides the theoretical base for the debates on appropriate technology.

Thus, a national innovation system encompasses the concentrations, dispersions and the interrelatedness of diverse 'knowledge stocks'. The determinants, and measures, of these stocks include scientists and engineers who generate and adapt technological change as well as the skills contents of the non-homogeneous labour force which determines the economy's capacity to implement innovations in the processes of production. The system is defined by the nature of interrelationships within the institutional web which contains and generates these stocks. The distribution of knowledge stocks is considered across economic sectors and across institutions, be they public or private. Thus the prime goal of the analysis of these systems is to understand the complex interactions among the various institutions within a specific system and among the different stocks of knowledge but this task necessarily has to account for history.

The most problematic aspect of the national innovation system concept is that its specificity is bound to the nation state. However, the modern international mosaic of nation states, as sovereign political units, is quite recent, with components of different vintages and in many parts of the world, not limited to third world countries, liable to often violent attempts at redefinition. Consequently there are two perspectives, the intra-statal and the supra-statal, from which the proposition of the nation state as the boundary defining unit of an integral and distinct innovation system can be questioned. While the introduction of specificity in the analysis of technological and economic development is a virtually undisputed breakthrough in the analysis of innovation, the identification of the nation state as the defining unit of containment of specificity has been quite a contentious issue.

At the intra-statal level Cooke, Uranga, and Etxebarria (1997) point out, following Gellner (1983) and others, that a state can be and often is composed of several nations, each with its own ethnic specificities, however these may be defined, often exhibiting distinctly different institutional networks, and even different economic and industrial development paths. Cooke et al. examine different possible configurations of institutional relationships between specific regions within a country and the state apparatus. It is this relationship that determines the relative autonomy of different localised innovation systems within the borders of the nation state. This critique obviously opens up the analysis of innovation systems to a higher degree of specificity than that accounted for by the national innovation system concept. It also makes the specific characteristics of intra-statal innovation systems contingent on the legal and implicit relations of power between the local and the central political authority.

Critiques that are based on a consideration of the supra-statal economy, on the other hand, tend to erode the significance of specificities in innovation systems. The relatively recent phenomenon of accelerating globalisation, arising from a combination of the rapidly growing ease of the transnational coordination of production processes and the increasing incidence of systemic technological programmes, has inadvertently posed a significant source of critique for analyses which emphasise a localised contextual determinateness.[7] This phenomenon has revitalised neoclassical approaches through the positing of a convergence of the diverse technological paths taken by nations towards some general mode which represents 'best practice' regardless of setting.[8] The adoption and, indeed, the appropriation of 'Western' technology by the successful Asian economies and the collapse of the Soviet Union are cited as evidence of this apparently indelible principle of an overriding rationality.

In his rebuttal of this critique Nelson proposes that national borders still impart a specificity to the nature of technological capabilities, i.e. that there is a range of 'factors' that enter into the generation and absorption of innovations which are, at best, imperfectly mobile, and at worst perfectly immobile, across national boundaries. This premise is a necessary implication of the assumption of tacit knowledge and its corollary of core capabilities. Kozul-Wright (1995) reinforces this proposition by arguing that the fact that the concentration of the seats of transnational corporations still lies in first world economies attests to the continuing grounding of core competencies in specific locations due to the significant role played by the non-transferable determinants of their underlying capabilities sets. Maskell and Malmberg (1999) reinforce this proposition by arguing that in a global context where competitive advantage is increasingly being located in knowledge[9], context specific core capabilities are rapidly becoming the main determinant of firm and country differentiation.[10] This also forms the basis of the refutation of the positive welfare implications that have been attached to globalisation by the neo-liberal agenda. Given the inability, and disinterest, of private corporations in respect of altering the structural underpinnings of the various countries over which their operations are spread, the globalisation phenomenon can easily, if unaccompanied by targeted initiatives by host countries, act as a reinforcing mechanism *vis-à-vis* current international inequalities.[11] In fact, as Miyoshi (1997) argues, global investment patterns in developing economies are often contingent on sets of conditions which lock host economies into specific structures which limit the possibilities for development.[12] From this follows Kozul-Wright's second argument that national policies still have a significant role in determining the context, the set of rules, explicit or implicit, within which transnational corporations operate. This leads to his apparently perverse proposition that the removal of national differences in the operational setting may actually increase the uncertainties which corporations face, thus leading to a reduction in cross-border investment. Finally, Kozul-Wright argues that unless the appropriate global regulatory regime to ensure convergence is established, the costs of globalisation for the weaker economies, or for sections within them (the erosion of the bargaining power of trade unions is the most obvious example), will continue to escalate. The negotiation of a global regulatory framework which assures a modicum of equality in development prospects will only occur through the strong lobbying of individual nation states or groupings of states which belong to roughly similar development categories.

The reading of the implications of the formation of economic blocs for the enduring relevance of nation states is contentious. On the one hand this phenomenon may be seen as an intermediate phase towards a fully integrated global economy, since within these blocs the barriers posed by national frontiers to trade and to resource mobility are eroded. The counter argument points to the difficulties experienced by the most successful bloc, the European Union, along its long tortuous path to unification as a testimony to the fact that generally this phenomenon is still in its early stages. The obstacles encountered in the attempts to integrate former Soviet satellite countries into the EU strengthen this assessment. These difficulties have reinforced the argument that effective economic unification requires a high degree of similarity in the development levels of the member countries, in order to offer reasonable prospects for convergence. From this flows the implication of the likelihood that economic blocs will actually intensify the North-South division of power, through the elimination of rivalries among the member countries in the industrial centres, without a corresponding increase in the power wielded by blocs representing weaker nations. There are therefore strong asymmetries at a global level in the nature and power of economic blocs that are emerging which will further reinforce the global imbalance of economic power.

The conditions for viability thus emerge from an understanding of the core components of the national innovation system concept. The particular definition of innovation that is adopted as relevant immediately identifies, through its emphasis on specific aspects, the context within which it is being applied. The less viable the innovation system the more all-embracing the definition of innovation has to be. This relationship between the broadness of definition and the stage of development of innovation systems also applies to institutions. In less-developed systems the implicit institutions which govern economic relationships cannot be assumed as given and suitable for the attainment of viability. Finally, the foundation on which the integrity of the national innovation system concept is premised is the effectively uncontested nation state.

The Prospects for National Innovation Systems in Sub-Saharan Africa

When we come to Sub-Saharan Africa, we find that in most cases the basic requirements for the establishment of viable innovation systems are absent in most of the countries in this region. In the first case, and perhaps the *primum causa* of the economic woes, there is the history of the development of nation states within the region. Most of the modern African states were an ersatz colonial creation whose subsequent independence was mostly bought at the cost of devastating wars of liberation. The first decades of the post-colonial period saw this part of the world turn into a cold war arena, with competing models of economic and civil governance imposed on ill-suited settings and with often extremely authoritarian regimes propped up by one or the other of the cold war superpowers. It is a region which is still devastated by major areas of armed conflict and of countries whose economic structure is still ravaged by recent wars. South Africa which stands alone as satisfying most of the preconditions for the attainment of viability is still burdened by the economic structure inherited from apartheid.[13] Even where the legal legitimacy of particular African states is well established and stable the ability of governments to devise and implement economic policy is severely circumscribed by crippling foreign debt burdens and the strictures of structural adjustment programmes imposed by international financial institutions. The end result is the current situation of a subcontinent composed of too large a number of nation states whose internal legitimacy is still being contested and the implications of this instability for individual innovation systems are ominous.[14]

A fluid national context obviously inhibits the development of an appropriate institutional framework within which technological capabilities can grow. In most cases the economic power relations which are now established as the foundation on which those formal institutions that affect the generation of human capabilities are built are still unsettled in most economies in the region. The relationship between labour and capital in most Sub-Saharan economies is often skewed with high unemployment rates, weakened labour unions and the labour cost requirements of foreign investors depleting workers of any bargaining power within the market.

In fact if one were to identify the principal bottleneck in an institutional framework in most economies it would be human capital formation. In a number of countries current wars and the devastation of past wars have caused an impoverishment and a degradation of the human capital base. Human capital development has been further inhibited by the combination of crippling effects of foreign debt and the imposition of structural adjustment programmes on the fiscus. The fiscal constraints imposed by these programmes severely limit the ability of the state to provide the set of basic needs and safety nets which are essential for a long-term human capital formation process. The IDRC (1999) study on measures to promote cooperation in science and technology within the Southern African Development Community (SADC) provides a number of crucial insights into the possibilities for the development of regional innovation system(s) within Sub-Saharan Africa. The findings of an earlier report[15] noted in this study identify the three main sectoral research characteristics among SADC countries as: '1. [T]he commercial sector undertakes substantial research through institutes owned by specific agricultural industries often funded through producer levies. 2. The mining sector is dominated by multinationals whose research is pursued offshore and whose local geological exploration activities frequently bypass the existing state geological surveys. 3. Industry on the other hand tends to be importers of technology and generally operate at the lower end of value addition' (IDRC 1999: 13, numbers added). The spread of HIV/AIDS and other diseases such as TB, malaria and cholera, all of which are strongly correlated to poverty, have intensified the erosion of the skills base of the regional labour force. Wars and economic conditions have generated massive refugee flows and a high rate of migration across national borders. All these factors have obviously considerably shortened the planning time horizon for human capital investment.

Given these fundamental obstacles, the prognosis for the establishment of viable national innovation systems is, in most cases, extremely poor. Indeed the entire region shows all the signs of being subject to a 'low-level equilibrium trap' where the development of specific innovation systems is often constrained by factors which are exogenous to the specific economy in question. These factors can be regional, such as wars and other forms of armed conflict as well as economic catastrophes which spill over across borders generating migrations of political and economic refugees. They can be extra regional as in the case of foreign debt servicing and the crippling effects of structural adjustment programmes on the ability of governments to establish the conditions for human capital formation. It is in fact useful to focus on human capital formation as the nexus of the symptoms of ailing innovation systems within the region and consequently as the prime yardstick of the effectiveness of policies which affect the prospects for systems. There is within the region a recognition that the obstacles to the development of innovation systems require a direct and overt direction by the state. Thus the recommendations of a report quoted in the IDRC (1999: 16) study points to a '..need for the state to display a more interventionist role in the underdeveloped countries than in the industrialised ones.[16] Implicitly what is being argued for is a set of national systems of innovation with the state actively laying down the regulatory, physical, human and social infrastructure.' Emphasis is placed on an overt interventionist role of governments in the fostering of broad based human capital development in order to increase absorptive capacity.

The Rationale for a Regional Approach to Innovation

The arguments which have been outlined in the previous section against the possibility of establishing viable innovation systems in most Sub-Saharan countries can cut both ways when arguing the case for a regional innovation system. On the one hand it can be argued that the formidable constraints on attaining viability can be seen as truly insurmountable impediments to the formation of an integrated regional innovation system. Moreover, the history of the numerous economic integration initiatives in the region does not really augur well for regional approaches to the region's development. Overall, the effects of regional associations within Sub-Saharan Africa on economic upliftment have been minimal.[17] One of the main obstacles to integration is what may be termed the reversed north-south phenomenon within the region where all countries except South Africa are mainly primary sector exporters. South Africa occupies the position of the single regional economic superpower, potentially the new economic coloniser of countries that cannot trade with one another due to the low degree of complementarity in their production bases. This situation obviously reduces significantly the scope for mutual benefit which is usually the main motive power for economic integration. There are of course specific initiatives which have been successful. In the case of S&T there are cooperative programmes within SADC that are in operation and planned projects which have every chance of being successful. However, these are specific cases which, from an innovation system perspective, are symptomatic of a fragmented approach to the development of innovation capability.[18]

On the other hand the overall improbability of most countries' innovation system attaining viability raises the urgent need for an alternative approach. In spite of the impediments facing any integration initiatives there is still a clearly recognised need for economic cooperation and even integration within the Sub-Saharan region. The era of rapid globalisation has rendered regional cooperation necessary as a means for individual countries and regions to retain a voice and a degree of power in the emerging global economic order. It is thus rapidly becoming clear that, whatever the prospects, the integration of innovation systems is the only policy option for the development of viability as it has been defined in this chapter.

Of course empirical necessity does not of itself constitute a sufficiently valid theoretical case for the concept of a regional innovation system. Such a case can, however, be made. As discussed earlier, the paradigmatic breakthrough of the national innovation system concept lay in its introduction of specificity and localised knowledge into the analysis of innovation. The choice of the nation state as the unit of analysis was empirically determined and hence contingent and essentially peripheral to the theoretical core. An emphasis on the nation state is therefore tenable only with respect to specific historical contexts. The emergence of a European innovation system is testimony to that.

There is of course a unique feature of the progression towards a regional innovation system that is being proposed here for Sub-Saharan Africa. In the case of Europe the regional system evolved from already viable national innovation systems.

In the case of the Sub-Saharan region the creation of a viable regional innovation system would have to leapfrog the preliminary stage of viable national innovation systems. Again, on a theoretical level, given the provision for historical contingencies, there is no contradiction that arises in the inversion of the progression towards a regional innovation system.

The rationale for cooperation is well established. There is the promise of a regional market with its implications for economies of scale and division of labour. There is the enhanced potential for developing sound indigenous technology bases and for exploiting economies of scale and complementarities in innovation activities. There are the obvious benefits that can accrue from the free movement of capital, expertise, technology and, although this is a controversial issue, of labour across the region. Sub-Saharan Africa is a potential economic powerhouse, with vast untapped reserves of mineral resources and populations which have often been forced, through the very conditions which inhibit planned human capital development, to develop a high degree of entrepreneurship. The question that then emerges is what distinguishes a regional innovation system as a planning and negotiating context from the general economic integration framework.

There are obviously large overlapping areas between the two frameworks. However, the distinctions both theoretical and pragmatic (from an implementation perspective) are quite clear. Within a single innovation system supply side bottlenecks of all sorts can be addressed through complementary markets with a greater probability of success than could be achieved within the planning framework of a number of individual impoverished national innovation systems. The current initiatives in cooperation in S&T activities would be placed within a broader innovation context in order to increase the probability that the outcomes of these initiatives would translate into a higher level of technological absorptive capacity across the region. The process of integration of innovation systems would over time engender a reordering of relative core competences within the region. Obviously there would be an element of attrition with specific nodes of competences disappearing. If the integration process is to be feasible, this must be more than offset by the gains from the emergence of a sounder, and more diversified regional core competence base which over the long term is less needful of protection. The benefits of this process must obviously be spread across the region so that no country is a net loser through the transition. The potential gains from the integration of innovation systems will emerge from two complementary processes. The first is where new technological complementarity matrices emerge within the regional innovation context. The second would be the exploitation of economies of scale in innovation within the larger internal factor and commodity markets.

Conclusion: The Policy Framework for the Integration of Innovation Systems within Africa

The current policy framework for economic integration and development within Africa is the *New Partnership for Africa's Development* (NEPAD) document issued in 2001. This is the first comprehensive approach to the emergence of a regional innovation system across the continent. This policy is itself set within the broader context of the newly formed *African Union*. The NEPAD document (October 2001) makes a rather startling statement in its introduction (item 6) when it says

> The resources, including capital, technology and human skills, that are required to launch a global war on poverty exist in abundance, and are within our reach. What is required to mobilise these resources and to use them properly, is bold and imaginative leadership.

Whatever the empirical basis for the statement about the availability of resources, this statement is remarkable in that, in a manner that recalls Hirschman, it identifies political will and decision-making capacity as the crucial scarce resource in the development process. Further on in the introduction the basic requirement for the resolution of this restraint is seen as democracy and a civil society. This shortcoming is referred to in a number of instances but is captured succinctly in item 23, again in the introduction, where it is stated categorically that '[t]oday, the weak state remains a major constraint to sustainable development in a number of countries'. In the discourse on policy this statement opens up a hitherto unexplored theoretical space in the approach to integration in Africa. It addresses for the first time the fundamental judicial, political and social preconditions for the establishment of a viable regional innovation system. Beyond that, this policy initiative is broad enough to form the type of integrated policy framework that is required to draw together what have up to now been fragmented, and hence largely ineffective, policies for a long term sustainable growth path.

Notes

[1] These three terms are used distinctly and specifically. 'Reproduction' refers to the replication of existing systems, 'growth' implies the expansion of systems, and 'evolution' indicates the process of mutation of systems.

[2] Within the constraints of this chapter I will not deal specifically with the temporal aspect of the evolution of innovation systems, of the streams of events, accidental or necessary depending on the particular historiographic approach that is adopted, that determine present states.

[3] The language that I use in referring to innovation systems carries implications of anthropomorphising, of endowing institutional formations with human (or at least organic) characteristics. This biological metaphor, deriving from the evolutionary account of technological change, is diametrically opposite to what Mittermaier (1986) refers to as the 'mechanomorphism' implicit in neoclassical economics.

4 Nelson (1993: 4-5) defines an innovation system as '... a set of institutions whose interactions determine the innovative performance ... a set of institutional aspects that, together, plays the major role in influencing innovative performance'.

5 Johnson (1988: 280) defines institutions in broad terms as the '... sets of routines, rules, norms and laws, which by reducing the amount of information necessary for individual and collective action make society, and the reproduction of society, possible'.

6 'In exercising power, individuals employ the resources available to them ... by accumulating resources of various kinds, individuals can augment their power ... While resources can be built up personally, they are also commonly accumulated within the framework of institutions, which are important bases for the exercise of power' (Thompson 1995: 13).

7 '... the globalizing tendencies of modern social life have rendered territorial delimitation increasingly problematic. Particular nation-states are increasingly embedded in networks of power ... which extend well beyond their boundaries and which limit, to an extent which varies greatly from one country to another, the room to manoeuvre of ... national governments. ... there are a range of issues concerning, for instance, the activities of transnational corporations, problems of pollution and environmental degradation, the resolution of armed conflict and the proliferation of weapons of mass destruction which cannot be satisfactorily addressed within the political framework of the nation-state' (Thompson 1995: 253).

8 '... a new best-practice set of global rules for designers, engineers, entrepreneurs, managers and marketing agents is emerging ... The resulting product (of trans-national corporations) is ... a complicated bundle of inputs, produced in a variety of locations, assembled in home or host countries for sale in those countries or anywhere in the world' (Kozul-Wright 1995: 152, parenthesis added).

9 '... the knowledge-based economy is characterised by three elements: the growing importance of economic transactions focussed on knowledge itself; rapid qualitative changes in goods and services; and the incorporation of the creation and implementation of change itself into the mission of economic agents' (Maskell and Malmberg 1999: 167).

10 '... knowledge creation of even the most globally oriented firms or sectors is, at least to some extent, influenced by differences in the economic properties of their place of location. Firms are progressively stimulated by and dependent on localised technological capabilities in order to maintain and increase their competitiveness *precisely* because of the drive towards globalisation and the resulting homogenisation of formerly critical factors of production' (ibid.: 168, emphasis added).

'... a logical and interesting consequence of the present development towards a global economy is that the more easily codifiable (tradable) knowledge can be accessed, the more crucial does tacit knowledge become for sustaining or enhancing the competitive position of the firm. If all factors of production, all organisational blueprints, all market information, and all production technologies were readily available in all parts of the world at (more or less) the same price, the market process of competition between firms would dwindle' (ibid: 172).

11 Kozul-Wright offers the following sound rebuttal to the neo-liberal prescriptive non sequitur that the emergence of the trans-national corporation and the weakening of the nation-state in the era of globalisation will lead to the attainment of Pareto efficiency across national borders:

'The presence of these firms, in itself, is testament to discontinuities in the economic environment and there is no *a priori* reason for assuming that the activities of TNCs, any more than international trade, will by themselves remove inequalities between and within regions. ... weak states are more likely to establish weak, and ultimately unsustainable, growth paths and reinforce global segmentation' (Kozul-Wright 1995: 159-160).

12 Capital can move freely in time and space as long as the target area guarantees: (1) a stable political structure (often meaning dictatorship); (2) viable labour conditions (that is, cheaply trained or trainable labour plus the absence of unionism, feminism, and human rights); (3) reasonable infrastructure; (4) lower tax rates (favoured treatment by the host governments); and (5) indifference to the environment (Miyoshi 1997: 50).

[13] It can be argued that South Africa's liberation from apartheid came at least a decade too late to allow its first democratic government sufficient manoeuvring space in policy formulation. After the collapse of the Soviet Union most forms of government intervention in the market, apart from facilitation, became relegated, through a logical *non sequitur,* to a universally discredited model. This has severely limited the perceived set of policy options for the reconstruction and transformation of the country.

[14] Gellner (1983, 1996) argues that the coherence of a nation state requires a sort of forced amnesia of ethnic allegiances and of nationality. The carving up of Africa among the imperial powers without regard to most of the considerations that determine national borders seriously impairs the possibility of fulfilling this condition. However, even the notion of ethnicity and nationhood tends to be blurred in the region. If a nation is, *a la* Anderson (1983), an imagined community, there are few areas where the underlying imagination is more evanescent than in Sub-Saharan Africa. These factors certainly pose serious obstacles for the development of the stable nation state.

[15] Mshigeni (1994).

[16] Ogbu, Oyeyinka, and Mlawa (1995).

[17] There are currently ten economic blocs within Sub-Saharan Africa. They range from small three country units such as the East African Community to the large groupings such as ECOWAS, COMESA and SADC. Individual countries often belong to a number of blocs.

[18] It is relevant to in this case to look at the economic indicators for those SADC countries that show the highest GNP per capita figures (IDRC 1999: Table 1). These are the Seychelles, Mauritius, Botswana and South Africa, in that order. If we compare the first three countries with South Africa, which lies at the bottom of this range, we find little immediate evidence for the relevance of formal S&T policy to the development process.

References

Anderson, Benedict (1983), *Imagined Communities*, New York: Verso.

Cooke, Phillip, Mikel G. Uranga, and Goio Etxebarria (1997), 'Regional Innovation Systems: Institutional and Organisational Dimensions', *Research Policy*, Vol. 26: 475-491.

Gellner, Ernest (1983), *Nations and Nationalism*, Oxford: Basil Blackwell.

Gellner, Ernest (1996), *Conditions of Liberty: Civil Society and its Rivals*, London: Penguin Books.

International Development Research Centre (IDRC) (1999), *Framework for Regional Cooperation in Science and Technology*, Cape Town: Edunet Consulting.

Johnson, Björn (1988), 'An Institutional Approach to the Small Country Problem', in Christopher Freeman and Bengt-Åke Lundvall (1988), *Small Countries Facing the Technological Revolution*, London: Pinter Publishers.

Mshigeni, Keto (1994), *Study on Science and Technology*, Gaborone: SADC mimeo.

Kozul-Wright, Richard (1995), 'Transnational Corporations and the Nation State', in Jonathan Michie and James G. Smith (eds.) (1995), *Managing the Global Economy*, Oxford: Oxford University Press.

Maskell, Peter and Anders Malmberg (1999), 'Localised Learning and Industrial Competitiveness', *Cambridge Journal of Economics*, Vol. 23, No. 2: 167-185.

Michie, Jonathan and James G. Smith (eds.) (1995), *Managing the Global Economy*, Oxford: Oxford University Press.

Mittermaier, Karl (1986), 'Mechanomorphism', in Israel M. Kirzner (ed.), *Subjectivism, Intelligibility and Economic Understanding: Essays in Honour of L. Lachman*, New York: New York University Press.

Miyoshi, Masao (1997), 'Sites of Resistance in the Global Economy', in Keith Ansell-Pearson, Benita Parry, and Judith Squires (eds.) (1997), *Cultural Readings of Imperialism: Edward Said and the Gravity of History*, London: Lawrence & Wishart.

Nelson, Richard R. (1993), *National Innovation Systems: A Comparative Analysis*, New York: Oxford University Press.

Ogbu, Osita M., Banji O. Oyeyinka, and Hasa M. Mlawa (1995), *Technology Policy on Africa*, Ottawa: IDRC.

Schumpeter, Joseph A. (1954), *Capitalism, Socialism and Democracy*, London: George Allen & Unwin.

Thompson, John B. (1995), The Media and Modernity: A Social Theory of the Media, Cambridge: Polity Press.

12

Innovation Systems and Endogenous Development

A Perspective of Asia for Africa

Shulin Gu

Introduction

Globalisation and technological progress might bring about possibilities wider than ever before in ensuring the freedom of people from want. However development performance is not for celebration. In the past two decades differences in economic growth among regions and countries were widened. Although East Asia and the Pacific Region grew at 6 per cent a year and quadrupled per capita income in 1975-99, growth in the Arab States and Latin America and the Caribbean has been slower at less than 1 per cent in the period. Most devastating has been the development of Sub-Saharan Africa (SSA), where already low incomes have fallen: in 1975-99 GDP per capita growth averaged 1 per cent (UNDP 2001: 10-11). The extent of the stagnation and reversals in SSA in the 1990s is striking and not even seen in previous decades.

Beyond the statistics, the most serious is perhaps the overwhelming pessimistic atmosphere. It is so pervading that almost anyone, as Muchie and Lundvall (2001) point out, who writes on Africa from any academic direction seems to assert the 'failure' of Africa. The pessimist cognitive orientation has affected not only international but also African researchers; negative portrayal of Africa's capacities and possibilities repeatedly appear in analyses and reports. What is wrong – the package of structural adjustment projects (SAP)? The lens that observers and analysts use when looking at Africa? Or, was Africa doomed to be hopeless?

This chapter explores implications of Innovation Systems (IS) as an analytical lens and policy tool supportive for Africa to make innovation systems and build competences in the era of globalisation. As a researcher with Chinese background, my discussion will be limited, based on experiences and lessons in Asia especially in China. Particularly I address three questions: (1) How important is learning and competence building for development? (2) Is the learning model and development path exclusively single? (3) What kind of international assistance is needed, provided that the international community has recently in called on the rich countries and international organisations to take concrete actions for the 'priority countries' (of the total 59 priority countries, 38 are in SSA (UNDP 2003))? We end the chapter by a summary: what Asian Experiences are to be referred to and what are not.

How Important is Learning and Competence Building for Development?

First of all, let's talk about the current dominant theory for development; see how it causes misleading, and then explore what was the essence for successful catching-up in Asian NIE.

The current dominant view of development is neo-classic, its hallmark being the aggressive World Bank and IMF Structural Adjustment Programmes (SAP). This theory focuses on full utility of existing resources, assumes a central role to the market for effective resource allocation, or 'getting the prices right' (Kruger 1995). This view of development, provides useful references though, seems to be away from handling the essential changes associated with development. Earlier popular, the classic development theory also gave no attention to institutional change and little guidance to technological capability building.[1]

As Behrman and Srinivasan (1995) and Rodrik (1995) indicate, the result of SAP is roughly that: a) it led to so-called (static) efficiency of resource allocation by the reduction of price-cost margins in import-competing sectors (where previous distorted competition status misled the allocation of benefits). It also led to technical efficiency where existing incumbent firms whose TFP (Total Factor Productivity) increased (while in some cases the incumbent firms went to bankruptcy); b) but the impact of trade reform on structural rationalisation has not been apparent, for change in industrial structure relates to intra-firm restructuring as well as entry and exit of firms; and more seriously c) investment in R&D has declined. The decline of investment in agricultural research and extension (R&E) services (Enos 1995; Friis-Hansen 2000) was even more noticeable. Many extension workers lost their jobs. Investment in agricultural infrastructure and its maintenance was cut down. Multinational seed companies went in to substitute for previously state or quasi-state seed agents, with the coverage of quality seeds provision reached to no more than a small proportion of demand (2-4 per cent), the majority of farmers could not access to or not afford for such provision.

These are typical illustrations of policies based on static cost-efficiency calculation. It ignores the necessity to invest in institution and technology and infrastructure for long-term development. Particularly the commitment that is mainly put to international sources for agriculture is mistaken, for international provision could not fit with diverse and sometimes unique local endowments, not to say about heightened factor price of this supply that is prohibitive to local people. As a result, agricultural production got worse. The already started agro-industry innovation system, rudimental though, is being dismantled with few exceptions. This is the part that the international value chains have interests in and incorporate. This not only is the root cause for formidable poverty but also has weakened the basis for long-term development in SSA where more than 70 per cent of population is still rural residents. Ruttan (2001) has strongly argued that a minimum level of agricultural R&D is one of the necessary conditions for the poorest countries in order to sustain development. Commentators (e.g. Rodrik 1995: 2971-2972) have concluded that 'getting prices right, in itself and of itself, will be insufficient to make Bolivia or Ghana grow at Korean rates'.

But what is the essence for successful development in Asian NIEs (Newly Industrializing Economies, like South Korea and Taiwan)? The essence is undertaking change. Developments there proceed in association with changes in technology, institution, structure and attitude, through learning (Amsden 1989; Kim 1997; Mytelka 1998; Wade 1990). Development will not happen at all if it really falls in an equilibrium state by effective allocation of currently possessed resources, which would be only a 'low equilibrium track'. Korea for example was in 1960 an economy with the comparative advantages in raw materials and agricultural products. By 1999 Korea became one that is comparative in sophisticated manufactured goods such as semiconductor and automobile. The rapid structural upgrading was not automatically achieved. It was the result of the proactive government policies that created a policy environment for capital investment including capital investments that were temporarily disadvantageous while structurally important. Meanwhile enormous investment was made in education, training, and various infrastructures and in acquisition of technology. Furthermore behind the policy and investment, there were extraordinary efforts made by people – managers, workers, engineers, they often worked seven days a week, 14-16 hours a day to absorb and exercise upon external sourced technology.[2] Elsewhere I call such learning as 'enhanced' (Gu 1999b). Efforts for narrowing the gaps with the economic frontier have to be via purposeful 'enhanced' efforts, because the restraints in industrial structure and competence structure could not be removed merely by means of effective allocation of currently possessed resources like the neo-classic view taught. They have to create competences and foundations for higher ordered structure. Surely, political stability was one of the necessary conditions for the enhanced learning, and openness to international knowledge flows was another one. With these necessary conditions, Korea mobilised their resources to learn, started from the place where per capita GDP was only about US$50 – not richer at all than the above mentioned 59 priority countries. The essence for successful development is undertaking change through enhanced learning. It is intrinsically an endogenous process.

Table 1 Korea's Top Ten Exports, 1960-1999 (US$ million, per cent)

	1960			1980			1999		
	Item	Amount		Item	Amount		Item	Amount	
1	Iron ore	5.3	(13.0)	Textiles	5041	(28.8)	Semiconductor	18852	(13.1)
2	Tungsten ore	5.1	(12.6)	Electronics	2004	(11.4)	Automobile	11169	(7.8)
3	Raw silk	2.7	(6.7)	Iron and steel products	1570	(9.0)	Textile fabrics	9111	(6.3)
4	Anthracite	2.4	(5.8)	Footwear	908	(5.2)	Ship	7489	(5.2)
5	Cuttle fish	2.3	(5.5)	Ship	610	(3.5)	Petrochemical products	7017	(4.9)
6	Live fish	1.9	(4.5)	Synthetic fibres	571	(3.3)	Iron & steel products	6908	(4.8)
7	Natural graphite	1.7	(4.2)	Metal products	401	(2.3)	Electronics home appliances	6361	(4.4)
8	Plywood	1.4	(3.3)	Plywood	352	(2.0)	Textile products	5827	(4.1)
9	Rice	1.4	(3.3)	Fish	352	(2.0)	General machine	5486	(3.9)
10	Bristles	1.2	(3.0)	Electrical goods	324	(1.9)	Plastic products	27	(2.0)

Source: The Bank of Korea, Economic Statistics Yearbook, reproduced from Choi (2000).

Stiglitz (2001) recognises clearly the deficiency of conventional development thoughts. He argues for a new development agenda as the following. Innovation Systems and underpinning Evolutionary Theorem of Growth and Development should make great contributions to the new development agenda.

> The seeming disappearance of development economics as a separate discipline some quarter century ago could not have come at a more inopportune time [...] In the last two decades, there has been a growing awareness of the limitations of the competitive paradigm [...] Much of the theoretical and empirical work in *developed* countries has focused, for instance, on agency theory, the new industrial organization, finance, and R&D. Yet, in this same period, the reigning paradigm in development economics was the Washington consensus, which ignored these considerations, despite the fact that they are even more important to developing countries.
>
> [...] A new development agenda thus must center around (i) identifying and explaining key characteristics of developing countries, [...] and exploring the macro-economic implications, e.g. for growth and stability; (ii) describing the process of *change*, how institutions, including social and political institutions, and economic structures are altered in the process of development [...]. (Stiglitz 2001)

The notion of innovation systems (Freeman 1987; Lundvall 1992; Nelson 1993; OECD 1999), and underpinning evolutionary theory of growth and development (Nelson 1995; Nelson and Winter 1985) are developed from the Schumpeterian tradition. It emerged in the 1990s, in response to accelerating paces of change, aiming at incorporating insights from innovation studies into investigation and explanation of growth and competitiveness of large social systems. One may ask: is this notion and theory relevant for developing countries, provided that either Schumpeter or the present-day pioneering researchers have their focus on advanced economies? The answer is affirmatively 'yes'. Change is as central in development of developing countries as in advanced market economies, perhaps even more so for the former ones while in somehow different ways. As economic historians (e.g. Fei and Ranis 1997; Ohkawa and Rosovsky 1973) observed against Japan and the Asian NIEs, development of a developing country involves historical transition, or 'transition growth', from a pre-modern pattern towards a modern pattern. In transition growth, the complexity comes from combined fundamental institutional development, which increases the sophistication of division of labour, and intensified learning to master 'modern' science and technology-based innovation.

The notion of innovation systems and the underpinning evolutionary theorem may serve a reference framework for a 'paradigm shift' of thoughts about development and policy-making. Innovation concerns more than R&D. Innovation policy has to be a part of economic policy, while up to now more than often the two chunks of policies in developing countries are separated from each other, because of the status of development economics that Stiglitz points out. IS also provides an observation lens and analytical tools that assume a central importance to learning and competence building for sustainable dynamics of development. I believe only with this could intellectual works on Africa become informative in support for positive loops of learning and experimentation, no longer as merely failure-telling.

Is Learning Model and Development Path Exclusively Single?

How to deal with successful experiences worked out somewhere else? What is the value of Asian NIEs successes? Is there a single model for learning and competence upgrading?

An evolutionary view nullifies any single set of models or policies that succeeded in some countries as to be universally applicable to all the others. In line with IS, one of the most outstanding findings in international comparison is that innovation systems are country-specific (Nelson 1993, 1996; OECD 1999). Specifics to countries are not only in specialised patterns of science and engineering base, in advantageous areas of innovation and international trade, but also in policy institutions and measures and supporting institutions. This finding of country specifics might be surprising to those who are preoccupied with conventional views. To them the world is universally homogenous, while this is natural from an evolutionary lens. The complex and dynamic interactions between technology and institutions cast the specifics of innovation systems, as well as distinct learning models and plural development paths.

We have summarised the experience of Korea: enhanced learning in association with high level of investment in capital equipment, infrastructures and human development under intensive exchange with the world market. Taiwan shares with

Korea the basics. However, looking into learning mechanisms with which the two economies climb up the competence ladder, one finds that they are distinct to each other (Gu 1999b). Characteristically the Korean IS has large firm structure; in contrast, small firms dominate in Taiwan. Hence Korea created a model of individual large firms-based learning, and Taiwan a model of small firm networks-based. As a result of technology-institution interactions, distinctive areas of learning were selected. By the end of 1990s Korea developed international comparative advantages in large mechanical systems like automobile and ships and scale-processed components like semiconductor (see Table 2); while Taiwan has the advantages in small systems like computer motherboards and mice, image scanners, monitors, keyboards, simple CNC machine tools. Furthermore, policy priorities and supporting institutions differ while both fitted to their respective technology-institution frameworks. More comparisons in learning and IS between Korean and Taiwan are listed in the following table.

Table 2 Learning in the Korean and Taiwanese IS

Learning and NIS Characteristics	Korea Individual large firms-based learning	Taiwan Small firm networks-based learning
Learning mechanism	Cyclic reverse travelling up the capability ladder	High entry and forward and backward linkages
Community of learning practice	Individual firms	A group of firms
Priority of S&T and industry policy	'Picking the winners' to give direct support	Focused on infrastructure and neutral business regulations
Supporting institutions	Provided by firms themselves under the close alliance with the government	Network, technological infrastructure; market friendly regulatory institutions
Location of R&D	80 per cent at private firms	50 per cent in public institutes
Technological strengths	'Mass' technology, large systems	'Niche' technology, small systems

Source: Gu (1999b)

How about China? What characterises the mechanisms behind its development in the 1980s and 1990s? China has since the end of the 1970s been undertakinig market reform, meanwhile achieved rapid growth and structural upgrading. Elsewhere we (Gu 1999a: 308-310; Gu and Steinmueller 1996/2000) summarised the learning in the period of economic transition as 'recombination learning'. Several parallel processes intertwined to assist and reinforce each other: (1) the stimulation of market reform and trade liberalisation that produced new incentives and induced the reallocation of innovative capabilities; (2) the re-organisation of previously accumulated capabilities in production, design, testing and R&D in novel and productive ways to meet the challenges of market reform and trade liberalisation; (3) the intensive learning devoted to identifying and filling major gaps in the competences inherited from the previous system; and (4) the efforts aimed at institutional restructuring that support these developments. Compared with the experiences in Korea and Taiwan, this learning has been characterised by large scale of institutional adjustment and restructuring, and in subtle re-deployment of accumulated technical assets in support of intensive acquisition of foreign technology. In terms of targeting market of production and learning, both domestic and international markets were important. In this connection, 'export orientation' should not be fixed as a doctrine. Unfortunately, it has been used more or less like a doctrine for Africa. Historically, the domestic market was extremely important for the rise of United States in the turn of the 19th and 20th century. For agricultural production in SSA, I assume domestic consumers should be the major clients, since export-agriculture accounts only a small part of the output while a not small proportion of people there are in hunger.

Up to the second half of the 1990s, symptoms increasingly manifested the fact that the space of development created by reforms and re-combination learning is just about exhausted. And the accession to WTO aggregated one more element to the need that China and its innovation system move into a new period of development. The WTO regulations ruled out the trade and industry policies that Korea and Taiwan took in the 1970s and 1980s, and elevated the costs for acquiring foreign technologies, these together were the major means for their learning, a learning that rested heavily on 'imitation' by repeated acquisition of foreign technology. China has to do more of endogenous innovation soon, if China is to continue the momentum of development in the coming years. The Decision by the Central Committee of Communist Party and the State Council (1999) reflects this necessity.[3] Favourable conditions include (Lu and Mu 2003) a certain level of technological capability, which has been re-organised in the past decades, the large domestic market, which embraces diverse demands/tastes (some rather advanced some preliminary while many unique), and the general trends of 'open architecture' and 'modular revolution' (e.g. Langlois 2001) that make technological knowledge more migratory. However this requires a lot of efforts rather different from what has been undertaken in the past decades, in development strategy and policy, business management, and IS construction. Likewise, many Asian NIEs are undergoing changes in innovation policy and innovation systems. South Korea, for example (Choi 2000; Kim 2000; Youn, Kwon, and Chung 2000), has since the financial crisis in 1997 urged *chaebol* to transform their large and multi-branched structure into core business-based smaller companies. The 1997 Korean Special Act for the Promotion of Venture Business made the policies supportive for small initiatives. Hence the large firm structure of Korea is in alternation, and the pace to endogenous innovation seems in accelerating.

What the above says is that development paths and learning models are country-specific (also sector-specific and region-specific). Even for the same country, development approach and the pattern of learning change over time. There is no such thing as *the* East Asia model for success, except the very importance of making change and managing change. Sticking to doctrines produces failures. For example SAP takes 'basically the same set of policies' to implement in 'a wide range of very different situations and countries with entirely inadequate concern for these differences throughout Africa' (Friis-Hansen 2000). Successful stories could spring out only with creativeness and experimentation rooted in the soil. And the IS approach provides insights and tools for looking into the 'micro-foundations' for learning and development, as I have applied a little in the above analysis of learning.

What Kind of International Assistance is Needed?

In order to end human poverty especially in SSA, the international community, based on the historic Millennium Declaration adopted by 189 countries at the UN Millennium Summit in September 2000, has called on the rich countries and international organisations to take concrete actions (UNDP 2003). International funds flowing in will increase considerably in the near future. What kind of international assistance is needed? The answer is: respect for recipient national ownership and an emphasis on policy capacity building.

From an evolutionary perspective each policy programme, with or without international resources involved, is an initiative to some kind of social innovation. In a dynamic world, where information is imperfect; choices are vague; and outcomes of a change involve uncertainty, the policy maker stands in the situation just like what a firm's manager faces in his initiation of a technological change (Metcalfe and Georghiou 1998: 80). He has to accept a considerable amount of indeterminacy and uncertainty in the consequence of the policy initiative, must adjust the policy by learning about the work of the system over time. Aoki and his associates (Aoki 1996; Aoki, Kim, and Okuno-Fujiwara 1996) argue that particularly for radical systems transformations involved in economic development, a positive, learning policy process is indispensable. Following a major reform initiative, various agents, organisations and individuals must make adjustments to cope with the disturbance and to re-build complementary relations with each other. Such institutional adjustment takes its own way, with outcomes often not as initially expected – 'unexpected fit'. It is impossible for a reform program to be designed perfectly, or even roughly properly in advance.

Elsewhere I have discussed the adaptive policy process responsible for the unprecedented transformation of the R&D system in China during the 1980s and 1990s (Gu 1999a: Part I). Initially in 1985, the design for transforming the R&D system, which was developed in the period of central planning, focused on the 'technology market' solution. It was soon recognised as inadequate as both buyers and sellers of technology felt difficult in transaction of technology for a number of reasons. As a response, in 1987 reform policy began to promote merger of R&D institutes into existing enterprises or enterprise groups. This policy was only partially applicable at the start. Huge gaps between the merging parties were hard to be filled up immediately. In the next year 1988 reform policy adopted the practice created by scientists and engineers, launched the Torch Programme to encourage spin-off enterprises, and this initiative received active responses. By the end of the 1980s, reform policy furthered to embrace transformation of R&D institutes on a whole institute basis, which was in effect legitimating the actual progress, already apparent to many industrial R&D institutes from their adaptations during a number of years. In this way by the early 1990s solutions for transformation enlarged as multiple and practical as possible, all the mentioned approaches played a part.

These are root reasons for respect for the recipient's national ownership of international funded programmes: national experts and managers are in the context, aware of the needs and possibilities; they are the actor in interaction with their systems, and they will be the carrier of learning results.

Unfortunately thus far external experts have more often than not controlled international assistant programmes. For example, external experts wrote the SAP programme (Friis-Hansen 2000). Participation of local policy makers was absent or minimal. This obstructed information about local conditions from being incorporated; it also demoralised extorts and officials of the recipient country in work with policy experimentation; and thereby adaptive policy process broke up. Realising the problems of agricultural productivity decline and the deterioration of natural resource base, analysts question the viability of the current programmes of agriculture modernisation in SSA that SAP pushes, a modernisation that is based on high external inputs and with export orientation (Barrett, Aboud, and Brown 2000; Friis-Hansen 2000; McMillan, Rodrik, and Welch 2002). It is observed that there are alternative 'low external input sustainable agriculture' (LEISA) approaches in practice. Such LEISA approaches integrate pest management and soil fertility management, encourage participatory conservation and use of plant genetic resources. The LEISA approaches are accordingly well in correspondence to the accumulated knowledge of people and suitable to local factor endowments – at low monetary expenditure and with higher labour intensity. Some international organisations and local NGOs are carrying out experiments on the approaches as well. While likely appropriate to the particular biophysical and socioeconomic conditions in SSA, LEISA are not reflected in the SAP documents. Not surprisingly, a programme designed and implemented as such could go nowhere but to failure; and the analysis upon similar stance could be nothing but failure-telling. As Friis-Hansen (2000) observed, confronted with wide dissatisfaction, some analyses 'feel that privatisation and liberalisations have not been pushed far enough'; and some others 'believe that it has been pushed too far too fast'. However, viewed from the perspective of IS, the story of LEISA might have indicated ways towards successes and the ways have been at the ground. These are ways that are not Asian, not Japanese or American, but must be African.

Local policy makers, on the other hand, have to be capable in negotiation and implementation with international organisations and donors in order to let the donated programmes better match actual needs, like Asians do who used to be big receivers of international assistance in the 1960s to 1980s. Weak policy capacity was argued as one of the reasons for bypassing local agents, while the fact is that the weaker the capacity the recipient policy institution is, the greater the need that both international and local partners to stand in engagement in capacity building. A recent work by UNDP calls for a new paradigm for capacity building. It recognises that capacity matters as much for development as do economic policies. Knowledge underpinning the capacity cannot be transferred but must be learned. Ten Principles are raised to make capacity development work. I cite them in the following with some reiteration (UNDP 2003: 151):

Ten principles for national stakeholders and external partners in search of promising approaches to building capacity:

- *Think and act in terms of sustainable capacity outcomes.* Capacity development is at the core of development. Every action should be analysed to see whether it serves this end.
- *Don't rush.* Capacity development could not be achieved by sudden external pressure or a quick fixation in short-term.
- *Scan globally, reinvent locally.* External experiences can serve as reference. Policy institutions and capacity have to be locally rooted and accumulated through experiencing.
- *Use existing capacities rather than create new ones.* National expertise, and social and cultural capital are the ground for the society-embodied policy governance and competence to grow.
- *Integrate external inputs with national priorities, processes and systems.* Key for external inputs to effectively work is to handle the interface of it with national needs and possibilities. Where national systems are not strong enough, they need to be reformed and strengthened, not bypassed.
- *Establish incentives for capacity development.* Distortions in public employment are major obstacles to capacity development. Ulterior motives and perverse incentives need to be aligned with the objective of capacity development.
- *Challenge mindsets and power differentials.* Capacity development is not power neutral, and challenging vested interests is difficult. Establishing frank dialogue and moving to a collective culture of transparency is essential to overcoming these challenges.
- *Stay engaged in difficult circumstances.* Weak capacity is not an argument for withdrawal or for pushing external agendas. People should not be hostage to irresponsible governance.
- *Be accountable to ultimate beneficiaries.* Even if governments are not responsive to the needs of their people, external partners need to be accountable to their ultimate beneficiaries and help make national authorities responsible. Approaches need to be discussed and negotiated with national stakeholders.
- *Respect values and foster self-esteem.* The imposition of alien values can undermine confidence. Self-esteem is at the root of ownership and empowerment.

A Summary: What Asian Experiences are to be Referred to and What are Not

We have briefly discussed experiences in Asia. One of the best ways to comprehend experiences in Asia is to acknowledge from the perspective of learning, but not some single pieces of policy. As pioneer authors on capability building analysed, every move ahead that they have made entailed investment in knowledge, skills, and institutional structure in association with investment in physical production means. We also demonstrated that development paths differ among Asia NIEs, the paths were not fixed even in themselves. Fundamental changes are now under way in cope with both problems accumulated from the past and challenges and opportunities newly emerged.

One of the strengths of Asian innovation systems is that they have the governments with strong position in promoting development. Proactive government policies distinctively drove the developments there by means of mobilising national resources for the acquisition of technology and the establishment of a modern manufacturing basis. Another strength in Asia is in well-built educational, scientific and technological infrastructures, and in entrenched industrial capacity. Throughout the period, a catching-up mentality emerged; it was gradually consolidated with initial achievements, and it, once spread in people's mindsets, became a part of the strength of the innovation systems as well. The 'state-assisted' approaches to development in Asia as outlined have its cultural and historical reasons. And such approaches have exerted both positive and negative impact on the developments in Asian NIEs.

Taking positive experiences in Asia, Africa needs to develop efficient institutions of governance (Rodrik 1999; UNDP 2003), in order to lend a framework for a 'system' to operate upon endogenous factors. With rather different pre-modern cultural and societal traditions, Africa has to find out own ways for the development of governance institutions and innovation systems, distinctively suitable to their historical and social backgrounds. From the viewpoint of IS, the role of government is not a counter-power of market, rather, it serves a coordination mechanism including that for the development of market institutions.

The state-assisted capitalist approaches were not without costs, among which are tremendous ecological damage, a crisis in agriculture in quite some Asian countries, and the over-preference for high and 'big' technologies which leads to dependence to various extents on foreign technology while omitting traditionally accumulated knowledge especially with natural endowments and local ecological systems. Inequality and disparity between urban and the rural and between advanced and backward areas grew. These problems were largely associated with centralised top-down policy making. In terms of industrial strategy, Asian NIEs might have overwhelmingly pursued manufacturing capacity at the expense of rural development and ecological wealth. Africa should develop policy institutions and innovation systems more participatory and democratic, to avoid the drawbacks of Asia. The ambitious industry-centred strategy that Asia took needs to be reviewed. Africa could not, and should not, wholly imitate it. Africa needs to find out ways of more balanced development in which the agriculture and rural economy keep pace with the development of modern manufacturing sectors.

In this chapter we have exemplified the power of Innovation Systems. The analytical lens of IS focuses on change. It looks for explaining change through investigation on interactions between institutions and technology at the micro-foundations of systems. In so doing, it turns our attention towards the people and the society, namely the earth from where development is to be started, and in where the resources for development – the skills and knowledge, and the hopes and pains, all are stored for mobilisation. IS must be understood and taken as a useful intellectual instrument for Africa to create its future.

Notes

[1] Based on a general production function, the classic view insists that physical capital investment gives the major contribution to economic development, expressed in the Harrod-Domar model (Fei and Ranis 1997; Hayami 1997). Institutions are not considered, and technology is conceived homogenous and approximated in the intensity of capital. Policy recommendations are generally towards that the savings rate determines the capital investment rate and hence the growth rate. Although enormous modifications have been made later on, problems remain from the basic conception of economic development; and its influence remains strong (Evenson and Westphal 1995).
[2] One case of enhanced learning is at Hyundai for skills and competences for automobile. It is reported by Kim (1997).
[3] It declares for 'enhancing technological innovation, developing high technologies and promoting commercial production of S&T achievements'. For the full document, refer to:
 http://www.most.gov.cn/t_a3_zcfgytzgg_a.jsp.

References

Amsden, Alice H. (1989), *Asia's Next Giant: South Korea and Late Industrialization*, New York: Oxford University Press.

Aoki, Masahiko (1996), 'Unintended Fit: Organizational Evolution and Government Design of Institutions in Japan', in Masahiko Aoki, Hyung-Ki Kim, and Masahiro Okuno-Fujiwara (eds.): *The Role of Government in East Asian Economic Development: Comparative Institutional Analysis*, Oxford: Clarendon Press: 233-253.

Aoki, Masahiko, Hyung-Ki Kim, and Masahiro Okuno-Fujiwara (eds.) (1996), *The Role of Government in East Asian Economic Development: Comparative Institutional Analysis*, Oxford: Clarendon Press.

Barrett, Christopher B., Abdillahi Aboud, and Douglas R. Brown (2000), *The Challenge of Improved Natural Resource Management Practices Adoption in African Agriculture: a Social Science Perspective*, Paper prepared for the workshop on 'Understanding Adoption Processes for Natural Resource Management Practices for Sustainable Agricultural Production in Sub-Saharan Africa', Nairobi: Kenya, July 3-5, 2000, http://www.cnr.usu.edu/research/crsp/icraf.pdf.

Behrman, Jere and T. N. Srinivasan (1995), 'Introduction to Part 9', in Jere Behrman and T. N. Srinivasan (eds.), *Handbook of Development Economics*, Amsterdam: Elsevier: 2467-2496.

Choi, Youngrak (2000), *Paradigm Shift in Korea's Science and Technology Policy*, presentation at the International High-Level Seminar on Technological Innovation, co-sponsored by the Ministry of Science and Technology, China, and United Nations University, September 5-7 2000, Beijing, China.

Enos, John L. (1995), *In Pursuit of Science and Technology in Sub-Saharan Africa*, London and New York: Routledge in association with the UNU Press.

Evenson, Robert E. and Larry E. Westphal (1995), 'Technological Change and Technology Strategy', in Jere Behrman and T. N. Srinivasan (eds.), *Handbook of Development Economics*: 2209-2299.

Fei, John C. and Gustav Ranis (1997), *Growth and Development From an Evolutionary Perspective*, Oxford: Blackwell Publishers.

Freeman, Christopher (1987), *Technology Policy and Economic Performance: Lessons From Japan,* London: Pinter Publishers.

Friis-Hansen, Esbern (ed.) (2000), *Agricultural Policy in Africa after Adjustment*, Copenhagen: CDR Policy Paper Centre for Development Research.

Gu, Shulin (1999a), *China's Industrial Technology, Market Reform and Organizational Change*, London and New York: Routledge in association with the UNU Press.

Gu, Shulin (1999b), *Implications of National Innovation Systems for Developing Countries: Managing Change and Complexity in Economic Development*, Maastricht: UNU/INTECH Discussion Paper, No. 9903, November.

Gu, Shulin and W. Edward Steinmueller (1996/2000), *National Innovation Systems and the Innovative Recombination of Technological Capability in Economic Transition in China: Getting Access to the Information Revolution*, Maastricht: UNU/INTECH Discussion Paper, No. 2002-3.

Hayami, Yujiro (1997), *Development Economics: From the Poverty to the Wealth of Nations*, Oxford: Oxford University Press.

Kim, Linsu (1997), *Imitation to Innovation: The Dynamics of Korea's Technological Learning*, Boston: Harvard Business School Press.

Kim, Linsu (2001), 'The Dynamics of Technological Learning in Industrialisation', in *International Social Science Journal*.

Kruger Anne O. (1995), 'Policy Lessons from Development Experience Since the Second World War', in Jere Behrman and T. N. Srinivasan (eds.), *Handbook of Development Economics*, Amsterdam: Elsevier: 2497-2550.

LU, Feng and Ling MU (2003), 'Endogenous Innovation, Capability Building, and Competitive Advantages: The Development of the Laser Video-Disc Player Industry in China and Policy Implications' (in Chinese), forthcoming in, *Journal of China Social Science (zhongguo shehui kexue)*.

Lundvall, Bengt-Åke (ed.) (1992), *National Systems of Innovation: Towards a Theory of Innovation and Interactive Learning*, London: Pinter Publishers.

McMillan, Margaret, Dani Rodrik, and Karen H. Welch (2002), *When Economic Reform Goes Wrong: Cashews in Mozambique*, RWP02-028, John F. Kennedy School of Government, Harvard University.

Muchie, Mammo and Bengt-Åke Lundvall (2001), *Proposal To Danida – Innovation Activities for the Ecological Industrialisation of Africa: A Comparative Study of the Relationship of the National System of Innovation to the Agro-industrial and Resource-based Sectors of Egypt, South Africa, Kenya, Nigeria, and Ghana*, Memeo.

Metcalfe, J. Stanley and Luke Georghiou (1998), 'Equilibrium and Evolutionary Foundations of Technology Policy', in *OECD STI Review*, No. 22: Special Issue on New Rationale and Approaches in Technology and Innovation Policy, Paris: OECD: 75-136.

Mytelka, Lynn K. (1998), Learning, Innovation, and Industrial Policy: Some Lessens from Korea, in Michael Stoper, Stauros Thomadakes, Lena Tsipouri, and Tavros B. Thomadakis (eds.), *Latecomers in the Global Economy*, London: Routledge.

Nelson, Richard R (1995), 'Recent Evolutionary Theorizing About Economic Growth', *Journal of Economic Literature*, Vol. 33, March: 48-90.

Nelson, Richard R. (ed.) (1993), *National Innovation Systems: a Comparative Analysis*, Oxford: Oxford University Press.

Nelson, Richard. R. and Sidney. G. Winter (1985), *An Evolutionary Theory of Economic Change*, Cambridge: Harvard University Press.

OECD (1999), *Managing National Innovation Systems*, Paris: OECD.

Ohkawa, Kazushi and Henry Rosovsky (1973), *Japanese Economic Growth: Trend Acceleration in the Twentieth Century*, Stanford: Stanford University Press.

Rodrik, Dani (1995), 'Trade and Industrial Policy Reform', in Jere Behrman and T.N. Srinivasan (eds.), *Handbook of Development Economics*, Amsterdam: Elsevier: 2925-2982.

Rodrik, Dani (1999), *The New Global Economy and Developing Countries: Making Openness Work*. Washington, D.C.: Overseas Development Council.

Ruttan, Vernon W. (2001), *Technology, Growth and Development: An Induced Innovation Perspective*, Oxford: Oxford University Press.

Stiglitz, Joseph E. (2001), *An Agenda for the New Development Economics*, Draft paper prepared for the discussion at the UNRISD meeting on 'The Need to Rethink Development Economics', 7-8 September 2001, Cape Town: South Africa, Geneva: United Nations Research Institute for Social Development, http://www.unrisd.org.

UNDP (2001), *Human Development Report 2001: Making New Technologies Work for Human Development*, Oxford: Oxford University Press.

UNDP (2003), *Human Development Report 2003: Millennium Development Goals: A Compact among Nations to End Human Poverty*, Oxford: Oxford University Press.

Wade, Robert (1990), *Governing the Market, Economic Theory and the Role of Government in East Asian Industrialization*, Princeton: Princeton University Press.

Youn, Moon-Seob, Yong Soo Kwon and Sungchul Chung (STEPI, Korea) (2000), *Government Policy for the Promotion of New Growth Industries*, presented at the 'Joint German – OECD Conference: Benchmarking ISRs', Berlin: October 16-17, http://industry-science-berlin2000.de/pres/kwon.pdf.

PART IV

Innovation Systems and the African State

13

The Role of Government in Shaping the National System

of Innovation

The Case of South Africa Since 1994[1]

Sunil Mani

Introduction

The national system of innovation perspective is a convenient framework to understand the process of innovations occurring in an economy and especially within the manufacturing sector. In this framework, the economy is decomposed into various components such as the government, independent research institutes, firms and the higher education system, which supplies human capital to firms, research institutes, and the government. The success of an innovation system depends very much on how closely knit the relationship between the various components are. Needless to add, innovation policy instruments and institutions play a very important role in cementing the relationship between various components. An application of this framework to the situation in a specific country is very useful in identifying the systemic failures that hamper the generation of innovations. Public policy can then be applied to correct for such systemic failures. In the present chapter we survey the role of government with respect to shaping the innovation system of a large developing country, which has a rather unique history.

The Case of South Africa

The South African economy is an outlier on the African continent. It is the most developed country on the continent, and with a per capita income of US$3000, it is more similar to Malaysia than to its immediate neighbours. However, the economy has not grown at all during the 1990s, but has experienced only violent fluctuations. But in the arena of science and technology, South Africa can report a number of achievements. It is the only African country to have been awarded a sizeable number of patents in the US. On average, inventors from South Africa secured 57 patents in the US during the period 1985–2001, thus making it one of the top innovative countries in the developing world. Since 1994, the new government has been completely revamping the S&T administrative apparatus in the country and has been putting in place a whole host of essentially research grants to encourage innovation by three important components of its national system of innovation, namely the business enterprise, the government research institutes and the higher education sector.

The Manufacturing Sector in South Africa

The manufacturing sector in the country accounts for about 20 per cent of the total economy. There are two distinct dimensions of the economy that merit our attention. The first is the performance of the sector, and the second is the structure of the manufacturing sector in terms the number and size distribution of firms, as well as in terms of the industry-wise distribution of firms.

PERFORMANCE OF THE MANUFACTURING SECTOR

In terms of physical performance, the sector did very badly during the 1990s. For instance, the manufacturing value added decelerated during much of the 1990s, and especially during the last two years. However, according to Schneider (2000), the deceleration in manufacturing output started as far back as the 1970s. The gross fixed investments in the sector have also shown sharp year-to-year fluctuations.

STRUCTURE OF THE MANUFACTURING SECTOR

Two dimensions of structure are considered: firstly, in terms of the degree of concentration, and secondly, in terms of the industry-wise distribution of firms. In terms of the degree of concentration, large firms dominate the sector. Kaplinsky and Manning (1998) found that: (i) in 1992, the six largest conglomerates were estimated to control approximately 87 per cent of issued capital on the Johannesburg Stock Exchange; (ii) large South African firms were responsible for between 70.9 per cent and 75.3 per cent of total manufacturing employment; and (iii) over 80 per cent of industrial output in South Africa emanates from large enterprises. This high degree of concentration, coupled with protection from external competition, meant that the demand for innovations is very low. In terms of distribution of industries, the manufacturing sector is fairly spread out, with the food processing industry accounting for the largest share.

Policy Outcomes

Given the fact that the innovation policy of the new government is very recent (post 1996), it may be too early to measure the end result of the policies. However, what is attempted here is to trace the trends in three standard indicators: namely (i) investments in R&D; (ii) patent granted to South African innovators inventors in the US; and (iii) exports of high technology products.

Investments in R&D

Figure 1 Gross Investments in R&D, 1983-2000

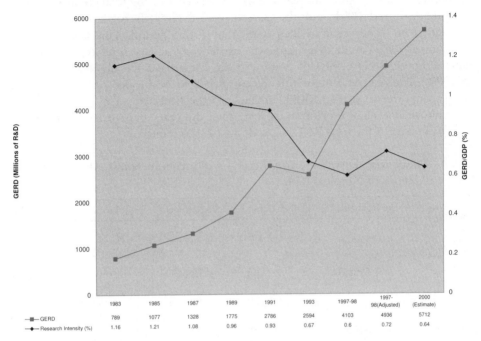

	1983	1985	1987	1989	1991	1993	1997-98	1997-98(Adjusted)	2000 (Estimate)
GERD	789	1077	1328	1775	2786	2594	4103	4936	5712
Research Intensity (%)	1.16	1.21	1.08	0.96	0.93	0.67	0.6	0.72	0.64

Source: Foundation for Research Development (1996) and South Africa. Department of Arts, Culture, Science and Technology; DACST (2000); National Advisory Council on Innovation (2002).

Gross investment in nominal R&D (despite the possible massive underestimation in data) shows a dramatic increase in nominal terms, although the R&D intensity of the country has been showing a secular decline over time (Figure 1). According to DACST (2000), the fall in the R&D intensity (in 1997-98) is specifically due to the so-called 'dramatic withdrawal of government from funding related to defence R&D. Recent defence data indicate that defence receives 25 per cent (in real terms) from what it used to receive in the beginning of the 1990s and no measures have been adopted to compensate for that withdrawal of R&D funding'. If indeed this is the case, how does one rationalise the reported increase in nominal GERD in 1997-98 (even with underestimation)? The fall in the research intensity has more to do with the relative rates of growth of GERD and GDP, especially when the GERD registered a strong growth in 1997-98. The estimates for 2000 shows a further fall in over all research intensity. In order to understand this overall trend, I now analyse the sector-wise financing and performance of R&D.

About one half of R&D investments in the country are expended by the private industrial sector (Table 1): the share could even be higher if the revisions are made for the possible under-estimation of this sector. Government accounts for the next highest category. However according to a recent study, cited in National Advisory Council for Innovation (2002), R&D expenditures of the 17 largest corporations in the country declined from 1.46 per cent to 0.97 per cent of the total company budgets between 1997 and 2001 and the percentage of research work outsourced had increased from 7 per cent to 26 per cent over this period. This finding is quite interesting as I shall argue that despite the existence of at least three types of research grants in-house R&D efforts of South African firms have actually decreased during the 1990s. The real reason for this decline needs to be located in the shortage of skilled manpower to perform R&D activities at the firm level. The higher education sector has seen a major erosion in its share over the years and currently accounts for only 10 per cent of the total R&D expenditure of the country.

Table 1 Sector-wise Distribution of R&D Investments in South Africa, 1983-1997 (Value in Millions of Nominal Rands)

	Private Industrial	Higher Education	Government	Non Profit	Abroad	Total
1983	394 (51)	121 (16)	247 (32)	7 (1)		769
1985	445 (41)	240 (22)	382 (35)	11 (1)		1,0 77
1987	549 (41)	262 (20)	509 (38)	10 (1)		1,3 29
1989	744 (42)	407 (23)	613 (35)	10 (1)		1,7 75
1991	1,304 (47)	554 (20)	903 (32)	25 (1)		2,7 86
1993	1,411 (53)	338 (15)	818 (31)	27 (1)		2,5 94
1997	2,052 (50)	410 (10)	1,354 (33)	Nil	287 (7)	4,103

Note: Figures in parentheses indicate percentage share of the total.
Source: Foundation for Research Development (1996); DACST (2000).

Traditionally, the private sector financed its own R&D investments in their entirety (Table 2). However, during the 1990s, the share of government increased significantly, reaching 10 percent in 1991 (an aberration?) and then subsequently declining to 5 per cent. Similar data for the latest survey are not available.

It is thus interesting to note that the importance of government in funding private industrial R&D is very much on the increase in South Africa since the 1990s. The separation of private industrial R&D between local and foreign companies is not available.

Table 2 Financing of Private Industrial R&D in South Africa, 1983-1997 (Percentage Share)

	Own Funds	Government	Other External Funds*
1983	99.22	0.49	0.29
1985	99.16	0.77	0.07
1987	99.31	0.67	0.01
1989	97.87	2.02	0.12
1991	89.34	10.31	0.35
1993	95.11	4.55	0.34
1997	NA	NA	NA

Note: * This indicates funds from the higher education and non-profit organisations.
Source: Foundation for Research Development (1996).

PATENTING PERFORMANCE OF SOUTH AFRICAN INVENTORS

Table 3 The Top 15 Most Emphasised Patents by Inventors from South Africa, 1995-2001

Technology Class	Technology	Cumulative Number of Patents Granted, 1995-2001
Class 210	Liquid Purification or Separation	37
Class 514	Drug, Bio-Affecting and Body Treating Compositions	28
Class 340	Communications: Electrical	28
Class 405	Hydraulic and Earth Engineering	21
Class 015	Brushing, Scrubbing and General Cleaning	17
Class 604	Surgery (Medicators and Receptors)	16
Class 606	Surgery (Instruments)	16
Class 052	Static Structures (e.g., Buildings)	13
Class 429	Chemistry: Electrical Current Producing Apparatus, Product and Process	11
Class 473	Games using Tangible Projectile	11
Class 209	Classifying, Separating and Assorting Solids	11
Class 335	Electricity: Magnetically Operated Switches, Magnets, and Electromagnets	8
Class 423	Chemistry of Inorganic Compounds	13
	Total for the above 15	230
	Cumulative total of all technology classes 1995-2001	791

Source: USPTO (n.d.).

An examination of the patenting behaviour of inventors from South Africa shows that of the 284 patents that were granted to inventors from South Africa during the period 1995–2001, more than 75 per cent were awarded to individuals and the rest to domestic and foreign firms. Domestic firms include two governmental agencies, namely the Atomic Energy Corporation with ten patents and the Council for Scientific and Industrial Research (CSIR) with nine patents. Only three foreign companies have secured patents based on research done in South Africa. Of which the most important is British Technology Group (BTG); however, these are very likely to have been purchased from other South African individuals or companies. In terms of technology classes (Table 3), there is a mix of both high and medium technologies. Some of the patents are in fact in non-manufacturing areas such as medicine (surgery).

Thus my analysis shows that the research intensity of the country as a whole has actually declined with fluctuations. The number of patents granted to South African inventors has also declined sharply, and most of the patents are actually taken by individuals and government organisations. This is very significant, as much of the R&D is actually performed by the enterprise sector despite the fact that the growth rate of this sector has actually been decelerating in terms of domestic value added. It is against this background that one analyses the innovation policy of the country.

EXPORTS OF HIGH TECHNOLOGY PRODUCTS
High technology exports (excluding armaments) from the country have been growing at a rate of 20 per cent per annum, but the level of exports is very low when compared to the US, Singapore and Malaysia (Table 4). However, the high technology intensity of South Africa's manufactured exports is slowly increasing.

Table 4 Performance of South Africa with respect to High Technology Exports, 1992-1997 (Value in thousands of US$)

	High Tech Exports	Share of High Tech Exports	Ratio to the U.S.	Ratio to Singapore	Ratio to Malaysia
1992	418,341.009	4.83	0.004	0.019	4.09
1993	443,801.791	4.73	0.004	0.016	3.29
1994	479,766.897	4.91	0.004	0.012	2.50
1995	696,737.044	5.77	0.005	0.013	2.74
1996	725,660.030	5.70	0.005	0.013	2.76
1997	977,890.117	7.59	0.006	0.016	3.31

Source: Mani (2000).

Policy Instruments and Institutions

INNOVATION POLICY IN HISTORICAL PERSPECTIVE
In terms of evolution of the policy on innovation, there are essentially three phases: phase one covers the period up to 1994, phase two covers the period 1994-2002 and phase three covers the post-2002 period. In this study, we are concerned only with the latter two phases: Marais (2000) summarises the S&T policy and systems in the country during the first phase. The major impetus for restructuring the innovation system was initiated by an IDRC report entitled 'Towards Science and Technology Policy for a Democratic South Africa' released in 1993 and commissioned by the

African National Congress (ANC). This report led to a Green Paper on Science and Technology in January 1996. The paper contained the main points for discussion and was subjected to a wide consultation process within and outside the government and especially among the various stakeholders. These consultations led to the drafting of a White Paper on Science and Technology entitled 'Preparing for the 21st Century', which was approved by the government on 4 September 1996. In essence, this document constitutes the official science and technology policy of the country. This policy commits government to attaining excellence in the use of science and technology in maintaining a cutting-edge global competitiveness and addressing the urgent needs of disadvantaged citizens (defined as those who are less able to assert themselves in the market). The framework for the new policy is that of a National System of Innovation (NSI). The NSI is described as a 'means through which the country will seek to create, acquire, diffuse and put into practice new knowledge that will help the country and its people to achieve their individual and collective goals'. However, the term 'NSI' is described in rather a textbook fashion, and the 'White Paper' does not delve into any of its indicators in empirical terms. No attempt is made here to subject this comprehensive document to a detailed review. However, there is an excellent review of it by Kaplan (1999). According to him, the White Paper has not operationalised the term 'National System of Innovation' in any precise manner and, moreover, the goals and objectives outlined in the Paper were very broad and general and consequently these objectives could not be easily translated into an operational plan of action. Suffice it to say that on a number of crucial issues (such as human resource development, promoting linkages between various components of the NSI and science and technology infrastructure), the Paper suffers from a sort of profound vagueness. To illustrate, though the Paper states that collection and maintenance of statistical indicators of the NSI is the primary responsibility of the government, the country does not even have good quality data on such commonly employed indicators such as R&D expenditure. Regarding human resource development, in all fairness, one will also have to take into account two other policy documents, namely (i) 'The Skills Development Strategy for Economic and Employment Growth in South Africa' of 1997 and (ii) The Education White Paper 3, dealing with 'A Programme for the Transformation of Higher Education', also published in 1997. The White Paper has thus effectively set the background for promoting innovation in the country.

However after six years of implementation of the strategy contained in the White Paper, as indicated by our discussion on policy outcomes, it was once again felt that the health of the innovation system was far from being satisfactory. An explicit manifestation of this understanding resulted in the launch of yet another policy initiative in 2002 titled 'National R&D Strategy'.[2] The strategy has three core objectives: (i) enhanced innovation; (ii) science, engineering and technology, human resources and transformation; (iii) an effective government, science and technology system and infrastructure. An important and welcoming feature of the new policy initiative is an explicit recognition of one of the serious weaknesses of the innovation system, namely the shortage of good quality human resources for performing R&D projects. The new strategy then proceeds to remedy the situation.

This shows that although the White Paper of 1996 is supposed to have adopted a NSI perspective it was only in form and not in content. The systemic failure of the higher education system (coupled with emigration of scientists) to improve enrolment rates especially at graduate and postgraduate levels in science and engineering was not addressed by this policy. This is now sought to be corrected by the new policy initiative.

From the point of view of the NSI, the following two components merit our attention:

(i)   policies affecting the supply of technically trained personnel in particular;

(ii)   policies leading to the issuing of various types of fiscal incentives, especially research grants;

POLICIES AFFECTING THE SUPPLY OF TECHNICALLY TRAINED PERSONNEL

South Africa has a severe shortage of technically skilled personnel to engage in R&D. This is best captured by the statistic on the density of research scientists and engineers in R&D (RSE). The RSE of the country, which was 33 per 10,000 labour force in 1990 (Foundation for Research Development 1996) sharply declined to 16.3 by 1997-1998 (DACST 2000). This decline is mainly due to the almost 21 per cent decline in the number of researchers (on a full-time equivalent basis).

The decline is greatest in the government and the higher education sectors, while there has actually been an increase in the business sector. Further according to the NACI (2002) there has been a 16 per cent drop in the number of private sector researchers over the period 1998-2002. In fact, even within a short period of four years, between 1992 and 1996, the science and engineering workforce as a percentage of the workforce in the formal sector reduced from 3.6 to 2.3 per cent. There are three relevant factors here. First this reduction can be explained by the reduction in the enrolment for university-level courses in the natural sciences and engineering (Table 5). In fact, the enrolment in science and engineering courses has been virtually stagnant during the period since 1991. Second, is the emigration of technically trained personnel, more popularly known as the 'brain drain'. This issue has been analysed in some depth in a recent report by Crush et al. (2000). Perhaps this explains the reduction in the number of researchers in the higher education and government sectors. Estimates of emigration over the period 1989–1997 amount to over 200,000. Given the strict controls on immigration by the Ministry of Home Affairs, the net migration has been increasing over time. Third, is the ageing scientific population of the country: according to estimates 50 per cent of the scientific output is by scientists over the age of 50 during the late 1990s while it was only 18 per cent during the early part of the decade. While the White Paper is silent on the issue of making available more scientists and engineers for doing R&D, the more recent *R&D strategy* document is more explicit about the shortage of this vital resource (NACI 2002). Most of the strategies that are implied in other governmental policies (for instance, the skills development strategy) are aimed at making available or improving the skills levels of workers for production activities rather than for research. This shortage of sufficiently qualified human resources is thus an important element of discord in the national system of innovation. This brings to the fore the important point that despite using such terms as 'NSI', an actual application of the real idea of an NSI is wanting in South Africa.

Table 5 University Enrolment for Degrees* in South Africa, 1985-1997

	Total Enrolment for Degrees*	Total Enrolment in Natural Sciences and Engineering**	Percentage Share
1985	179,127	34,326	19
1986	194,723	35,166	18
1987	207,690	35,820	17
1988	224,923	37,171	17
1989	238,643	38,660	16
1990	247,602	39,298	16
1991	267,196	41,007	15
1992	278,265	42,088	15
1993	284,403	42,484	15
1994	295,231	44,240	15
1995	320,197	47,532	15
1996	318,762	47,505	15
1997	307,767	47,813	16

Note: * This include the following levels: (a) First degree; (b) Honours; (c) Masters; and (d) Doctoral. ** This includes the following disciplines: a) Mathematical Sciences; (b) Physical Sciences; (c) Life Sciences; (d) Agricultural Sciences; (e) Engineering; and (f) Architecture.
Source: For the data up to 1993: Foundation for Research Development (1996); and for the data from 1994–1997: Unpublished data from National Research Foundation.

PROVISION OF FISCAL INCENTIVES FOR INNOVATION
There are essentially ten different government-supported financial schemes that encourage innovation-related activities within the enterprise sector, each of which addresses the financing gap at a specific stage in the innovation chain. None of the schemes are comprehensive enough to cover all the components of the innovation chain. The schemes can be divided into three categories:

(i)   Research grants: Innovation Fund; Technology and Human Resources for Industry Programme (THRIP); Support Programme for Industrial Innovation (SPII); and Partnership in Industrial Innovation (PII).
(ii)   Development finance especially for enterprise development: Venture Capital; Development Finance; Seed Capital; Feasibility Study Scheme.
(iii) Tax Incentives for R&D.

I now discuss the more important of these schemes in greater detail:

The *Innovation Fund* grant scheme is an important outcome of the White Paper and was initially administered by the Innovation Trust, which functions under DACST. But recently the fund has been shifted to the National Research Foundation. According to DACST, it is a programme of support that addresses problems 'serious enough to impede socio-economic development or affect the county's ability to compete in products and services'. The funds are disbursed through competitive bidding and are intended for large-scale collaborative projects that should involve a significant component of R&D and generate new knowledge leading to novel products, processes or services. The projects have to be completed within a period of three years, and there have been three rounds of selections since 1998-1999. A total of 57 projects totalling R325 million have been funded in five broad areas, namely, crime prevention, information society, information and communications technologies, biotechnology, and manufacturing in general (referred to as value addition) (Table 6). Most of the projects are in manufacturing *per se*, and most of the projects have been in either the SETIs or the universities. The private sector accounted for about a third of the amount granted in the first round, but its share was virtually zero in the second round and 28 per cent in the third round. Similarly, the CSIR alone accounted for about a third of the grants in the first round, but this again has came down significantly over the three rounds. In short, the Innovation Fund is, by and large, targeted at governmental agencies, though one is not arguing that this has been a deliberate strategy of DACST. However, given the large size of the grants under the Innovation Fund and the small size of the number of researchers, it is very likely to crowd out other grant schemes, especially the THRIP scheme.

Table 6 Profile of Innovation Fund Grants during the First three Rounds (Value in Thousands of Rands)

Field/Area	Number of Projects	Amount Granted	Average per project
1. Value Addition	22	132,583 (41)	6,027
2. Biotechnology	12	76,904 (24)	6,409
3. Information Society	11	64,037 (20)	5,827
4. Information and Communication Technologies	5	26,198 (8)	5,240
5. Crime Prevention	7	25,390 (8)	3,627
Total/Average	57	325,112	5,704

Source: Unpublished data provided by DACST.

Since all the projects are for a period of three years, even the projects that were sanctioned under the first round would only just be nearing completion, and therefore it is rather too early to evaluate the efficacy of this programme. However, some comments can be made on the management of the programme. First, there is enough information available on the existence of this Fund and the selection criteria are explicitly laid out and appear to be transparent. In order to measure whether the competition for Innovation Fund grants had increased over time, I define two rates. The first is called the rate of success and is computed by taking the share of projects which are successful during each round as a percentage of the total (adjusted) applications received during each round. An increase in this ratio would mean that the competition for the funds has decreased over the course of the rounds. The second ratio is the rate of frivolous projects and is computed by taking the percentage share of outright rejects in the total number of applications received during each round. The ratios have been worked out and indicate that while the competition for Innovation Fund grants has decreased significantly the quality of proposals received has also has deceased. This appears to be a worrisome fact. Since the number of researchers is limited and the projects are large, in terms of the quantum of grants, there is always a tendency for more mediocre proposals to get funded. It is also a matter for concern that the cost of administering the Innovation Fund is not readily available.

The *Technology and Human Resources for Industry Programme* (THRIP) is another research grant scheme funded by the Department of Trade and Industry (DTI) and managed by the National Research Foundation (NRF). The scheme was introduced in 1992–1993, but real allocations under the project commenced only in 1994–1995, and the real take off took place only from 1996–1997 onwards (Table 7).

Table 7 Progress under the THRIP Funding (Value in Millions of Rands)

	Available for Funding	Actual Funding	Success in Funding (%)
1993	1.5	0	0.0
1994	8.17	0	0.0
1995	14.84	2.98	20.1
1996	11.86	5.99	50.5
1997	27.33	22.72	83.1
1998	46.7	46.34	99.2
1999	75.28	71.2	94.6
2000	99	97.4	98.4

Note: The success in funding is defined as that percentage of available funding that has actually been allocated to projects.
Source: Unpublished data from National Research Foundation.

In other words, the scheme had a rather long gestation period. Representatives of all agents in the NSI, namely industry, the SETIs and the higher education sector, guide the working of this scheme. Judged by the above, it is a really a unique project in that it seeks to link the various components of the NSI such as the government, the higher education sector, the SETIs and the business enterprises. Moreover, it explicitly addresses the shortage of skilled manpower for research. Finally, the choice of technological focus for the activities to be supported by THRIP is left to the industrial participants and their partners. The mechanisms comprising THRIP as well as the selection criteria and other details are set out in DTI (1998). The funding under the scheme is approved for the duration of the project, up to a maximum of five years, subject to meeting all the conditions of the grant and achieving satisfactory progress. Funding of the projects is based on a matching support basis, and there are essentially two funding formulae. The first is referred to as 'R1 for R2', in terms of which THRIP will generally consider contributing a maximum of one rand for every two rands invested by industry in a project that otherwise satisfies the criteria for support. The second is called 'R1 for R1', in terms of which every rand contributed by industry is matched equally by THRIP. There has been a significant increase in the success (narrowly defined) in fund allocation over time.

Since the scheme has been in operation for over five years, it is instructive to analyse the tangible output that has resulted from this scheme. Five indicators of output are employed, namely three measures of human resource development and two measures of physical output of the research funded under the scheme: (i) number of students supported at university and technikon levels, (ii) number of researchers involved; (iii) number of researchers who were exchanged between universities and business enterprises, known as Technology Innovation Promotion through the Transfer of People (TIPTOP) candidates; (iv) number of both local and foreign patents secured; and (v) number of products or artefacts developed (Table 8).

Table 8 Actual Output of THRIP, 1995-1999

Indicator	1995	1996	1997	1998	1999
1. Number of students supported	246	1,053	1,589	1,711	2,131
2. Number of researchers involved	NA	NA	597	911 (309)	927 (296)
3. TIPTOP Candidates	NA	NA	12	46	46
4. Number of Patents (Local and Foreign)	8	19	39	49	36
5. Products and artefacts developed	NA	NA	116	242	290

Note: Figures in brackets indicate the number of team leaders.
Source: Unpublished data provided by National Research Foundation.

There are, of course some obvious limitations in the data. First, the data were all collected by the funding agency itself, namely the NRF. Secondly, the definition, and indeed measurement, of some of the indicators can be problematic. For instance, the number of students supported does not tell us whether the students actually graduated or not. Again, the number of products and artefacts developed does not indicate how many of these have actually been transferred to industry and are in commercial production at the moment. It holds good for the number of patents, as there is no information on how many of these are under actual exploitation. But then these data, though ideal, are really hard to come by. So going by the data presented in Table 8, THRIP is a really successful programme, as most of the indicators register considerable increases over the years. However, the ultimate acid test of the success of THRIP is its ability to increase the number of research scientists and engineers engaged in R&D. Its performance should be judged against increases in RSE over time. There was a further evaluation of the THRIP in 2001 by an independent review panel. Its report published in 2002 was generally positive, although the report did not have much quantitative data on impacts of the scheme.

The *Support Programme for Industrial Innovation* (SPII) is once again a DTI scheme, which is actually implemented by the Industrial Development Corporation (IDC). The scheme, which was introduced in 1993, is designed to promote technology development in the enterprise sector and consists of three separate schemes: the matching, partnership in industrial innovation, and the feasibility schemes.

There has been a phenomenal growth in the number of applications, but the rate of rejection has also come down, implying thereby reduced competition for the funds. In fact, the probability that an application will be successful in obtaining funding certainly increased in the late 1990s as opposed to the earlier period up to the mid 1990s. It is not clear whether this is due to: (a) relaxed criteria; or (b) a significant increase in the quality of the research proposals over time. A second dimension of the scheme is that only two-thirds of the total disbursements of R202.66 million (cumulatively 1993–2000) has been actually utilised by the applicants. A third concern, in terms of industry-wise distribution of the funds approved, nearly 60 per cent have gone towards the electronics industry (including software). The chemical and pharmaceutical industry is the next highest recipient with about 11 per cent, followed by the mechanical industry with about 8 per cent. It is now instructive to find out whether the scheme has been successful or not. Once again, the evaluation is conducted on the basis of a survey of the companies involved by the IDC. The companies are required to provide post-completion data for three years (after the completion of the project) to the IDC on a compulsory basis.

Three indicators have been collected, namely sales of the product developed and sold (both domestic and export sales), the number of new jobs created as a result of the manufacture of this new product and the average research intensity of the company involved (Table 9). Two caveats have to be borne in mind when interpreting this data. First, the coverage of firms has been only approximately 50 per cent. Secondly, because of the time lag between project funding and the realisation of sales, it is difficult to draw a direct causal link between the projects funded and the resultant level of economic activity.

Table 9 Performance of the SPII Recipients (Based on Matching Scheme Only)

Indicator	1997-1998	1998-1999	1999-2000
(i) Sales (in millions of Rand)	249.7	614.6	552.7
Domestic	159.5	432.2	289
Export	90.20	183.3	265.7
(ii) Employment created (in numbers)	860	3675	1395
(iii) Average Research Intensity of the Companies (in per cent)	4.3	11.6	12.7
Response Rate (in per cent)	NA	63	45

Source: Unpublished data from Industrial Development Corporation.

The data do not present a clear trend. However, it is interesting to note that, despite the lower response rate in 1999–2000, the average research intensity of the companies has shown an increasing trend. It must be stated that the research intensity is unusually high, as, according to the latest R&D survey the research intensity for the nation as a whole is only 0.64 (see Figure 1). However, the higher research intensity may also be a reflection of the fact that much of the SPII grants have gone towards the research-intensive electronics, chemical and machinery sectors.

The feasibility scheme was introduced only in June 1998. As mentioned before, the scheme aims to assist small, medium and micro enterprises to prepare a thorough feasibility study for a potentially innovative project. Theoretically speaking, the feasibility scheme is likely to encourage small businesses to initiate R&D projects that can then be funded under the matching scheme. The scheme is currently under review to improve its functioning. During its first year (1998–1999) two studies were completed and both resulted in successful matching scheme applications. In its second year of operation (1999–2000), the scheme attracted eleven applications, of which nine, accounting for R 0.24 million, were approved. Seven of these were completed during the year and six of these resulted in matching scheme applications. However, with the introduction of a 'competitiveness fund' by the DTI, the feasibility scheme is likely to be 'crowded out'.

The third scheme, 'Partnership in Industrial Innovation' is directly administered by the DTI. The scheme is intended to support large development projects, and there will be no limit on the amount of the grant other than budgetary constraints. However, there is absolutely no data whatsoever available on the functioning of this grant scheme.

Thus the three grant schemes (Innovation Fund, THRIP and SPII) all compete with one another in supporting industrial innovations in the country. Given the fact that the number of researchers that are qualified to engage in R&D is rather limited, the grant schemes run the risk of 'adverse selection'. Some of the schemes, especially THRIP, are designed to address this shortage in technically trained manpower.

Finally, the White Paper makes explicit reference to the existence of *Tax Incentives for R&D*. This shows that South Africa operates a simple tax deduction for R&D purposes and no tax credit schemes. Although the White Paper refers to the widespread use of these incentives, there are hardly any data on the number of enterprises that have claimed these incentives. In a recent paper, Kaplan (2000) has undertaken an empirical exercise in measuring the attractiveness of the country's tax regime with respect to R&D by computing what is called a B-Index.

On the basis of the value of the B-index (1.008), Kaplan reached the conclusion that the county's tax regime is not very favourable to R&D and, given the existence of some evidence about the efficacy of tax incentives in promoting R&D in developed countries in particular, he makes a strong case for extending and strengthening tax incentives for R&D in the country.

However, I am not sure whether one can make a case for improving tax incentives for R&D by analysing the level of the B-Index across space, as the B-Index is only an indicator of how favourable a country's tax regime is (with respect to R&D). It does not measure the effectiveness of tax incentives. A better measure of effectiveness is the elasticity of R&D expenditure with respect to a unit reduction in the cost of doing R&D. However, since South Africa does not have a tax credit scheme, it is impossible to compute any measure of elasticity. It is pertinent to note that the Katz commission was very much opposed to the idea of providing tax incentives to the private sector as it 'erodes the tax base and is difficult to administer'. Even in the developed market economies, over 60 per cent of the direct support for technology development in enterprises is in the form of research grants, while there has been considerable erosion in the share of tax incentives. Grants can be better administered to achieve specific targets.

Conclusion

South African policy makers have shown considerable sophistication in innovation policy formulation. Policies, especially the technology policy of the country, have been framed after considerable consultation with stakeholders and have been revised from time to time. The policies, as well as the institutions that support science and technology in general, have been subjected to detailed reviews. The country is also one of the few from the developing world to explicitly use the national systems of innovation approach. However, my analysis shows that this subscription to seemingly sophisticated terms and concepts is more in form than content. What has been attempted in this chapter has been a preliminary analysis according to which none of the instruments have effectively addressed, or are poised to address, the severe shortage of skilled manpower, not only for simple manufacturing but also for research.

Notes

[1] This is a summarised and updated version of the chapter on South Africa in Mani (2002): 173-213.
[2] See Department of Science and Technology, NACI (2002).

References

Crush, Jonathan, David McDonald, Vincent Williams, Robert Mattes, Wayne Richmond, Christian M. Rogerson, and J. M. Rogerson (2000), *Losing our Minds: Skills Migration and the 'Brain Drain' from South Africa*, Southern Africa Migration Project, Migration Policy Series No. 18, http://www.queensu.ca/samp/-publications/policyseries/policy18.htm.

Department of Arts, Culture, Science and Technology, DACST (2000), *White Paper on Science and Technology, Preparing for the 21st Century*, Pretoria: DACST, http://www.dacst.gov.za/default_science_technology.htm.

DTI (1998), *THRIP: 'Guide to Research Support'*, Pretoria: Department of Trade and Industry.

Foundation for Research Development (1996), *SA Science and Technology Indicators*, Pretoria: Directorate for Science and Technology Policy.

Kaplan, David (1999), 'On the Literature of the Economics of Technological Change: Science and Technology Policy in South Africa', *The South African Journal of Economics*, Vol. 67, No. 4: 473–490.

Kaplan, David (2000), *Rethinking Government Support for Business Sector R&D in South Africa: The Case for Tax Incentives*, Private Bag: Science and Technology Policy Research Centre, School of Economics, University of Cape Town, *processed*.

Kaplinsky, Raphael and Claudia Manning (1998), 'Concentration, Competition Policy and the Role of Small and Medium-Sized Enterprises in South Africa's Industrial Development', *Journal of Development Studies*, Vol. 35, No. 1: 139–161.

Mani, Sunil (2000), 'Exports of High Technology Products from Developing Countries: Is it Real or a Statistical Artifact?', Discussion Paper: 2001, Maastricht: Institute for New Technologies, United Nations University, http://www.intech.-unu.edu/publications/index.htm.

Mani, Sunil (2002), *Government, Innovation and Technology Policy: An International Comparative Analysis*, Cheltenham: Edward Elgar.

Marais, H. C. (2000), *Perspectives on Science Policy in South Africa*, Pretoria: Network Publishers.

National Advisory Council on Innovation, NACI (2002), *South Africa's National R&D Strategy*, http://www.naci.org.za/a05a.cfm?item=319

Schneider, Geoffrey E. (2000), 'The Development of Manufacturing Sector in South Africa', *Journal of Economic Issues*, Vol. 34, No. 2: 413–424.

USPTO (n.d.), http://www.uspto.gov/.

14

Technology, Knowledge, and Egypt's Competitiveness

Lobna M. Abdel Latif

Introduction

The importance to Egypt of expanding and diversifying its export base, particularly in manufactured products, has risen with its membership in the WTO and the European Partnership. Most of the advantages that Egypt has against major export-oriented countries in the developing world, in the form of quota and tariff free access for its resource intensive exports especially to the European markets, are on their way to disappear with the complete application of the GATT by 2004. Egypt will face a sharp increase in competition in labour-intensive exports from many Middle East and North African countries (MENA) as well as Asian countries. Competing in these low-technology products depends mostly on cost reduction, which requires high productivity growth to reduce the unit labour cost while maintaining growth in real wages. Alternatively, however more difficult and demanding, competition in low-technology products can depend on product differentiation, which requires high degree of mastering the production process (in terms of new design, components, materials, etc.). Otherwise, competition in these products will be at the expense of suppressing real wages and welfare in the economy. Also, as a technologically lagging economy, Egypt will find it difficult to compete against high technology firms in industrialised countries and also against medium technology firms in the newly industrialised countries.

This chapter explores the prospects of globalisation on the Egyptian manufacturing sector. It focuses on the gap between technology and knowledge as one of the major determinants of the performance of the sector. Section two provides the analytical background. The analysis shows that the gains from the globalisation process depend on the structure and the performance of the manufacturing sector, especially on the factors of technology and knowledge. The third section explores the structure of the manufacturing sector in Egypt and the resultant pattern of trade and competitiveness. Section four analyses the future prospects of the manufacturing sector under two scenarios: specialisation in labour-intensive products as compared to technology-intensive. The fifth section discusses some features of the national innovation system in Egypt and pinpoints some of the factors that are behind the gap between technology and knowledge (or skills) in Egypt's manufacturing sector.

Section six concludes with policy implications on the pattern of modernisation needed to accelerate the economy's integration with the world economy.

Analytical Background

Openness

The relation between openness and growth has been thoroughly investigated in the economic literature in a large number of empirical studies. In the studies aimed at providing support for the policy advice on openness and at identifying the key features that distinguish countries with high and low rates of economic growth (for instance Bosworth, Collins, and Chen 1995; Dollar 1991; Edwards 1993; Pritchett 1996; Singh 1994). A consensus saying that openness fosters growth has been reached. Openness provides more export opportunities and widens the base of national products, it is argued.

On the other hand, a growing body of empirical studies began to focus on and account for the differences in the level of gains from openness. Their major finding is that the levels of gains differ according to the differences in the economic structures of developing countries. Export structures, which are dominated by technology-intensive products, have better growth and welfare prospects than others do. Activities having high-technology content generally enjoy faster growing demand *vis-à-vis* low-technology activities. Therefore, over time it becomes difficult to sustain high export growth in the latter category of activities. To continue competing in these types of products, the economy should be able to continuously reduce the unit cost of its exports. Unless the competition strategy is accompanied by high growth rates of productivity, it would negatively affect the real wage level and the welfare of the economy. Also, low-technology products, by nature, have limited potential for productivity growth – contrary to those products with high-technology content. This means that competing in low technology products could be a good starting point, but the economy should rehabilitate itself towards the shift to technology-intensive products to guarantee a continuous increase in the level of welfare.

Therefore, openness is a necessary but not a sufficient condition to enjoy the benefits of trade. The economy might be trapped in Ricardian exports (i.e. land-resource-intensive exports) and simple-type Heckscher-Ohlin exports (factor-intensive exports) without being able to contribute to the growing trade in product-cycle exports (technology-intensive exports). Therefore, the interactions between the domestic features of the relevant economy and trade reform actions are considered a compound element that determines the magnitudes of gains from trade and openness. That is to say, domestic structures pave the way for the economy to maximise the benefits from trade. Successful management of this interaction would guarantee the economy's sustainable participation in the growing international trade. Thus, to be a fast integrator, depends on the economy's capabilities to produce – up to international levels – goods in accordance to consumer preferences and being able to join the evolving process of international production sharing beyond Ricardian exports. Hence, a diversified and well-developed domestic production base is a pre-condition to integrate fast and smoothly.

Capability vs. Capacity

These findings raise many questions as regards to the challenges the economy would confront if its structure of industry fails to generate high levels of productivity growth in traditional low-tech/labour-intensive branches, or could not make the shift towards high-tech products. The new literature on competitiveness stresses the important role of capability versus capacity as the main determinant for the structure and growth of manufactured output and exports. Traditional trade theories assume that once technology is imported, firms in developing countries can use it efficiently to generate the desired structure of output and exports. Thus, the role of the government in order to optimise the comparative advantage of the economy is to get prices right so that firms select the technology appropriate for the natural endowments of the economy (Lall 2000a).

On the contrary, the availability of equipments having high technology-content would not generate any comparative advantage to the country unless exports are flexible and responsive to the induced shift in the structure of output. To guarantee this, the domestic industry should be able to incorporate the imported technology into its knowledge fabric (Lall 2000a, b). This will only take place if the economy has the capabilities that match the imported level of knowledge and is able to move quickly on its learning curve. Therefore, even if two developing countries have similar endowments of labour and capital, their patterns of exports would differ subject to the learning curve that these economies have reached, and their capabilities to move along or to develop a higher one. Hence, their short-term structures of output may be close to each other, but their long-term structure would differ completely because sustaining the desired shift is subject to the degree of competitiveness. This again depends on the structure of the learning curve of the economy, which is the outcome of the degree of developments of the economy's national innovation system.

National Innovation System

One of the most crucial issues of trade liberalisation is its link to government intervention. As mentioned by Lundvall '...it is assumed that the most important resource in the modern economy is knowledge and, accordingly, that the most important process is learning' (Lundvall 1992: 1). The dynamic process of learning starts by the learning orientation stage (Kim 2000). In the early stage of industrialisation, firms in developing courtiers undertake the duplicative imitation of foreign mature products. They are to 'learn by doing', through examination of the learning curve of these mature products. Then, as industrialisation progresses, they shift their focus from duplicative imitation to creative imitation. Now they 'learn by research' to produce imitative products with new performance features, which is a must to sustain competitiveness in traditional products and also to export some medium-technology products. When a developing country catches up with advanced countries and reaches the technological frontier, major emphasis shifts from imitation to original innovation. Now the developing economy reaches the stage of 'knowledge-based' in the structure of its industry, as well as its exports, and is capable of exporting its own technology.

To proceed successfully in these stages, it is necessary to develop a conceptual framework for the process of learning, through which the process of institutionalisation of learning is to be built up. 'A national innovation system' is a system constituted by various actors in the economy who influence the technological capabilities– such as educational and training institutions, scientific bodies and organisations, industrial firms (in an organised form). It operates through various interactions among these actors. It is national to the extent that it has identifiable national and societal specificities. In order for this system to operate efficiently in the local initial conditions and to retain flexibility for further evolution in the learning process it is in need of influential government policies and support. Learning must be supported and embodied in an institutional context (Gu 1996) to guarantee the efficient work of the interactive linkages among the macro-sectoral (industry) and micro (firms). The work of government is indispensable to support the restructuring of institutions to be able to capture and interact with the modernisation process. Also, it has a distinctive role in compensating for inadequate market mechanisms and the modernisation and upgrading of technical capabilities.[1] In other words, the national innovation system is an approach that is organised by the government and incorporates all players related to the industry to facilitate and accelerate the industry modernisation process in a systematic manner.

Egypt's Manufacturing Sector

The Industrial Base

At the macro level, Egypt's manufacturing sector experienced some improvements towards the end of the 90's (1998), achieving the highest growth rates among all economic sectors, reaching 7 per cent on average in the years 96-98, from an average of 4 per cent in the years 1992-95. This was reflected in the growth of the industrial base, which more than doubled when compared with the outset of the 90's (1991). Also, the contribution of Egypt's manufacturing sector to value added at the national level reached its highest level over the 90's, amounting for 19 per cent in 1998/99.[2]

This level of performance is close to those of most of the countries in the MENA region. Yet, it is lower than the average of the developing countries (with 26.7 per cent contribution of manufacturing sector to GDP, on average during the period 1991-1998). It is also much lower than that of the newly industrialised countries in Southeast Asia. The manufacturing sector amounted to about 30 per cent of GDP in each of Korea, Malaysia and Thailand during the period 1991-1998.[3] This figure is the outcome of the high growth rates that the industrial bases in these countries had experienced over many decades (Abdel Latif and Selim 1999). Also, when looking at the group of countries in our sample, Egypt's industrial base is almost half that of Thailand or Indonesia, and two thirds that of Malaysia or Turkey, while those of China and Korea are beyond the comparison. However, the manufacturing output in Egypt is more than double that of Morocco and Tunisia.

Structure

Table 1 depicts the share of the major manufacturing branches in total manufacturing output. In general the four branches: 'food and beverage', 'textile and clothing', 'chemicals', and 'machinery and equipment' constitute the bulk share in manufacturing output in the sample countries, ranging from 45 per cent to 70 per cent, and having different domestic structures. These branches constitute a combination of resource-based, labour-intensive (low knowledge), and knowledge-intensive products.

Table 1 Structure of Manufacturing Output Generation (average 1991-95)

Country	31*	32	33	34	35	36	37	38	39
China	12.0	15.1	1.2	2.8	18.4	6.3	11.8	30.2	2.2
Egypt	25.3	13.7	0.6	3.3	29.1	6.7	8.4	12.9	0.1
Indonesia	22.6	19.6	9.8	4.7	15.2	3.5	6.0	18.0	0.8
Jordan	20.3	4.7	2.9	5.2	43.0	10.3	4.9	8.4	0.2
Korea Rep.	9.1	11.3	2.0	4.3	18.1	4.4	8.4	41.4	1.1
Malaysia	14.4	4.4	6.3	2.6	15.6	3.0	4.7	48.0	0.9
Morocco	35.0	17.5	2.0	4.4	18.2	6.9	1.6	14.4	0.1
Philippines	29.5	8.8	2.1	3.8	26.0	3.4	6.6	19.1	0.8
Thailand	..	12.8	1.4	..	13.0	5.2	4.9	25.3	2.0
Tunisia	23.9	24.5	3.7	2.7	23.2	7.8	5.6	7.4	1.1
Turkey	17.9	18.4	1.2	3.6	23.2	5.3	9.7	20.4	0.2

Note: *ISIC classification: Food, beverage, and tobacco (31) - Textiles, wearing apparel, leather, and footwear (32) - Wood and wood products, including furniture (33) - Paper and paper products, printing and publishing (34) - Chemicals (35) - Non-metallic minerals (36) - Basic metals (37) - Metal products, machinery and equipment (38) - Other manufacturing industries (39).
.. Figures are not available.
Source: Calculated from UNIDO, ISIC3 database, 1998.

The most developed structure of industrial production appears in Malaysia and Korea; with a high share of machinery and equipment that tends to approximate the levels of the industrial countries. China and Thailand are trying to catch up followed by Turkey, Philippines and Indonesia. Contrary to this, the most primitive structure appears in Jordan and Tunisia, followed by Egypt and Morocco. Specifically, the industrial structure in this group is heavily natural resource-based. This is reflected in the bias of the structure of manufacturing output towards chemicals, food and textile. Next to Jordan, the importance of chemical industries in Egypt is the highest over the sample. Most chemical products in industrial activities are classified to be knowledge (technology)-intensive products. However, the chemical products that Egypt specialises in are out of this category and are included in the group of primary products.

The share of products with technology-intensive content is lower than the anticipated one (the combined share of both engineering and chemicals) in the structure of manufacturing sector in all MENA countries and to a less extent in the case of the Philippines, due to the big share of resource-based products in the chemicals activity. Table 2 presents a detailed distribution for the three major contents (resource, labour and knowledge) of manufacturing output, regardless of the type of activity.[4] As it could be seen from the table, the importance of manufacturing products that have high knowledge-content in Egypt's manufacturing sector is almost half those in Korea, Malaysia and Thailand. Yet, it is almost, or close to those of other countries in the MENA region.

Table 2 Manufacturing Outputs' Structure (1994)

Country	Resource-based	Labour-intensive	Technology-intensive
China	..	..	..
Egypt	62.1	23.0	15.0
Indonesia	47.1	33.3	19.6
Jordan	62.7	15.8	22.5
Korea Rep.	30.2	28.8	41.0
Malaysia	36.6	14.6	50.2
Morocco	..	..	..
Philippines	59.2	18.2	22.6
Thailand	39.2	36.2	24.6
Tunisia	39.5	36.4	24.1
Turkey	43.1	37.4	19.5

Note: .. China is not available at the INDST-4 digit.
Source: Calculated from UNIDO INDST-4 digit. 1998

The share of high-tech products in the structure of Egypt's manufactured exports is lower than the share of these products in the structure of output, as shown in Table 3.[5] This means that technology content in the structure of production could not be channelled to the export structure. It also means that Egypt cannot compete internationally using the high level of imported technology. For example, Tunisia does not share with Egypt this phenomenon, even though industrialisation in Egypt has deeper historical roots and a wider industrial base. Yet, it did not evolve to a stage beyond that of Tunisia.

Table 3 Manufactured Exports' Structure (1994)

Country	Resource-based	Labour-intensive	Technology-intensive
China	12.2	56.3	31.5
Egypt	35.5	54.7	9.8
Indonesia	40.3	37.6	22.1
Jordan	..	..	..
Korea Rep.	6.8	27.5	65.7
Malaysia	21.8	11.4	66.7
Morocco	47.7	31.6	20.7
Philippines	21.5	30.9	47.5
Thailand	19.6	32.0	48.3
Tunisia	26.6	50.5	22.8
Turkey	20.1	58.7	21.2

Source: Calculated from PC/TAS database, 1998.

Competitiveness

This situation lead all writers interested in sustainable development to question the capabilities of the Egyptian economy to compete internationally. Where 'International competitiveness is the ability of a country to produce goods and services that meet the test of international markets and simultaneously to maintain and expand the real income of its citizens' (Gassmann 1996: 40). Measuring competitiveness encompasses two indicators: absolute efficiency which reflects the level of production costs relative to other countries, and relative efficiency which explains the pattern of international specialisation in production (ul Haque 1995). Comparisons of relative wage adjusted for productivity – measured in terms of unit labour cost – would give a simple, not comprehensive, technique to measure absolute efficiency.[6] For relative efficiency, comparative advantage techniques; revealed comparative advantage (RCA) is one of which, would help in exploring the relative indicator.[7]

However, with the growing tendency of construction of free trade areas, the globalisation process of the manufacturing sector will not affect only the international position of the sector, but will also have an impact on the domestic presence of the sector. In this case, local competitiveness of the sector is an important issue that ought to be explored.

WAGES AND PRODUCTIVITY

As it appears form Figure 1, Egypt has one of the highest unit labour costs among the compared countries. Despite the relatively low wage level per worker in the manufacturing sector as compared to those countries, the levels of productivity of manufacturing workers in Egypt are relatively very low (Figure 1b). Therefore, lower wages do not constitute any comparative advantage to Egypt's manufacturing sector.

It can also be seen from Figure 1 that, the bulk of the plot clusters in the bottom of the distribution, indicating a phenomenon of low wage-productivity equilibrium in the

Egyptian manufacturing. This case is further highlighted within international comparison, as Figure 1b shows Egypt lies behind most other countries, even though many of these countries are in a close stage of industrialisation as that of Egypt, such as Tunisia and Jordan.

REVEALED COMPARATIVE ADVANTAGES

When applying the RCA technique, according to the SITC classification in 1995, which is calculated from UNIDO, ISIC3 database 1998, to the figures of exports of each country in our sample.[8] The same phenomenon, of primary commodity bias in Egyptian trade, manifests itself when the structure of commodity groups in which Egypt has a revealed comparative advantage is explored. Most of the commodities that Egypt has RCA in are natural resource-intensive or labour-based industries. The relative highest commodity group in which Egypt enjoys the greatest number of commodities with RCA, in comparison to those of other countries of the same group, is the group of 'mineral fuels, lubricants and related materials', in which petroleum and its products are included. This RCA in petroleum and its products is not expected to be sustainable in the long run. And the lowest RCA prevailed in the relative structure of Egyptian merchandise is the group of machinery and equipment; the group of continuously increasing share in world trade.

Figure 1 Relation Between Wages and Productivity

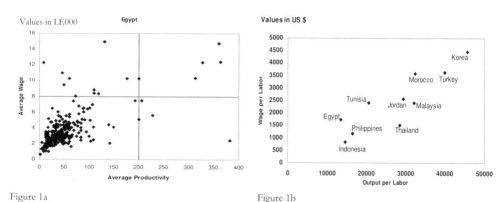

Figure 1a   Figure 1b

Source: Figure 1a author's calculations based on CAPMAS data; Figure 1b author's calculations based on UNIDO (1998).

Also, the first category in which Egypt has the highest number of commodities with RCA 49 is 'the manufactured goods classified chiefly by materials', in which the groups of 'textiles' and 'wearing apparel' are included. For these two groups, Egypt enjoys international competitiveness, yet in a lower degree concerning apparel than textiles (Sakr and Abdel Latif 2000). Therefore, the international competition confronting Egypt regarding these two groups is very high, it comes from almost all the sample (China has the highest RCA 109) which enjoy RCA in a wider range of goods related to these two groups (textiles and apparel).

DOMESTIC COMPETITIVENESS

The threat arising from import competition becomes quite tangible when the policy of economic openness is pursued at a high speed. In this situation, consumers' demands develop more rapidly than local firms' capacity. Domestic suppliers begin to loose their internal markets, allowing thereby a continuous increase in the foreign component of domestic consumption. If local firms fail to dominate their own domestic markets, they would presumably have little chance to compete internationally.

For example in the textiles sectors in Egypt, with the relaxation of tariffs on imports under the Egypt-EU partnership Agreement, effective rates of protection on textiles products would decrease (to negative rates by the end of the phases of tariff reductions due to the non-incorporation of agriculture and services commodities in the agreement). Prices of most of these products (either inputs, or final outputs) are much higher than the international prices, due to inefficiency in domestic production processes. Kheir-El-Din and El-Sayed expected that the combined effect of production inefficiency and tariff phasing out would reflect itself on a new mix in the components of producing ready-made garments in behalf of imported textiles (Kheir-El-Din and El-Sayed 1997). This shift towards imports will affect negatively the level of domestic production. It will also decrease the degree of inter linkages among the textiles sectors, which will reduce local competitiveness.

Therefore, this loss of local competitiveness in the Egyptian market could be generalised to many industrial branches. Many of local industrial outputs have higher prices than their international counterparts, due to inefficient practices in the production and pricing process, such as the presence of both technical and allocative inefficiencies and high mark up prices.

THE CAPACITY TO EXPORT

The importance of manufactured exports is relatively lower in Egypt than any of the compared countries. Moreover, the growth of manufactured exports is the lowest in Egypt *vis-à-vis* other countries, even in those countries that have industrial bases close to that of Egypt (Table 4).

Table 4 Merchandise Exports and Some Exports' Indicators

	Average merchandise exports (US$ Millions)		Average export growth (annual %)		Growth difference (%)	Manufacturing exports (% of merchandise exports)*	
	91-95	96-98	92-95	96-98		91-95	96-98
China	87.0	172.4	14.3	11.9	-2.4	80.4	85.3
Egypt	4.0	4.9	5.9	-1.2	-7.1	35.6	38.7
Indonesia	37.5	52.3	8.9	9.1	0.2	49.0	46.0
Jordan	1.4	1.8	9.8	4.3	-5.5	50.0	..
Korea Rep.	89.7	133.6	15.8	15.3	-0.5	93.0	91.3
Malaysia	49.7	75.6	15.6	4.1	-11.5	68.8	77.3
Morocco	5.5	7.0	9.0	-3.4	-12.3	54.0	49.5
Philippines	12.2	25.1	10.5	7.5	-3.0	47.8	86.7
Thailand	39.3	54.6	14.1	1.3	-12.7	69.8	71.0
Tunisia	4.3	5.6	6.5	4.5	-2.0	74.4	80.0
Turkey	16.9	32.1	10.5	17.1	6.6	71.2	75.3

Note: * Does not include SITC 68.
Source: Calculated from WB-WDI database, 2000.

In Search for an Appropriate Strategy

This critical situation raises many questions about how to proceed to deal with all these challenges. Is it better for the economy to specialise in producing and exporting labour-intensive products? As mentioned elsewhere, many countries were able to compete internationally and raise high levels of growth through this strategy (Cooper 1995). Proceeding in this strategy depends on the economy's capabilities to reduce the unit labour cost of its labour-intensive exports to practice price competition. If the skills and the productivity of the labour force are not growing fast, implementing the suggested export strategy will be at the expense of reducing real wages. If real wages are already very low to generate income levels above the poverty stage, then the poor employees in the manufacturing sector will subsidise the export strategy, having a high negative impact on welfare, which is likely the case of Egypt.

According to the Family Budget Survey of 1996, about 42 per cent of workers in manufacturing (public and private) in Egypt are below the poverty line (El-Laithy, El-Khawaga, and Riad 1999). Focusing on the private portion of the industry, the proportion of poor worker is more than 70 per cent (Abdel Latif 2001).

To compete through a cost reduction strategy, wages would be suppressed which would exacerbate the poverty problem. Also in the absence of an effective national innovation system, moving towards technology-intensive products might not work.

A good measure of the level of development of the technological capabilities of the industrial base is the intra-industry trade ratios (Yeats 1998).

Countries with relatively high technology content in their structure of exports have higher degrees of intra trade, for example in the category of exporting machinery and transport equipment, Korea has the highest SITC of 0.49. This means that they are engaged in importing and exporting the same trade categories, which reflects the complexity of their industrial products and raises the need to diversify the parts and components to comply with the increasing demand on differentiated products. Low figures of Egypt's intra trade ratios in the same category, having an SITC of 0.19 indicates a relatively simple structure of production processes (Abdel Latif 2001).

The National Innovation System of Egypt

Diagram 1 draws some of the features of Egypt's national innovation system. As it shows there is a specialised ministry for scientific research, which incorporates many specific councils and units for technology upgrading. Also acting upon the Ministerial Decree of the Minister of Industry on July 2 1983, many industrial specialised centres (such as Spinning and Weaving centre, Engineering designs centre) were established to be directly under the Minister's supervision. The Decree also spurred manufacturing public firms to establish its own R&D units.

All these units, centres and councils were linked to each other at the macro level. However, there is no complete information system that linked all these bodies. They were also not linked to the education system. Moreover, these bodies were governmental or public units serving other public units (the manufacturing public firms). Therefore, the whole system had all the disadvantages and inefficiencies of public production.

Recently in the 90's, universities were allowed to construct their own technological research centres. It was thought that those centres would support the technological upgrading of private manufacturing firms, especially since these firms were reluctant to cooperate with the public technology system (affiliated to the Ministry of Industry or to the Ministry of Scientific Research).

Technology research centres in universities being run by faculty members, were counted on to provide the missing link between the technology upgrading system and the education system. Yet achievements on both sides have not been much largely due to the absence of real competition inside the Egyptian markets and the existence of high tariffs rates on imported manufactures (especially finished goods). Therefore, most of private manufacturing firms were reluctant to make any technological upgrading or be involved in programs to upgrade the skills of their labour. Also, the lack of confidence in the public technology system and the whole education system, lead many firms – that are aware of the importance of skill and technology upgrading – to seek experts from abroad. This upgrading had no positive externalities through diffusion of technology and efficiency to other firms. Both backward and forward industrial relations are very weak among private firms in Egypt. Most firms are involved in producing final products and basically depend on imported components. However, some recent programs were developed to try to solve the problem of incomplete and ineffective national innovation system by addressing the technology and knowledge problems on a group or cluster basis (such as the Industrial Modernisation Program, under the auspices of the Egyptian Ministry of Industry).

Diagram 1 National Innovation System (Industry)

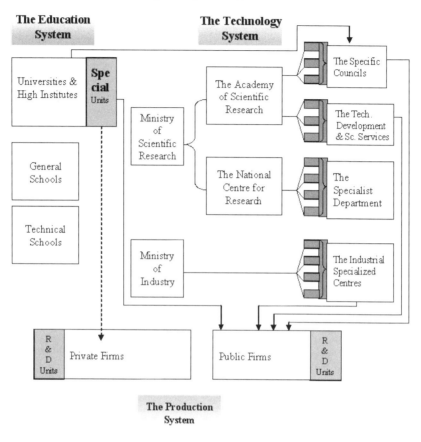

Source: The author, based on Amin (1995); Academy of Scientific Research (2000).

Concluding Remarks: Industry Modernization

The present chapter dealt with the prospects of the Egyptian industrial sector as it is facing a growing degree of globalisation. The analysis pinpointed the gap between technology and knowledge in the manufacturing sector and showed its trade implications. It is clear that the sector suffers from many problems that are reflected in its weak performance in the context of international trade. Building an export promotion program based on the cost reduction strategy will negatively affect the level of welfare of workers in the manufacturing sector, unless the program is integrated with a modernisation plan.

The common interpretation of the concept of modernisation is to import new machines with modern technology; therefore, productivity would be higher. However, it was proved that machines are produced to match the capabilities of their surrounding learning levels. In other words the degree of technology that is built in them postulates a certain level of knowledge and skills. If this match could not be made, the productivity of these machines will be lower than expected. In the case of Egypt, the productivity of imported capital is lower than the expected by about 50 per cent due to lower labour skills than those in the country of origin of the capital (Acemoglu and Zilbotti 1999). In other words, there is a waste of resources.

Some firms in the manufacturing sector suffer from obsolete technology, but this is not the core problem of the sector. One of the striking findings of the analysis in this chapter is the dichotomy between the industrial structure and that of exports. This makes industry in Egypt a good representative example of the consequences of the discrepancy between the level of technology that is built in machines and the level of knowledge that is built in the labour skills. The assembling industries in Egypt are the industries least capable of competing internationally or even in the local market if they are not highly protected. Their unit labour cost is one of the highest in the world, despite acquiring very modern technology. Yet, they do not have the knowledge to master this technology.

Then, what is meant here by modernisation? History of the evolution of industry indicates that the first step is to establish an industrial complex system, stemming from the desire of the industrial unit – on the micro level – to increase its efficiency in using economic resources (labour, capital, managerial, collective bargaining).

Therefore, the first step in modernisation should focus on the micro level. Parallel to this, the creation of an effective innovation system is crucial. Also, learning must be institutionalised. Moreover, the link between scientific bodies, laboratories, universities, technology centres, from one side and the knowledge building process on the firm level is essential to sustain the process of knowledge upgrading at the micro level. In other words, modernisation depends on the success of creating a mechanism to generate the learning process within firms on the one hand, and on linking the acquisition of technology and its diffusion with educational and training programs. The long-term development of this structure and its organic relations to firms is to play a large part in the process of technological upgrading.

Firms' performance is path-dependent, since the process of developing industrial capabilities needs time. And because time dimension is of essence, it is crucial for the government to intervene in a way that accelerates the process. Government actions are indispensable to create the enabling environment for a well-functioning national innovation system, to guide firms on how to acquire technology, as well as to support scientific bodies and education organisation to outsource firms with continuous stream of technology, information, and knowledge.

This modernisation is the only hope to enhance productivity and to be able to compete in labour-intensive products, on low-cost bases. After this the economy should be able to move to competition in differentiated labour-intensive products, with the evolution of the level of mastering technology over time.

One may argue: are all these actions needed just to compete in labour-intensive products, despite its low-tech content? The answer is yes. Competition is becoming stronger than ever before, which necessitates increasing the sophistication of the production of the labour-intensive products. Danish wooden furniture and Italian spectacle frames are two examples of low-tech exports. Yet the degree of tacit-knowledge is intensive (Mytelka 2000), and the search for new designs, and new compounds are endless in these two highly successful examples of exports of labour-intensive products.

Labour-intensive exports are currently the major fields of specialisation in Egypt, and will remain so for many years. Targeting a mere increase of these exports will negatively affect economic welfare. In order to be able to raise national income and economic welfare, building competitiveness for this type of exports is a must. Egypt has no other alternative, but to accelerate the process of restructuring its national innovation system and speeding up the evolution of knowledge capabilities to achieve the shift from learning by doing to learning by research. By then, the economy will be ready to enter the stage of innovation and competition in high-tech exports.

Notes

[1] For more details as regards to the market mechanism and its discourageable role of technical upgrading see Gu (1996).

[2] Ministry of Planning (2000).

[3] WB-WDI database, 2000.

[4] For details of this classification see Lall (2000a, b); Mayer and Wood (1999).

[5] Both terms: high tech products or technology intensive products are used for the same meaning. They are counterparts to both medium and high- tech products in Lall (2000a, b).

[6] Measuring absolute efficiency in an international context is quite a complicated task, due to the difficulty in transforming all data needed to comparable sets for all relevant countries (uL-Haque 1995: 4).

[7] Specifically, if x_{ij} is the value of country i's exports of commodity j and X_{it} is the country's total exports, then commodity j's revealed comparative advantage index is (Balassa 1965):

$RCA_{ij} = (x_{ij} / X_{it}) / (x_{jw} / X_{tw})$

Where the w subscripts refer to world trade totals.

[8] Based on a study conducted by Abdel and Selim (1999).

References

Abdel Latif, Lobna M. (2001), *Globalization of Industry and Modernization*, INP-UNDP.

Abdel Latif, Lobna M. and Kamel Selim (1999), 'Egyptian Industry and the Economy Speed of Global Integration', *Economic Studies*, No. 12, CEFRS, FEPS .

Academy of Scientific Research (2000), *Academy's Achievements in Twenty Years*, Academy of Scientific Research.

Acemoglu, Daron and Fabrizio Zilbotti (1999), *Productivity Differences*, Working Paper, No. W6879, January, National Bureau of Economic Research.

Amin, Mohamed (1995), *The Technology in Egypt's Industrial Sector: An Approach for Development*, Academy of Scientific Research.

Balassa, Bela (1965), 'Trade Liberalization and 'Revealed' Comparative Advantage', *The Manchester School of Economic and Social Studies*, Vol. 33: 99-124.

Bosworth, Barry, Susan M. Collins, and Yu-chin Chen (1995), *Accounting for Differences in Economic Growth*, Conference on 'Structural Adjustment Policies in the 1990's', October 5-6 1995, Institute of Developing Economics, Tokyo, Japan.

Cooper, Charles (1995), *Technological Change and Dual Economies*, Discussion Paper Series, No. 9510, INTECH, United Nations University.

Dollar, David (1991), 'Outward-oriented Developing Economies Really Do Grow More Rapidly: Evidence from 95 LDCs, 1976-1985', *Economic Development and Cultural change*.

Edwards, Sebastian (1993), 'Openness, Trade Liberalization, and Growth in Developing Countries', *Journal of Economic Literature*, Vol. 31, September.

El-Laithy, Heba, Ola El-Khawaga, and Nagwa Riad (1999), *Poverty Assessment in Egypt: 1991-1996*, Economic Research Monograph, Department of Economics, FEPS.

Gassmann, Hanspeter (1996), *Globalisation and Industrial Competitiveness*, OECD OBSERVER, No. 197, Dec.- Jan.

Gu, Shulin (1996), *Toward an Analytic Framework for National Innovation Systems*, Discussion Paper Series, No. 9605, INTECH, United Nations University.

Kheir-El-Din, Hanaa and Hoda El-Sayed (1997), *Potential Impact of a Free Trade Agreement with the EU on Egypt's Textile Industry*, Working Paper Series, No. 16, ECES.

Kim, Linsu (2000), *The Dynamics of Technological Learning in Industrialization*, Discussion Paper Series, No. 2007, INTECH, United Nations University.

Lall, Sanjaya (2000a), *The Technological Structure and Performance of Developing Country Manufactured Exports, 1985-1998*, QEH Working Paper Series, No. 44, University of Oxford.

Lall, Sanjaya (2000b), *Turkish Performance in Exporting Manufactures: A Comparative Structural Analysis*, QEH Working Paper Series, No. 47, University of Oxford.

Lundvall, Bengt-Åke (ed.) (1992), *National Systems of Innovation: Towards a Theory of Innovation and Interactive Learning*, London: Pinter Publishers.

Mayer, Jörg and Adrian Wood (1999), *South Asia's Export Structure in a Comparative Perspective*, IDS Working Papers, No. 91, UNCTAD.

Ministry of Planning (2000), *National Accounts*.

Mytelka, Lynn and Fulvia Farinelli (2000), *Local Clusters, Innovation Systems and Sustained Competitiveness*, Discussion Paper Series No. 2005, INTECH, United Nations University.

Pritchett, Lant (1996), 'Measuring Outward Orientation in LDCs: Can it Be Done?', *Journal of Development Economics*, Vol. 49.

Sakr, Mohammed and Lobna Abdel Latif (2000), 'International Competitiveness of Egypt's Textile Industry', *Economic Studies*, No. 14, CEFRS, FEPS.

Singh, Ajit (1994), 'Openness and the Market Friendly Approach to Development: Learning the Right Lessons from Development Experience', *World Development*, Vol. 22, No. 12.

ul Haque, Irfa (1995), *Trade, Technology, and International Competitiveness*, Washington: World Bank.

Yeats, Alexander J. (1998), Just How Big is Global Production Sharing, Working Policy Paper, No. 1871, The World Bank.

15

Towards an African National System of Innovation

Lessons from India

Angathevar Baskaran and Mammo Muchie

Introduction

India and Africa share a common history of colonisation and struggle for independence. India's independence influenced many national liberation movements across Africa. Further, both India and Africa have been facing some common post-independent political and socio-economic problems towards achieving development and modernisation. Because of this shared colonial history, India's economic development policies and its achievement have been keenly watched and to some extent followed in many African countries. After attaining independence, India was trying to find a right development model. Indian political leaders feared neo-economic colonisation by the Multinational Corporations (MNCs), as the memory of colonisation of India by the East India Company was still fresh. At the same time, the rapid industrialisation achieved by the Soviet Union through five-year plans appears to have made a big impression on them. As a result, Indian leaders, while following a Western democratic political system, decided to follow a 'mixed-economy' model where the public sector played a predominant role. India's science, technology and economic policies were tuned to achieve 'self-reliance' which until the mid-1980s was 'inward-looking'. This led to a very different experience compared to those of East-Asian countries such as South Korea and Taiwan. However, the experiences of the Indian innovation system over the last five decades are more relevant to Africa because of common socio-economic problems such as large rural population, illiteracy, health, food and poverty. In this chapter, first, we discuss the evolution of the Indian national innovation system (Phase I – inward looking and Phase II – outward looking) and then, we draw some lessons from India's experiences for making an African national system of innovation.

Phase I (1950s-mid 1980s): Inward Looking National System of Innovation

Figure 1 Three Major Features of the Indian Innovation System

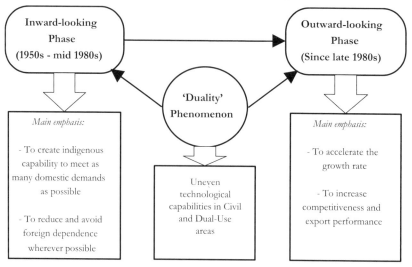

Figure 1 illustrates the three major features of the evolution of the Indian innovation system: (i) Inward looking Phase I; (ii) Outward looking Phase II; and (iii) the phenomenon of 'duality'. The Indian innovation system that evolved between the 1950s and mid-1980s was driven by two major factors; (i) 'blind faith' in science and technology; and (ii) an inward-looking policy of 'self-reliance'. Jawaharlal Nehru, India's first Prime Minister declared that 'science alone... could solve these problems of hunger and poverty' (Nayar 1983: 252). In 1958, the Scientific Policy Resolution committed the government 'to foster, promote, and sustain, by all appropriate means, the cultivation of science, and scientific research in all its aspects' (Nayar 1983: 288). Immediately before and after independence in 1947, India established a basic science and technology (S&T) infrastructure that included a network of public R&D organisations, universities, science and engineering institutions.

The principal policy objective behind India's industrialisation effort has been 'self-reliance'. Nehru said that India could not be economically or politically independent unless it strengthened its scientific and technological capacity (Eisemon 1984: 269). Indian leaders feared the domination and influence of foreign firms if free and unrestricted entry was allowed. Therefore, India's 'self-reliance' policy was defensive and inward looking rather than outward looking. India aimed to create local technological capabilities to meet mainly the domestic demands and reduce foreign dependency rather than developing an industry that should be competitive in the global market. This fundamental factor determined the shape and efficiency of the Indian system of innovation in Phase I. The 'self-reliance' policy influenced the development of other major elements of the Indian system of innovation – the incentive structures and supporting institutions, that is, financial markets, education systems and governments' macroeconomic and industrial policies. To achieve self-reliance, India implemented a number of measures such as industrial policy clearly defining the roles of private and public sectors, regulation of private investment through industrial licensing, regulation of foreign private investments, and regulation of technology imports to encourage indigenous research and development (Mascarenhas 1982: 4).

Until the mid-1960s India did not have a technology import policy and there was little concern about technological dependence. However, between the mid-1960s and mid-1970s, technology imports were regulated due to a number of factors such as problems of foreign exchange reserves, a need to regulate domestic competition, and to encourage the development of applied R&D institutions. The technology import policy (1965-68) was aimed at eliminating the advantages of the use of imported technology and to encourage import substitution. This led to scale-adaptation of imported technology to suit domestic market demands. This in turn led to the development of indigenous R&D capabilities and local machine tools and industrial equipment suppliers as little or no technical assistance was received from foreign technology suppliers (Cooper 1988: 117). The technology import policy has helped firms to become informed buyers and by the early 1980s, India has achieved a high level industrialisation through 'extensive collaboration for the import of foreign technology' (Mascarenhas 1982: 7). There were two major developments in the industrial sector. On the positive side, India has developed relatively a high level of indigenous technological capabilities to design and operate plants in a number of areas of capital and intermediate goods sectors (Mascarenhas 1982: 2). On the negative side, Indian firms hardly made major innovations to their products to establish a significant and sustainable export market. They mainly produced cheap and reliable products for the domestic market and a number of firms started in-house R&D to develop such products by adapting imported technology. By the early 1970s, most of the public R&D institutions made effort to catch up with research in developed countries and conducted research at the frontier level. Although they produced scientific knowledge and created a strong basic research base, often they did not contribute directly to help solve socio-economic problems of the country (Mascarenhas 1982: 2). The government's attempt to force firms to buy technology from public R&D institutions was given up in 1975 and by the early 1980s India started liberalising its policies towards import of 'new technology'.

Phase II (Since late 1980s): Outward Looking National System of Innovation

By the mid 1980s dissatisfaction with the performance of the economy started a shift towards an outward-looking national innovation system to achieve competitiveness and higher growth. This shift became clear when the industrial policy liberalisation was announced in 1991, which led to major changes in areas such as industrial licensing, foreign investment, foreign technology agreements, the public sector and Monopolies and Restrictive Trade Practices Act. The industrial policy – 1991 marked a clear shift from import regulating activity to export promotion activity. However, the impact of this liberalisation of the policy regime is not fully clear. While foreign technology import and investment has increased in the 1990s, there appears to be a small decline in domestic R&D investment of firms, particularly in the private sector (Goldar and Renganathan 1998; Kumar and Agarwal 2000). A number of foreign companies have established manufacturing operations in India since liberalisation. One of the significant developments is the opening up of R&D centres by MNCs in India and forging of collaborative relationships with Indian S&T institutions. The impact and benefit of this may only be clear after some years. Another development is the outsourcing of operations by foreign companies to India, mainly in the service sector. This seems to be increasing as this helps foreign companies to cut cost and enhance their efficiency, because of a high-skilled and highly qualified workforce available in abundance in India (Getty 2003). The nature of technological complexity and volume of outsourcing to India seems to be increasing and has already come under strong criticism by politicians and trade unions in the US, UK, and Germany (*BBC News*, 12 June 2003). In the era of the 'knowledge economy' it is an important development, as skills are as much valuable as technology and products. India was more used to 'brain draining' (Indians seeking jobs abroad) than 'brain seeking' (foreign companies employing skills in India). Although one might dismiss this as an attraction of 'cheap labour', this is not an insignificant development. However, it will be some time before the full implication of this development becomes clear.

In the area of export and competitiveness, progress appears to be slow in many industrial sectors, may be because of a gestation period required to shift from inward-looking to outward-ward looking. However, the IT sector, which emerged in the 1980s and 1990s as a major sector has witnessed significant export growth, particularly in software. There is a general perception that the availability of abundant skilled labour is the main reason for this. The answer is more complex than this. India missed the semiconductor revolution in the 1970s, due to protectionism and inter-departmental turf war. Compared to total world electronics production India's production was insignificant. For example, it was less that 0.5 per cent of the world production in 1982 (Commerce 1983: 1). The Indian electronics industry was found 'lagging far behind even the very small countries which joined the race much later' (Khandelwal 1981: 10). India learned valuable lessons and was careful not to repeat the mistake in the 1980s when the computer/IT revolution started. Since the early 1980s, that is, long before the major liberalisation in the 1990s, significant policy measures were taken to promote and expand the computer industry. The Computer Policy was announced in 1984 that removed capacity curbs, liberalised the licensing system and import duty to enable economies of scale and increase competitiveness (Commerce 1984: 845). The Electronics Policy 1985 noted that 'the software content of electronics is increasing and India is most appropriately placed to take advantage of this' (Bhojani 1985: 807). The computer industry was predominantly left in the private sector and a competitive environment was fostered. Soon, hundreds of firms in all sizes emerged. This subsequently appears to have established India as a leading player in the software market in the 1990s. By 1999-2000 India's software exports amounted to US$4.02 billion and it increased to US$6.3 billion in 2000-2001. India's success in this sector was mainly due to an intensive R&D effort by the companies and the presence of a strong basic research capability in the country (Government of India 2002). The liberalisation of policy regimes in the 1990s has demonstrated the potential of the Indian system of innovation in achieving a higher rate of growth despite its persistent weaknesses such as continuing problems in forging closer linkages between R&D institutions and firms.

Performance of the Indian System of Innovation

From the beginning, India put great emphasis on 'big-science' programmes such as nuclear and space programmes and they were allocated large resources. This affected the development of the S&T infrastructure in other areas such as agriculture and medicine until the late 1960s and also the funding of academic research in the universities (Krishna 2001: 6-7). Further, the promotion of 'big-science' programmes has led to a 'lopsided' or 'duality' phenomenon in the Indian system of innovation. That is, compared to most civil sectors, India has accumulated a high level of technological capabilities in complex dual-use areas (Baskaran 2000, 2001; Chengappa 2000).

India's innovation system often faced criticism because of its inefficiency that led to low rate of growth, its poor export performance, and relatively low quality of manufactured goods. These criticisms, although valid, either ignored or deliberately failed to take into account the context of the evolution of India's national innovation system. Particularly in the first phase, the principal objective of India's economic and S&T policy regimes was creating indigenous capabilities in the industry to meet as much domestic demand as possible, and thereby reducing or avoiding undue foreign dependence. Although ritual mentions were made in policy declarations about exporting, it was not the main driver of the Indian national innovation system in the first phase unlike the case of South Korea or Taiwan. Indian firms failed to export not because they were incapable, but because they 'prefer to exploit local markets where they have factor cost and marketing advantages' (Eisemon 1984: 272). Despite major flaws, there were significant achievements during the first phase of the Indian national innovation system. These included: (i) creation of S&T infrastructure and the expansion of higher education with great emphasis on basic research; (ii) development of indigenous capability to produce a range of goods which even today many developed countries are not capable of; (iii) implementation of the Green Revolution to achieve self-sufficiency in food grains; and (iv) creation of the scientific and industrial innovative potential to compete at the international market.

Considering that India fought three major wars, faced severe droughts and chronic foreign exchange constraints between the 1950s and 1980s, India's investment in S&T infrastructure and R&D expenditure was significant. Its R&D investment ranged between 0.8-0.9 per cent of GNP, which is comparable not only to developing countries like Brazil and China but also to some developed countries (see Tables 1 and 2). This created a vast network of basic S&T infrastructure. The number of research institutions at the time of independence in 1947 was 11. This increased to 63 by 1960 and 555 by 1980. The R&D units within firms were 13 in 1950, which increased to 400 in 1975 and 750 in 1981-82 (Subrahmanian 1990: 208; Eisemon 1984: 272). By 1980 the number of science and engineering graduates increased to 2.65 per thousand of population from 1.04 in 1960. There were 697,600 scientists and engineers and about 7 per cent of them were engaged in R&D activities. The growth in number of scientists, engineers and technicians (SET) per 1000 population in India by the 1990s, compared to other major countries, is also significant (see Tables 3 and 4).

An efficient innovation system is where technological accumulation and progress is also accompanied by higher growth performance of the industrial sector. During Phase I, the industry has witnessed significant growth, although 'the overall growth rate remained much below the plan targets and also below the achievements of several newly industrialising countries such as South Korea and Brazil' (Subrahmanian 1990: 205). Initial high growth rates gave way to stagnation since the mid-1960s. However, this changed since the mid-1980s when India started liberalising its industrial and technology policy regimes. Since then, the industrial growth ranged between 6.5 to 9 per cent (Government of India 2001: 303). The relative inefficient performance in Phase I appears to be largely because of rigid policy regimes.

The liberalisation in Phase II aimed to accelerate investment, growth, and employment appears to have produced mixed results. On the positive side, GDP growth was higher in the 1990s than in previous decades. The foreign currency reserves also increased from US$1 billion in 1991 to over US$45 billion in 2002. The debt service ratio was brought down from 35.3 per cent of current receipts in 1990-91 to 16.3 per cent in 2000-01. The external debt-GDP ratio has improved from 38.7 per cent in 1992 to 22.3 per cent in 2001. The literacy level has improved from 52 per cent in 1991 to 65 per cent in 2001 (Government of India 2002). However, after witnessing a high level growth in the initial period of liberalisation, the GDP has slowed down in the latter part of the 1990s. This trend has been repeated in the industrial sector growth (Government of India 2001: 303). Despite some inconsistent performances, the Indian system of innovation is refining and it is likely to perform with greater efficiency with increasing reforms to policy regimes. However, liberalisation of policy regimes may not be enough to realise the full potential of the Indian system of innovation. For this, fundamental changes of the institutions and research culture may be needed.

Table 1 Comparison of Major Economic Indicators and R&D Expenditure as Percentage of GNP Between India and Selected Countries (World-wide)

Country	Year*	R&D Expenditure (% of GNP)	Gross National Product (GNP) in US$ -billions (1999)	Per Capita GNP in US$ (1999)	GNP - Average Annual % Growth 1998-99	GDP - Average Annual % Growth 1990-99
India	*1998*	*0.82*	*442.2*	*480*	*6.9*	*6.1*
Argentina	2000	0.45	277.9	7,600	- 2.9	4.9
Brazil	2000	0.77	742.8	4,420	- 2.0	2.9
Canada	2000	1.84	591.4	19,320	3.8	2.3
China	2000	1.00	980.2	780	7.2	10.7
Egypt	2000	0.19	87.5	1,400	5.7	4.4
France	2000	2.15	1,427.2	23,480	2.4	1.7
Germany	2000	2.48	2,079.2	25,350	1.2	1.5
Israel	1999	3.62	---	---	--	5.1
Japan	2000	2.98	4,078.9	32,230	1.0	1.4
Madagascar	1995	0.18	3.7	250	5.5	1.7
Nigeria	1987	0.09	37.9	310	3.0	2.4
Republic of Korea	2000	2.68	397.9	8,490	11.0	5.7
Russian Federation	2000	1.00	332.5	2,270	1.3	- 6.1
Senegal	1997	1.40	4.7	510	5.1	3.2
Singapore	2000	1.88	95.4	29,610	5.6	8.0
South Africa	1993	0.70	133.2	3,160	0.8	1.9
Spain	2000	0.94	551.6	14,000	3.7	2.2
Sweden	1999	3.80	221.8	25,040	3.9	1.5
Tunisia	2000	0.45	19.9	2,100	6.2	4.6
Uganda	1999	0.75	6.8	320	7.7	7.2
UK	1999	1.87	1,338.1	22,640	1.7	2.2
USA	2000	2.69	8 351	30,600	4.1	3.4

Note: * Year relates to R&D Expenditure (% of GNP) only.
Source: UNESCO, Statistical Year Book 1999 and Science and Technology: Selected R&D Indicators (1996-2000), November 2002; World Bank, World Development Report 2000-2001.

Table 2 Comparison of Major Economic Indicators and R&D Expenditure as Percentage of GNP Between India and Selected African Countries

Country	Year*	R&D Expenditure (% of GNP)	Gross National Product (GNP) in US$ -billions (1999)	Per Capita GNP in US$ (1999)	GNP- Average Annual % Growth 1998-99	GDP- Average Annual % Growth 1990-99
India	*1998*	*0.82*	*442.2*	*480*	*6.9*	*6.1*
Benin	1989	---	2.3	380	5.1	4.7
Burkina Faso	1997	0.19	2.6	240	5.2	3.8
Burundi	1989	0.31	0.8	120	- 0.5	- 2.9
Central African Rep.	1984	0.25	1.0	290	3.7	1.8
Congo Rep.	1984	0.01	1.9	670	7.7	0.9
Egypt	2000	0.19	87.5	1,400	5.7	4.4
Madagascar	1995	0.18	3.7	250	5.5	1.7
Mauritius	1997	0.28	---	---	---	---
Nigeria	1987	0.09	37.9	310	3.0	2.4
Rwanda	1995	0.04	2.1	250	7.5	- 1.5
Senegal	1997	1.40	4.7	510	5.1	3.2
South Africa	1993	0.70	133.2	3,160	0.8	1.9
Togo	1994	0.48	1.5	320	2.1	2.5
Tunisia	2000	0.45	19.9	2,100	6.2	4.6
Uganda	1999	0.75	6.8	320	7.7	7.2

Note: * Year relates to R&D Expenditure (% of GNP) only.

Source: UNESCO, Statistical Year Book 1999 and Science and Technology: Selected R&D Indicators (1996-2000), November 2002; World Bank, World Development Report 2000-2001.

Table 3 Comparison of Number of Scientists, Engineers, and Technicians (SET) Between India and Selected Countries (World-wide)

Country	Year *	All R&D Personnel	Research Persons	Technicians	Support Staff	Year **	Research Persons/ million	Technicians /million
India	*1996*	*357,172*	*149,326*	*108,817*	*99,029*	*1994*	*149*	*108*
Argentina	2000	37,515	26,420	5,707	5,228	1995	660	147
Brazil	2000	78,565	55,103	21,914	1,548	1995	168	59
Canada	1998	139,570	90,200	31,380	19,560	1993	2,648	1,070
China	2000	922,131	695,062	---	---	1995	347	200
Egypt	1991	102,296	26,419	19,607	56,274	1991	459	341
France	2000	314,452	160,424	---	---	1994	2,583	2,873
Germany	1999	480,415	255,260	110,364	114,415	1993	2,843	1,472
Israel	1997	13,110	9,161	3,023	926	1984	4,828	1,033
Japan	1999	919,132	658,910	84,527	175,695	1994	6,293	827
Madagascar	2000	985	240	730	15	1994	12	37
Nigeria	1984	18,345	1650	9,696	6,999	1987	15	76
Republic of Korea	1999	137,874	100,210	26,160	11,504	1994	2,637	318
Russian Federation	1999	989,291	497,030	80,498	41,176	1997	3,587	600
Senegal	1996	78	19	29	30	1996	3	4
Singapore	2000	19,365	16,633	---	---	1995	2,318	301
South Africa	1993	60,464	37,192	11,343	11,929	1993	1,031	315
Spain	1999	102,237	61,568	40,670	---	1994	1,211	343
Sweden	1999	66,674	39,921	---	---	1993	3,706	3,166
Tunisia	1999	5,363	3,149	292	1,922	1997	125	57
Uganda	2000	1,187	549	330	308	1997	21	14
UK	1998	---	157,662	---	---	1993	2,413	1,017
USA	1997	---	1,114,100	---	---	1993	3,676	---

Note: * Year relates to All R&D personnel, Researchers, Technicians and Support staff columns only.
** Year relates to Research persons/million and Technicians/million.
Source: UNESCO, Statistical Year Book 1999 and Science and Technology: Personnel Engaged in R&D (1996-2000), November 2002.

Table 4 Comparison of Number of Scientists, Engineers, and Technicians (SET) Between India and Selected Countries (Africa)

Country	Year *	All R&D Personnel	Research Persons	Technicians	Support Staff	Year **	Research Persons/ million	Technicians/ million
India	*1996*	*357,172*	*149,326*	*108,817*	*99,029*	*1994*	*149*	*108*
Benin	1989	2,687	794	242	1,651	1989	176	54
Burkina Faso	1997	780	176	165	439	1997	17	16
Burundi	1989	814	170	168	476	1989	33	32
Central African Rep.	1996	19,500	---	---	---	1990	56	32
Congo Rep.	2000	217	101	111	3	1984	462	789
Egypt	1991	102,296	26,419	19,607	56,274	1991	459	341
Madagascar	2000	985	240	730	15	1994	12	37
Mauritius	1992	1,162	389	170	603	1992	361	158
Nigeria	1984	18,345	1,650	9,696	6,999	1987	15	76
Rwanda	1995	315	181	40	94	1995	35	8
Senegal	1996	78	19	29	30	1996	3	4
South Africa	1993	60,464	37,192	11,343	11,929	1993	1,031	315
Togo	1994	1,473	387	249	837	1994	98	63
Tunisia	1999	5,363	3,149	292	1,922	1997	125	57
Uganda	2000	1,187	549	330	308	1997	21	14

Note: * Year relates to All R&D personnel, Researchers, Technicians and Support staff columns only.
** Year relates to Research persons/million and Technicians/million.
Source: UNESCO, Statistical Year Book 1999 and Science and Technology: Personnel Engaged in R&D (1996-2000), November 2002.

Sharing India's Experience to Making an African System of Innovation

Hardly any research on the African system of innovation has been undertaken to date. There is, however, a need to elaborate the wider African as well as specific state, region and sector specific systems of innovations in Africa. The broad features, which can describe an African system of innovation are stated below.

- The specific aspect that needs changing to date is that the elements in the system that enter to constitute a specifically African NSI – both the technology and institutional dimensions – are mainly externally driven rather than having been endogenously propelled based on the interactions of national social-economic arrangements and the different knowledge systems, which exist in African societies. While India has mixed planning and the market to establish a largely self-reliant system of innovation, knowledge and technology to Africa have been diffused from external sources. India has the innovation system (lopsided and dualistic it may be) to internalise external knowledge into its system of production.

- The external dimension remains critical in innovation, learning, and accumulation of knowledge, the building of competencies and capabilities in organisation, product, process, techniques, market and management in the continent. India has managed to internalise these into its national context. The ability to choose from the world technological shelf in order to build local capabilities that can absorb, adapt and diffuse knowledge has been successfully carried out in India.

- Africa's research environment including its science and technology system has been dominated by foreign sponsorship.

The R&D expenditure as a proportion of the gross national product (GNP) for the continent as a whole was a mere 0.28 per cent in 1980, while Asia spent 1.40 per cent of its GNP on R&D, and North America 2.23 per cent. By 1990 the situation has worsened in Africa by R&D dropping to 0.25 per cent while in Asia it had increased to 2.05 per cent and in North America to 3.16 per cent. The structural adjustment impacts have not changed the science and technology system in Africa (Inos 1995). Overall Government support to R&D is lowest in Africa. On the whole countries in Africa spend nearly a tenth of the percentage of GDP that industrialised countries expend on R&D.

India has created local R&D and has invested in science and engineering to create both the human resources and physical resources internally.

- Africa has not put in place mechanisms for intellectual property and patents for inventions and innovations despite the setting up of two regional organisations (Organisation African de La Propriety Intellectual (OAPI) based at Yaounde, Cameroon set up as early as 1962 and the African Regional Intellectual Property Organisation (ARIPO) based in Harare, Zimbabwe (established in 1976)). The system of intellectual property continues to protect foreign patents rather than stimulate and furnish incentives to African inventors and innovators. The AU may make new efforts, but the problem still exists. India is active in WTO negotiations to create a national intellectual property regime that protects not only Indian inventors and innovators but also those from the developing world. Only a small number of countries in Africa such as South Africa play a significant role like India. In the recent trade round in Cancun, India played a lead role in the group of 21 articulating a position in favour of the farmers of the developing world against the heavy agricultural subsidies from Europe and USA.

- Many Governments in Africa have established a science and technology policy machinery assisted by UNESCO and foreign consultants, but the utilisation of science and technology for bringing about a structural transformation of the economy and society remains to be undertaken. State support to R&D has yet to grow and supplant the disproportionate donor funding it is projected to receive for the foreseeable future. India relies on a national science and technology system and funds its own science largely from its own resources.

- The private sector source's contribution to innovations is either from the in-house R&D departments of major multinational companies and/or from purchases in the form of capital and turnkey projects. The African centred R&D development and the link with production needs yet to be developed and increased. India's private sector is very active in innovative activity.

- While there is no problem in learning from outside the weakness of linkages between the formal and informal institutions, private and public institutions, and the indigenous and exogenous technological innovations dissipates the external input.

- It has been claimed that the market, state, production and business and learning systems do not often work in concert. Institutions, structures of production, and infrastructure have weak techno-economic networks. Inter-African communication linkages are still to be forged. India also has a lopsided and uneven national system of innovation.

- The science and technology system in African countries is mainly donor driven and much R&D requires donor input. For example in Senegal, between 30-40 per cent of scientists are French nationals (Tiffin and Osotimehin 1992: 44). This unduly injects donor influenced terms of references, priorities and donor preferences into the African system of innovation more than in any region in the world. About 80-90 per cent of the recurrent R&D budget is said to be devoted to personnel emoluments (Tiffin and Osotimehin 1992: 44). Local researchers are severely disadvantaged in research agenda setting with respect to donors. The pattern of assistance is said to be skewed to favour the learning of expatriate personnel more than domestic researchers. India has largely overcome this dependent situation.

- The scientific and technological human resources in Africa is said to be below the critical threshold necessary to provide effective and innovative leadership in R&D. India has trained skill labour which it is exporting to even the advanced countries.

- Many African researchers are said to be outside and those inside work for external actors and agencies. Indians work for both national and outside firms.

- It is claimed that there is no African research university comparable to the level and distinction of the major European and American centres of learning and research perhaps except a few universities in South Africa. India has a relatively well-functioning higher education system.

- The policy environment in facilitating linkages and techno-economic networks is said to be largely unreliable. The domains of state, market, and civil society remain weakly linked where the actors and activities emerging from them seem to sustain weak learning and innovation techno-economic networks. The Indian system has a stronger techno-economic network relative to Africa, though it may still not be coherent enough in relation to the developed Western economies.

Conclusion

The major lesson that African states can learn from India is the fact that India has established a national system of innovation capable of internalising knowledge from the outside. Africa is still not in a position to develop a strong system of innovation. It relies far too much on outside knowledge and technology and medium of transaction. This has to change. It can change if Africans build capacity by strengthening the African Union.

Most African economies rely on mineral and mono-crop export. They lack a strong industrial structure. India has evolved an industrial sector that can compete internationally.

Many African countries rely for about 80 per cent of their export earnings on primary commodities, and their share of manufactured exports continue to be very low compared with India. India's industrial growth was positively growing not fluctuating as in African countries. This is because India has successfully diversified its economic structure.

India has in Bangalore its own Silicon valley, whilst Africa's infrastructure-electricity and telecommunications network has been growing very slowly by international standards. In Sub-Saharan Africa except for South Africa, the average main telephone lines per 100 people is only 0.5 (9.5 in South Africa). African countries are universally lacking in locally manufactured computer hardware and in-local-language software, so the use of computer and Internet is not popular in the continent. Factory automation and computerisation are uncommon in African industries, and this is preventing the industries from becoming internationally competitive. Most African countries rely on the importation of IT technology and products. So far, more than half of African countries have some form of e-mail service and a gateway to the Internet. The pattern of dependence persists: For example:

> At present, e-mail sent from one Tanzanian computer to another must often transit Western Europe or the US before being redirected, greatly increasing costs for Tanzanian service providers and therefore customers. (Africa Business 2003: 43)

The main lesson that Africa must learn from India is to create a unification-nation within which a system of innovation can be embedded regardless of the imperfection of this system.

References

Africa Business (2003), *Special Report: IT in Africa*, August/September.

Baskaran, Angathevar (2000), 'Duality in National Innovation Systems: the Case of India', *Science and Public Policy*, Vol. 27, No. 5: 367-374.

Baskaran, Angathevar (2001), 'Competence Building in Complex Systems in the Developing Countries: the Case of Satellite Building in India', *Technovation*, Vol. 21, No. 2: 109-121.

Bhojani, Rajesh (1985), 'Electronic Policy: a Package of Surprise', *Commerce*, Vol. 150, No. 3857, April 27: 807-808.

Chengappa, Raj (2000), *Weapons of Peace: The Secret Story of India's Quest to be a Nuclear Power*, New Delhi: HarperCollins Publishers.

Commerce (1983), 'Growth of Electronics', *Commerce*, Vol. 147, No. 3771, December 1: 1-2.

Commerce (1984), 'Towards a Computer Revolution', *Commerce*, Vol. 149, No. 3835, December 1: 845.

Cooper, Charles (1988), 'Supply and Demand Factors in Technology Imports', in Ashok V. Desai (ed.), *Technology Absorption in Indian Industry*, New Delhi: Wiley Easter Limited.

Eisemon, Thomas O. (1984), 'Insular and Open Strategies for Enhancing Scientific and Technological Capacities: Indian Educational Expansion and its Implications for African Countries', in Martin Fransman and Kenneth King (eds.), *Technological Capability in the Third World*, London: Macmillan.

Getty, Sarah (2003), 'BT: Call Centres Better in India', *Metro*, July 17, London.

Goldar, Bishwanath N. and V. S. Renganathan (1998), 'Economic Reforms and R&D Expenditure of Industrial Firms in India', *Indian Economic Journal*, Vol. 461, No. 2: 60-75.

Government of India (2001), *Handbook of Industrial Policy and Statistics*, http://eaindustry.nic.in/handout.htm.

Government of India (2002), *Economic Survey 2001-2002*, Ministry of Finance, http://indiabudget.nic.in/es2001-02/general.htm.

Inos, John L. (1995), *In Pursuit of Science and Technology in Sub-Saharan Africa: The Impact of Structural Adjustment Programme*, London: Routledge.

Khandelwal, K. K. (1981), 'The Electronics Industry: Aspects and Prospects', *Commerce*, Vol. 142, No. 3648, May 16: 10-13.

Krishna, Venni V. (2001), 'Changing Policy Cultures, Phases and Trends in Science and Technology in India', *Science and Public Policy*, Vol. 28, No. 3: 1-18.

Kumar, Nagesh and Aradhna Agarwal (2000), *Liberalisation, Outward Orientation and In-house R&D Activity of Multinational and Local Firms: a Quantitative Exploration of Indian Manufacturing*, paper presented at Tokyo Conference, organised by GDN, December 10-13.

Mascarenhas, Reginald C. (1982), *Technology Transfer and Development: India's Hindustan Machine Tools Company*, Boulder: West View Press.

Nayar, Baldev R. (1983), *India's Quest for Technological Independence: Policy Foundation and Policy Change*, Vol. 1, New Delhi: Lancers Publishers.

Subrahmanian, K. K. (1990), 'India's New Policy Approach to Multinational Corporations and Technology Transfers', *Asian Journal of Economics and Social Studies*, Vol. 9, No. 3: 203-212.

Tiffin, Scott and Fola Osotimehin (1992), *New Technologies and Enterprise Development in Africa*, Paris: OECD.

16

National Innovation Systems, Development and the

NEPAD

Peter Gammeltoft

Economic development is a process of gradual capability and competence building; a process based on interactive and institutionally embedded learning in broader systems of firms, governments, research centres, universities, consultants, and other entities.[1] These systems can tap into stocks of global knowledge and technologies, assimilate and adapt it to local circumstances, and create new knowledge or technologies.

Such broader production systems are conceptualised in several different ways in the literature, e.g. Lundvall et al.'s 'national innovation systems', Richard Whitley's 'business systems', and Sanjaya Lall's concept of 'industrial technology development'. These approaches have primarily been developed to address nationally based institutional systems in advanced economies.

Both the ontological premises and the policy implications of these systemic approaches depart distinctly from the conventional orthodoxy on economic development as articulated in the 'Washington Consensus' and its later derivatives. This chapter explores which policy implications the adoption of such a systemic view might have for the New Partnership for Africa's Development (NEPAD).

Introduction

NEPAD is a new continent-wide development programme with the long-term objectives of 'poverty eradication, sustainable development, demarginalisation of Africa in the globalisation process and promotion of the role of women in all activities'. Even though issues of political governance such as peace, security, democratisation and human rights appear to dominate the rhetoric on NEPAD, the most concrete efforts and progress are so far concentrated in the more tractable domain of economic and corporate governance. Given that about half the population of Africa, or 340 million people, are living on less than one dollar a day and that life expectancy is only 47 years in sub-Saharan Africa, economic development is indeed an urgent task. Regional integration too is not an unreasonable aspiration: it makes little economic sense for an economy smaller than that of France to be divided into 54 separate states.

From an immediate observation, it might appear inappropriate to apply frameworks, which have predominantly been developed in the context of *economically advanced* countries and which identify themselves with *national* systems to a *regional* initiative in a *developing* region: NEPAD implies elevating the focus from national to regional and subregional systems and surely the concerns of countries in the South are different from those in the North.

However, historically, and in fact still, nation states have been the main political, economic and institutional vehicles of economic development. As such, the frameworks with a national focus have been tied to the principal institutional vehicle of development, not to the nation state *per se*, and might equally well be applied to a regional initiative were such an initiative to assume any primacy.

What the relevance to developing countries is concerned, it is common in developing countries for production systems to be fragmented. This originates in part from short-lived and alternating development strategies and in part from the selective and also impermanent penetration by global production chains of local production systems (Gammeltoft 2001, 2003a). The studies, which have documented the all to predictable fragmentation of innovation systems in developing countries, are already legio. So in this respect there may be limits to the analytical value of the national institutional frameworks. But the frameworks' prescriptive value is only so much the greater: in most developing countries the further progress of production systems is contingent on increased institutional integration and diversification of production – of broader, deeper and tighter innovation systems.

NEPAD does, at least in rhetoric, reflect the requisite perspective, resolve and long-term focus to address this. African development is critically conditioned from outside the continent, e.g. by the unsustainable debt burden, stagnating or falling foreign investments, declining post-cold war aid budgets, and lingering protection and subsidies in Northern markets.[2] However, progress depends equally critically on the configuration and development of internal institutional systems and NEPAD endeavours to address exactly such internal issues: it is promoted as the African side of a new deal between aid recipients and donors under which donors, in return, are expected to alleviate the external constraints.

To properly understand economic development processes, a systemic framework needs to be applied and a range of well-developed economic approaches do offer such frameworks. In the following we will concentrate on a body of literature dealing with the role of 'industrial policy'. In the account here we will predominantly relate to East Asian development experiences. After an account of the background and constitution of the NEPAD initiative, the conclusion explores the potential policy implications for NEPAD of adopting a systemic approach to development.

The Institutional Embeddedness of Production Systems

Even though enterprises are the primary drivers of growth, productivity and technological development, economic development involves much more than individual enterprises: enterprises do not emerge and succeed individually but in the context of wider production systems. Work in a wide range of disciplines such as economic sociology, economic geography, economic history, industrial organisation, business studies, and development studies informs us of the many ways in which economic activities take place and are structured by various humanly-constituted and devised structures. In a profound way, these approaches challenge the neoclassical perception of perfectly informed individual agents interacting instantly and costlessly through perfect markets and address the various ways in which economic activities are embedded in and influenced by a wider social context. There is also an implication that these issues significantly influence the *performance* of firms, beyond efficient markets, a stable and predictable legal system, macroeconomic stability, etc. (Hall and Soskice 2001; Hollingsworth, Schmitter, and Streeck 1994).

Applying a systemic perspective also has implications for the role of policy in development: the need for government intervention in the presence of certain market failures is generally accepted. More contested is the need for broader government activism to bring various markets and actors operating in them into existence and guide their development. There is also a general consensus as to the significance of macroeconomic stability, human resource development, high savings and investment rates, and export orientation, but if we consider government policies and programmes targeted more specifically at economic, industrial and technological advance, we move into highly contested grounds.

Even if one accepts that interventions are warranted in order to remedy information failures and co-ordination problems between whatever activities may already be present in an economy, economic *development* goes beyond this and depends on uncertain, long-term, and complex learning processes of a cumulative and path-dependent nature on the part of a diverse range of economic actors. If one accepts that such fundamental differences exist between processes of dynamic growth and static allocation, this strongly influences the extent and type of policy intervention conceivable.

Industrial Policy

There is a broad and diverse literature, which addresses the rationales, scope, requirements, and techniques of 'industrial policy'. The activism of governments of the successful East Asian states has been amply documented in the literature and we will illustrate the issues involved in industrial policy by drawing especially on the East Asian experiences.[3] We consider four broad areas of industrial policy and the techniques associated with them: the creation and nurturing of markets and agents, industrial organisation, the institutional infrastructure, and the regulatory framework.

AGENTS AND MARKETS

Markets are not universal, spatially and temporally uniform entities but are institutionally constituted and vary between contexts. Both markets and the economic agents operating in them may need to be created and nurtured. The Korean government's promotion of the *chaebol* is a prominent case in point. The large domestic business groups were a means to economise on limited local entrepreneurial and financial resources and to internalise deficient markets for capital, skills, information, and entrepreneurship. By design, a symbiotic relationship between government and the *chaebol* was created in which government was able to generate investment opportunities and the *chaebol* subsequently respond to them.

Specialised technological agents such as engineering firms, intermediate-goods producers, and capital goods suppliers, may act as repositories of technological capabilities and diffuse technology between firms. Flows through such intermediaries are often far more important than those directly between competing firms (Dahlman, Ross-Larson, and Westphal 1987). To nurture such complementary activities, governments have in the past set specific targets for machinery, parts, and raw materials that should be localised but such 'local content' policies have become disallowed under the WTO TRIMs agreements. Other possible measures are tax incentives, preferential financing, loan guarantees, and R&D subsidies.[4]

Countries have frequently targeted particular *infant industries* for certain periods, among the reasons being that advance in various 'base industries' such as information technology, new materials, and biotechnology may influence strength in downstream industries, so that countries cannot afford to let these sectors be exclusively controlled by foreign firms. A converse argument is that demand from strong downstream industries may be necessary to develop upstream component industries. More generally, technological linkages between firms may require that whole groups of activities are promoted as infant industries, since this will allow learning processes in individual companies to be co-ordinated.

Public enterprises have been common in activities where social benefits considerably outweigh private, in capital and technology intensive areas, and areas, which for one reason or another are considered nationally 'strategic'. Beyond the capabilities developed in these enterprises themselves, technology may be diffused into private industry through the linkages they form and through labour mobility.[5] In Korea, the government announced *procurement plans*, which induced activities in particular areas, and at the same time provided secure income to companies in the process of undertaking risky investments in other areas such as semiconductors.

INDUSTRIAL ORGANISATION

Besides promoting particular markets and agents there may also be a role for government in shaping the way agents interact and the way industrial activities are organised. As a production system develops, more and more advanced activities will be undertaken locally, and linkages between activities become more complex. In some instances, firms may not initiate activities even though they would be socially beneficial because of externalities, inability to appropriate the benefits, or information or co-ordination failures.

Even neoclassical economists who recognise that products and technology are not perfectly tradable emphasise the need to facilitate the flow of information and to coordinate investment decisions. Based on case studies of fifteen countries, Nelson (1996) finds that many countries encourage co-operation between private firms in R&D.

Government may need to encourage the establishment and use of various *supporting industries*, as referred to above. Pack and Westphal (1986) address the issue that previously transferred technology may be incompletely mastered and productivity therefore reduced due to insufficient diffusion of knowledge about production engineering, inadequate product specialisation among firms making similar products, and an insufficient extent of subcontracting. The absence of *subcontracting* is likely to result in different firms internalising the same activity, all operating it below full utilisation. If too many firms produce the same product, the advantages of specialisation may be foregone, a risk which may be reduced by government intervention, e.g. by encouraging rationalisation cartels among private industries. Export processing zones and industrial districts have been important in providing companies with physical and institutional infrastructure and facilitating cooperation between foreign and local companies and among local companies themselves.

INSTITUTIONAL INFRASTRUCTURE

Beyond extending support to individual firms, the institutional infrastructure may also function as a repository for accumulation of capabilities, and a channel through which information and also manpower can diffuse between firms. The heavy investment in *human resources* in general and technical training in particular is usually highlighted as one of the most important prerequisites for the rapid economic development in East Asian nations. The overseas training and hiring of returnees are also frequently cited as important. Policy recommendations may include encouraging industrial training by subsidies to or levies on firms; to increase enrolment rates with a focus on technical fields; to gear training to emerging technological needs; and to get industry involved in the management of training and education institutions.

Besides human resource development, a well developed local *S&T infrastructure* can induce the choice of socially appropriate techniques, improve the terms of technology imports, and stimulate capability development in local productive enterprises and specialised technological agents. Studies suggest that the main economic benefit from research activities is not the formal output as such but the resulting supply of scientists and engineers, their skills and network engagements (Bell and Pavitt 1993), and this testifies to the importance of labour migration between supporting institutions and firms. Government laboratories may spearhead the development of new technologies, but generally policies directed at the diffusion and application of technology, bringing industries up to world practice or spreading knowledge about new developments, can be more effective than the subsidisation of major breakthroughs. Since individual companies may not be able to appropriate the benefits of information gathering related to technology acquisition and absorption, and since such gathering is associated with large fixed costs, government may induce industry-wide efforts, possible with some compulsion to curb free riding.

The rationales for more mundane *industrial extension services* are the same as those for the more specialised activities related to science and technology. A well-functioning metrology, standards, testing, and quality assurance (MSTQ) system is central to the upgrading of local firms and to facilitating both local co-operation and international marketability.[6]

It has been argued that *government-business deliberation councils* contributed significantly to the economic success of some of the East Asian countries (World Bank 1993). They are fora which bring together various stakeholders, government, business, labour, consumers, academia, and the press to discuss policy, market trends, exchange information in general, and formulate visions for future development.

Due to the inherent uncertain and long-term nature of scientific and technological activities, *finance* poses a special problem.[7] In Korea, the government established various funds aimed at supporting activities, which traditionally have difficulties raising capital, such as technological development, small technology start-ups, R&D, equipment modernisation, and plant automation. It also took steps to establish a venture capital industry, primarily based on public firms.

REGULATORY FRAMEWORK

Various features of the regulatory framework, beyond the basic 'rules of the game' have a bearing on processes of technological development. Government may intervene to increase *technology transfer* or improve the terms under which it is conducted. Government may stimulate the participation of local agents in the transfer and absorption of imported technological packages by providing subsidies and fiscal incentives for local involvement, guide or subsidise TNCs to enter targeted activities or conduct R&D locally (Singapore), or encourage TNCs to conduct higher value-added activities locally through investments in education/training and through upgrading of local suppliers, infrastructure and support institutions.

Developing countries licensees are often disadvantaged *vis-à-vis* foreign licensors. Governments may impose limits on royalty payments or be able to achieve favourable changes in the terms of licensing agreements, e.g. through information dissemination or through their ability to control the access of licensors to the domestic market. Korea and Japan carefully screened licensing agreements, particularly to avoid export restrictions.[8]

Another universally recognised factor underpinning the economic growth of East Asian nations is their early push towards *exports*, which imposed dynamic incentives upon firms for upgrading and efficiency and provided them with learning opportunities (Hobday 1995; Nelson 1996).[9] In Korea and Taiwan, while the domestic market was protected from foreign imports and investments, this was combined with fierce *domestic competition*. A special mechanism to combine competition and protection discussed in the World Bank's 'Miracle Study' (1993) was the conduction of 'contests': in Korea and Japan, firms were encouraged to cooperate and the number of competing firms was kept down to be able to attain scale. The risk of inefficiency and collusion was reduced by requiring firms to compete for government-controlled scarce resources, particularly credit, foreign exchange, licenses to initiate or expand activities, and import protection. These favours were then granted according to export performance and international competitiveness.

Foreign direct investment is an important source of technology, but there are risks inherent in relying too heavily on foreign technology since TNCs tend to exploit static comparative advantages and retain advanced activities elsewhere. Interventions may encourage TNCs to 'deepen' local production and conduct more dynamic and complex activities locally. This might take the form of changing incentives to encourage local technological activity (as in Singapore) or restricting foreign entry and encouraging and supporting local companies to develop R&D and other technological capabilities themselves (Korea and Japan).

At the early stages of development, lenient *intellectual property right laws* facilitate local imitation of foreign products and processes, but later on when local companies themselves become able to undertake development work, lax laws may discourage local development efforts. Accordingly, one could envision systems of flexible and variable protection, contingent on the industry or activity in question and its state of development as being developmentally superior.

Governments commonly formulate economic *development plans*, which typically reflect ambitions to shift from a low to a high technology growth path by increasing local value added in production and design and taking on more complex industrial activities. Besides determining areas of direct government action, such plans constitute part of the incentive structure influencing the direction and intensity of private efforts.

SUMMARY OF ISSUES ADDRESSED IN INDUSTRIAL POLICY
Table 1 summarises the primary entities addressed by the literature on industrial policy. For a fuller account and a comparison with Lundvall et al.'s 'national innovation systems' (Lundvall et al. 2002), Richard Whitley's 'business systems' (Whitley 1999), and Sanjaya Lall's concept of 'industrial technology development' (Lall 1992, 1993), see (Gammeltoft 2001).

Table 1 Main Issues Addressed in Industrial Policy

Policy Area	Techniques
General policies (incentives)	Competition
Technology policies	Technology transfer
	Export push
	FDI regime
	IP protection
	Development plans
	Government-business deliberation
Labour and skills	HR development
Information and technical support	S&T infrastructure
	Industrial extension
Economic agents	Capable firms
	'Specialised technological agents'
	Infant industries
	Public enterprises and procurement
Industrial organisation	Supporting industries, subcontracting, SMIs
	Inter-firm cooperation, e.g. in R&D, rationalisation
Finance	Credit
	Foreign exchange
	Venture capital

NEPAD: Background and Purpose

We now turn to looking more specifically at the New Partnership for Africa's Development (NEPAD). NEPAD is the latest grand attempt to set out a continent-wide development programme.[10] Its stated long-term objectives are poverty eradication, sustainable development, demarginalisation of Africa in the globalisation process, and promotion of the role of women in all activities (NEPAD 2001). NEPAD emerged at the joint conference of Africa's ministers of finance and economic planning in May 2001. It is strongly promoted as African in origin and Africa-driven, though this characterisation is not uncontested (Gammeltoft 2003b).

NEPAD covers four broad areas which are defined as prerequisites for the success of the programme: Peace and Security; Democracy and Political Governance; Economic and Corporate Governance; and Sub-regional and Regional Approaches to Development. The stated aim of the initiative is to achieve the overall 7 per cent annual growth necessary for Africa to meet one of the Millennium Development Goals (MDGs): halving poverty by 2015. It is estimated that Africa needs a yearly transfer of US$64 billion to meet this target, a tall order compared to the current transfer of US$10 billion. The envisioned sources of transfer are increased debt relief, aid, investment and market access.

Even though the NEPAD has met with a variety of criticism, it is nevertheless widely deemed as the best chance the continent has had for years of ensuring that its concerns are heard widely in the international community. It has the necessary boldness and simplicity to move African concerns up the agenda of the G8 and the OECD and has already succeeded in doing so.

The Emergence of NEPAD

Regional economic integration has long been an aspiration in Africa: the 'Lagos Plan of Action' for economic integration was adopted in 1977 and in 1991 the 'Abuja Treaty', which planned the gradual establishment of an African Economic Community over a period of 34 years. Subregional integration has also been proceeding, with moves towards a monetary union in West Africa, the revitalisation of the East African Community of Kenya, Tanzania and Uganda, and other initiatives.

The primary champions behind the NEPAD have been the presidents of South Africa, Nigeria, Senegal, and Algeria:[11] Thabo Mbeki, Olusegun Obasanjo, Abdoulaye Wade, and Abdelaziz Bouteflika. In Algiers in May 2001, three preceding and similar initiatives were merged into one: The Millennium Partnership for Africa's Recovery Programme (MAP), the Omega Plan, and the Compact for African Recovery. The July 2001 summit of the Organisation of African Unity (OAU) in Lusaka mandated an implementation committee of 15 heads of state to manage it, in October its secretariat was established in South Africa, and it was officially launched in July 2002.

At the core of the NEPAD lie two important and novel concepts: the notion of 'enhanced partnerships' and the African Peer Review Mechanism (APRM), which hold the potential to substantially transform the aid relationship. The 'enhanced partnership' is put forth as a principle of joint responsibility and mutual accountability: rather than donor-imposed conditionalities, which have proved ineffective and burdensome in the past, 'enhanced partnerships' are to represent common commitments by African countries and donors to a set of development outcomes defined by African countries, whereby donors pool funds, guarantee them for an extended period and channel them through budgetary processes, which are then jointly monitored on the basis of outcomes.

The notion of 'enhanced partnership' sets NEPAD apart from the African Union[12]: whereas the AU has no criterion for membership except being located in Africa, participation in 'enhanced partnerships' is contingent upon meeting certain standards of governance and economic management. This introduces a politically very sensitive element of discrimination among African countries.

The African Peer Review Mechanism (APRM) is similar to the peer review in the OECD, which is regarded as a successful means of identifying and promoting appropriate practices: African countries are to monitor each other's progress according to a set of benchmarks and the scheme specifies an ambitious range of targets for conflict prevention, good governance, poverty reduction and disease control, including HIV/AIDS. The expectation is that this will introduce a pressure on badly performing countries to do better.

The aspiration is that the integrity and standard of the reviews will be sufficiently high for the donors to abandon their own monitoring process and accept the outcomes of the APRM. Reviews are not intended to be imposed on countries; rather countries should step forward for review themselves. One of the greatest incentives is the promise of debt relief. If the mechanism is implemented, a small group of well-performing countries are likely to step forward for review and the reward will be entry into 'enhanced partnerships' under which they will receive increased aid and investment.

NEPAD and the Donors

The progress of the NEPAD is obviously critically dependent on its reception by the donor community, and it has generally been received with applause and enthusiasm. NEPAD is also generally well-aligned with what was dubbed the 'Monterrey Consensus' at the Financing for Development Conference in Monterrey in March 2002, which *inter alia* stresses that more aid should be channelled to the 'good performers'; that developing countries should assume more responsibility for their own development processes; and that international trade and foreign direct investment are the primary drivers of development.

In Monterrey an additional US$12 billion per year were pledged for global development assistance by 2006. Shortly afterwards, at the G8 Summit in Kananaskis, Canada, the G8 agreed that half or more of this money *could* go to Africa, to those countries that 'govern justly, invest in their own people and promote economic freedom'. If implemented, this would represent an increase by almost 50 per cent of the US$10 billion development assistance to Africa in 2002.

Also in Kananaskis, where African leaders were allowed to attend for the first time (Stefanski 2002), the G8 adopted an Africa Action Plan in response to NEPAD. It is clear from the Africa Action Plan that the G8 countries will not decide *en-bloc* to transfer aid and other resources to African nations. Rather, each of the G8 countries reserves the right to decide with which African countries they want to partner and when and how. Individual G8 countries decide 'on the basis of measured results' which countries they would want to form partnerships with but in doing so they will be informed by the results of the peer review process NEPAD is putting into place. Yet, the G8 welcomes partnerships based more on African priorities.

The United Nations and the OECD have adopted NEPAD as the basis upon which to build future relations with Africa.[13] But the US Millennium Challenge Account (MCA) set up in November 2002, may seem at odds with the thrust of NEPAD: the MCA is US$5 billion of new yearly aid money to be dispensed to those countries, which score highly on a specific benchmark system measuring good governance, economic liberties and delivery. By acquiring recipients to comply to a set of pre-specified criteria this appears to run counter to the principles of indigenous aid agendas and African peer reviews.[14]

Recent Developments and Progress

Which concrete progress has the NEPAD made then, in the two years it has been in existence? In his speech at the latest meeting of NEPAD's 20-member Heads of State and Government Implementation Committee meeting in May 2003, President Mbeki criticised the commitment of African leaders to the NEPAD: only seven or eight leaders were present at the meeting and only 10 governments had ratified the protocol to establish a peace and security council to intervene in conflicts on the continent.

During the first year of NEPAD's existence, the international financial community met it with diplomatic enthusiasm but did not come up with any meaningful funding. At Evian however, a number of concrete targets for increase in donor funding were set (G8 2003), and commitments were made to a number of the infrastructure projects proposed by NEPAD (Battersby 2003). By June 2003, 41 out of 124 projects on a 'high-priority shortlist' had been initiated (Fabricius 2003). NEPAD must also be credited with having achieved to push Africa visibly up the international agenda, e.g. at Monterrey, Kananaskis, Evian, and the 2003 World Economic Forum Africa Summit in Durban in June.

When evaluating the progress it is also important to note that the different components of NEPAD are accorded different priority: the economic and corporate governance peer review mechanism is central in defining 'enhanced partnerships' and in unlocking increased development finance and is accorded the highest priority. The other governance and peace and security components of the initiative are still in preliminary stages of discussion (de Waal 2002). Peer reviews of *political* governance, when and if it is implemented, will be a wholly new practice that has not been tried anywhere in the world.[15]

In the past year, six prominent Africans have been appointed to the peer review unit, which is scheduled to start its work in July 2003. But with only 15 of the 53 member countries of the AU having signed up for peer review, the enthusiasm for what could be the core of the NEPAD concept does appear limited.[16]

The NEPAD Strategy and the Role of S&T

Having accounted for the general features of NEPAD, we now turn to its specifics: which concrete areas does the strategy target and which measures does it propose to advance the political, economic, social, and cultural progress of the continent?

The NEPAD Strategy

The programme of action in the base document of NEPAD (NEPAD 2001) identifies the following priority areas:

A Conditions for Sustainable Development
A1 The Peace, Security, Democracy, and Political Governance Initiatives
A2 The Economic and Corporate Governance Initiative
A3 Sub-Regional and Regional Approaches to Development

B Sectoral Priorities
B1 Bridging the Infrastructure Gap
B2 Human Resource Development Initiative, Including Reversing the Brain Drain
B3 Agriculture
B4 The Environment Initiative
B5 Culture
B6 Science and Technology Platforms

C Mobilising Resources
C1 The Capital Flows Initiative
C2 The Market Access Initiative

In the implementation plan the initiative proposes that some programmes be fast-tracked due to the need to sequence and prioritise, viz. communicable diseases (HIV/AIDS, malaria and tuberculosis), information and communication technology, debt reduction, and market access. The strategy duly recognises the shortcomings of a project-based approach to development but finds that a number of crucial individual projects need to be implemented, e.g. in agriculture, private sector development, and infrastructure and regional development.

The Role of Science, Technology and Innovation

We will now turn to looking more specifically at science, technology and innovation. Which role is it envisioned to play and which importance is it accorded with in the NEPAD process? Academics and researchers, representatives of sub-regional bodies and government officials met to consider these exact issues at a NEPAD Workshop on Science and Technology in Johannesburg in February 2003. The aims of the workshop were to develop a NEPAD framework for science, technology and innovation, define priorities, align national strategies and strengthen co-operation.

The declaration adopted by the workshop spells out a number of agreed areas of concern related to the African science and technology system (NEPAD 2003): lack of information on research activities on the continent; insufficient co-operation across national boundaries; weak linkages between scientific institutions and industry; underestimation of the potential of science and technology to address poverty issues; outward mobility and loss of African scientists; low quality of science education; and R&D expenditures below one per cent of GDP for most African countries.

A number of concrete measures are recommended to address these weaknesses, e.g. instituting a monitoring and evaluation system of S&T in Africa, also to inform policy; 'mainstreaming' of science and technology into the existing NEPAD sectoral programmes (health, agriculture, education, environment, governance, infrastructure, security, and investment and trade); creation and strengthening of centres of excellence; establishment of regional research and innovation programmes focusing on human development needs in for example space science, desertification, biotechnology and information technology.

The declaration also recommends that a special Forum on Science and Technology be established within NEPAD to push science and technology up the political agenda. The Forum should be made up of African science ministers and presidential advisors and supported by a panel of experts from science and industry and charged with identifying priority areas for African S&T. The declaration appeared to shy away from more sensitive issues such as intellectual property rights.

Conclusion and Recommendations

How do the industrial policy issues and lessons discussed in the preceding sections apply to NEPAD's economic policies? First of all, when exploring the applicability of lessons from East Asia on the NEPAD we should not overlook the obvious: that the so far weakly institutionalised pan-African initiative is not a strong East Asian state. History is littered with examples of governments intervening in counterproductive ways retarding technological development, efficiency, export growth, and structural change (Lall 1992; World Bank 1991). Lacking of such capacities, it is safer to adhere to a neutral policy regime or apply broader functional interventions, leaving it to market forces to sort out the best enterprises and technologies (Lall 1992; Pack and Westphal 1986).

However this is not the same as saying that the requisite institutional capacity could not or should not be developed. The World Bank recognises that state activism under the right circumstances leads to faster and more equitable growth (World Bank 1993, 1998). Developing such capacity would have to take into account that the political culture in Africa is generally very different from that in Japan, Korea, Taiwan and Singapore. Peter Evans (1995) identified 'embedded autonomy', i.e. East Asian states' concurrent embeddedness in and separation from the business community and wider society, as crucial for the success of their activist economic policies. African states generally function according to neopatrimonial principles and tend to be much less professionalised and less institutionally separated from society than their East Asian counterparts (Chabal 2002).

The most immediate and obvious recommendation is therefore that any success of NEPAD's economic policies is contingent on further institution building and, based on what we saw concerning the initiative's progress so far, broadening and deepening of political commitment.

If we assume that NEPAD was to develop as envisioned, which policy recommendations could our presentation imply? Among the most urgent issues in terms of African economic development are upgrading from agricultural produce and extraction industries and diversification of manufacturing production and this could be addressed by an array of the techniques discussed. Export orientation should be attempted from early on and techniques to encourage technology transfer applied; any preferential treatment should be tied to performance requirements, e.g. export performance or productivity increases; attention should be paid to attaining scale economies, and subcontracting and specialisation promoted; the processes should be supported with appropriate demand-driven extension services, designed and managed with active business participation.

We saw that some techniques applied in the past are no longer permissible under the new international trade regime. It would be worthwhile then to evaluate more systematically how NEPAD might exploit the WTO provisions for 'special and preferential treatment' for the least developed countries. Export subsidies and selective government procurement are still permitted in most African countries. Furthermore, we discussed two specific institutional techniques which might also be applied in the context of NEPAD: 'government-business deliberation councils' to enhance the quality of policy deliberation and the commitment to their subsequent implementation, and the use of 'contests' in public procurement as a mechanism to combine co-operation and competition.

We saw that economic development in East Asia benefited from the externalities flowing from local planning, monitoring and evaluation of government policies. Short of leaving the administration of the new US aid initiative, the Millennium Challenge Account (MCA), entirely to African states, it is likely that the wider the reliance on local planning, monitoring and evaluation, the greater the benefit for African states. Such local reliance could very well work through the African Peer Review Mechanism (APRM) and the enhanced partnerships.

The APRM has the potential to become an important instrument to strengthen innovation systems on the continent: peer reviews of good practice can stimulate policy learning (Dalum, Johnson, and Lundvall 1992) and thus provide a stronger institutional foundation for the promotion of innovation, competence building and industrial upgrading.

Notes

[1] A more in-depth version of this chapter is published as a working paper at the Department of Intercultural Communication and Management, Copenhagen Business School.

[2] It is often argued that a major reason for Africa's predicament is its insufficient integration into world markets. However Amin (2002) observes that Africa is in fact more integrated into the global economy in terms of the share of its external trade in GDP than other regions. Accordingly, it is not the extent but the modality of integration, which should be of concern. Measures to improve it would include upgrading beyond agricultural produce and extraction industries and strengthening and diversifying manufacturing production.

[3] Unless otherwise specified the case material in this section is especially based on (Amsden 1989, 2001) and (Wade 1990).

[4] Selective techniques are generally no longer permitted under the WTO. However what subsidies is concerned, countries with less than US$1,000 per capita GNP are exempted from the prohibition on export subsidies and have a time-bound exemption from other prohibited subsidies.

[5] Under the plurilateral agreement on government procurement administered by the WTO, local and foreign companies should be treated equally. However, so far only 28 out of the 146 WTO member countries are signatories to the agreement.

[6] A common problem with publicly provided R&D and extension services seems to be that they are often supply-driven, do not correspond to industry needs, and are of inadequate quality. Various mechanisms can be applied to secure the relevance and reach of such efforts, e.g. requiring them to be more demand-

driven, requiring that part of the budget is covered by fees; conducting joint public/private projects to secure relevance and reach; securing private sector input in management and operations; and conducting applied technological work rather than basic science.

[7] 'Anglo-Saxon' capital market-based financial systems are usually taken to favour short-term profit-oriented investments, whereas in credit-based financial systems, often associated with Germany and Japan, creditors tend to be more engaged in long-term growth-oriented investments and there are closer associations between financial institutions and firms (Whitley 1999).

[8] Korea's restrictive policies towards FDI and foreign licenses induced companies to acquire foreign technology in the form of capital goods and turnkey plants. A slight overvaluation of the local currency and tariff exemptions on imported capital goods facilitated these forms of transfers.

[9] In Korea, while the small domestic market was protected to foster infant industries, from early on government pushed and pulled companies to compete in export markets to obtain economies of scale. This also imposed stringent cost and quality requirements on the exporters, and companies were brought in contact with foreign OEM buyers from whom technology was transferred.

[10] But it is not the first: the Financial Times quotes a G7 official for saying that there have been some 18 Africa development initiatives in the last 20 years (Beattie and Lamont 2002). For brief reviews of some earlier initiatives, see (Akinrinade 2002; Stefanski 2002). Proponents argue that leadership and African ownership as well as a new set of circumstances makes this initiative different.

[11] In addition to these countries, Egypt is often included as a fifth country behind the initiation of NEPAD.

[12] The African Union (AU) replaced the Organisation of African Unity (OAU) and was officially launched in July 2002.

[13] UN General Assembly resolution A/RES/57/2 adopted NEPAD as the general framework around which the international community including the United Nations system should concentrate its efforts for Africa's development.

[14] Instead of dispensing the yearly US$5 billion MCA funds through international organisations and USAID, the Bush administration has set up a new organisation, the Millennium Challenge Corporation, which is envisioned to enter into business-like contracts with individual African states. Furthermore, the administration intended to only apply Congress for an initial allocation of US$1.7 billion and it is likely that only US$7-800 million will eventually be allocated.

[15] The track record of democratisation in Africa is not altogether bad: If Nigeria is included, 17 of Africa's 54 nations are now considered fully fledged or emerging democracies, compared with around four at the end of the 1980s. In three years elected governments in three countries, Senegal, Ghana, and Kenya, have handed over power peacefully after being voted out and this is a rare occurrence in Africa. But in Zimbabwe political repression has become worse, Ivory Coast is balancing on the verge of breakdown, and Africa's first military coup in more than three years took place in March in the Central African Republic.

[16] South Africa, Algeria, Ethiopia, the Democratic Republic of Congo, Ghana, Kenya, Mozambique, Nigeria, Rwanda, Uganda, Mali, Cameroon, Gabon, Burkina Faso, and Senegal.

References

Akinrinade, Sola (2002), 'NEPAD: the New Partnership for Africa's Development', *The Conflict, Security & Development Group Bulletin*, No. 15, May-June.

Amin, Samir (2002), 'Africa: Living on the Fringe', *Monthly Review*, March.

Amsden, Alice H. (1989), *Asia's Next Giant: South Korea and Late Industrialization*, Oxford: Oxford University Press.

Amsden, Alice H. (2001), *The Rise of 'The Rest': Challenges to the West for Late-Industrializing Economies*, New York: Oxford University Press.

Battersby, John (2003), 'Now Let's See What Africa Can Do – Mbeki', *The Star*, Johannesburg, 4 June.

Beattie, Alan and James Lamon (2002), 'African Initiative Struggles in Hope for a Fresh Start: Alan Beattie and James Lamont Explain the Problems Facing a New Partnership', *Financial Times*, 14 February.

Bell, Martin and Keith Pavitt (1993), 'Technological Accumulation and Industrial Growth: Contrasts between Developed and Developing Countries', *Industrial and Corporate Change*, Vol. 2, No. 2.

Chabal, Patrick (2002), 'The Quest for Good Government and Development in Africa: is NEPAD the Answer?', *International Affairs*, Vol. 78, No. 3: 447-62.

Dahlman, Carl J., Bruce Ross-Larson, and Larry E. Westphal (1987), 'Managing technological development: lessons from the newly industrializing countries', *World Development*, Vol. 15, No. 6: 759-775.

Dalum, Bent, Björn Johnson and Bengt-Åke Lundvall (1992), 'Public Policy in the Learning Society', in Bengt-Åke Lundvall (ed.), *National Systems of Innovations: Towards a Theory of Innovation and Interactive Learning*, London: Pinter Publishers.

de Waal, Alex (2002), 'What's New in the 'New Partnership for Africa's Development?', *International Affairs*, Vol. 78, No. 3: 463-75.

Evans, Peter (1995), *Embedded Autonomy: States and Industrial Transformation*, Princeton: Princeton University Press.

Fabricius, Peter (2003), 'Nepad Gathers Steam as Priority Projects Get under Way', *Cape Times*, South Africa, 12 June.

G8 (2003), *Implementation Report by Africa Personal Representatives to Leaders on the G8 African Action Plan*, Web site of the G8 2003 Evian Summit.

Gammeltoft, Peter (2001), *Embedded Flexible Collaboration and Development of Technological Capability: a Case Study of the Indonesian Electronics Industry*, Ph.D. dissertation, Roskilde: International Development Studies, Roskilde University.

Gammeltoft, Peter (2003a), 'Embedded Flexible Collaboration and Development of Local Capabilities: a Case Study of the Indonesian Electronics Industry', *International Journal of Technology Management*, Vol. 26, No. 7.

Gammeltoft, Peter (2003b), *The Institutional Embeddedness of Production Systems and the New Partnership for Africa's Development (NEPAD)*, Working Paper, Department of Intercultural Communication and Management, Copenhagen Business School.

Hall, Peter A. and David Soskice (2001), *Varieties of Capitalism: the Institutional Foundations of Comparative Advantage*, Oxford: Oxford University Press.

Hobday, Michael (1995), *Innovation in East Asia: The Challenge to Japan*, Aldershot: Edward Elgar.

Hollingsworth, J. Rogers, Philippe C. Schmitter, and Wolfgang Streeck (1994), *Governing Capitalist Economies: Performance and Control of Economic Sectors*, New York: Oxford University Press.

Lall, Sanjaya (1992), 'Technological Capabilities and Industrialization', *World Development*, Vol. 20, No. 2: 165-86.

Lall, Sanjaya (1993), 'Policies for Building Technological Capabilities: Lessons from the Asian Experience', *Asian Development Review*, Vol. 11, No. 2.

Loxley, John (2003), 'Imperialism and Economic Reform in Africa: What's New About the New Partnership for Africa's Development (NEPAD)?', *Review of African Political Economy*, No. 95: 119-128.

Lundvall, Bengt-Åke, Björn Johnson, Esben Sloth Andersen, and Bent Dalum (2002), 'National Systems of Production, Innovation and Competence Building', *Research Policy*, Vol. 31, No. 2: 213-231.

Nelson, Richard R. (1996), 'National Innovation Systems: a Retrospective on a Study', in Giovanni Dosi and Franco Malerba, *Organization and Strategy in the Evolution of the Enterprise*, Basingstoke: Macmillan.

NEPAD (2001), *The New Partnership for Africa's Development (NEPAD)*, OAU, Abuja, Nigeria, October.

NEPAD (2003), *Workshop on Developing a Science and Technology Framework for NEPAD*, Johannesburg, 17-19 February.

Pack, Howard and Larry E. Westphal (1986), 'Industrial Strategy and Technological Change: Theory versus Reality', *Journal of Development Economics*, Vol. 22, No. 1: 87-128.

Stefanski, Bogdan (2002), *Preliminary Analysis of the New Partnership for Africa's Development (NEPAD)*, Africana Bulletin, No. 50: 189-208.

Wade, Robert (1990), *Governing the Market: Economic Theory and the Role of Government in East Asian Industrialization*, Princeton: Princeton University Press.

Whitley, Richard (1999), *Divergent Capitalisms: The Social Structuring and Change of Business Systems*, Oxford: Oxford University Press.

World Bank (1991), *World Development Report: the Challenge of Development*, New York: Oxford University Press.

World Bank (1993), *The East Asian Miracle: Economic Growth and Public Policy*, Washington: World Bank.

World Bank (1998), World Development Report 1998: Knowledge for Development, New York: Oxford University Press.

PART V

Innovation in African Manufacturing and Services

17

Manufacturing in Sub-Saharan Africa and the Need of a

National Technology System[1]

Sanjaya Lall and Carlo Pietrobelli

There is an increasing concern about national 'competitiveness' among policy-makers in many countries. Equally shared is the agreement on the importance of industrial and, most of all, technological dynamism for competitiveness.

In developing countries industrial and technological performance is closely linked to their capacity to use technologies efficiently. This reflects the fact that they are seldom 'innovators' in a narrow sense, but they crucially need to be able to acquire the foreign technologies relevant to their competitiveness, absorb them, adapt and improve them constantly as conditions change.

Following this notion of innovation and technical change, we develop a concept of National Technology System, that builds upon, but differs in important respects, from the concept of National Innovation Systems. This paper contributes to this debate by specifically focusing on Sub-Saharan Africa (SSA). In this region, competitiveness is worsening, and deficiencies in the science and technological infrastructure seriously constrain industrial performance.

The paper uses detailed and original microeconomic evidence on scientific and technological infrastructure in support to industry in a sample of Sub-Saharan African countries, to conclude that, in spite of continuing liberalization and openness, this represents a fundamental weakness for African industry.

Introduction

'National technology systems' are the developing world's counterpart to the 'national innovation systems' in industrialized countries, the discussion of which is now a major element in the literature on technology policy there. The idea that innovation occurs in a 'system' – a set of interacting enterprises, institutions, research bodies and policy makers that engage in technological activity, share in knowledge spillovers and often engage in collective action – is now widely accepted.[2] The evolutionary literature, in particular, stresses the uncertain nature of the innovative process and the central importance of continuous interactions between agents (Nelson 1993). These interactions are *systemic* in the sense that the same elements recur in all economies and have a coherent set of predictable interactions. Thus, an analysis of the system and the strengths and weaknesses of its elements can be useful to policy makers.

Most developing countries do not create new technologies and so do not have 'innovation systems', in the usual sense of creating new knowledge at the frontier. However, they do have national systems within which they import, absorb, adapt and improve upon new technologies. Such technological effort is vital to their growth and competitive success, and it has systemic elements similar to those of innovation systems in advanced countries. While all such systems pertaining to knowledge creation and diffusion suffer from market failures, technology systems in developing countries are more likely to be prone to such failures in that markets and supporting institutions are less developed and information networks more confined. Moreover, technology systems in poor countries are set in trade and industrial policy regimes that are quite different from those in rich countries.

It is important to note, however, that technology development in industrial latecomers is not a trivial or automatic process. Even countries that import all their technology have to undertake significant, costly and risky effort to use the technology efficiently (Section 3).[3] This needs an efficient technology system that is able to offset some of the inherent market and institutional weaknesses in these countries. It is thus important for development policy to analyse the features and constraints of these technology systems. It is more important for the least industrialised countries that tend to suffer the greatest competitive weaknesses and consequently find themselves facing the most severe problems as they open their economies to global competition.

This paper analyses technology systems in five countries in Sub-Saharan Africa (SSA). These countries are at different levels of industrial development and so illustrate different sets of institutional problems. Ghana and Uganda are among the earliest liberalisers in the region, the former with an established industrial sector and the latter with a very small one. Zimbabwe is the most industrialised country in the region after South Africa (at least until its recent problems). Kenya is the next most industrially developed country in East Africa, while Tanzania is one of the weakest. Section 2 provides some background on the region.

Background

The poor industrial performance of SSA is well known. Much of the industrial sector has been state-owned, oriented to the local market and technologically backward. Despite liberalization and a cheap labour force (now probably among the lowest paid in the world), it has failed to build a competitive edge in export markets. It has attracted very little of the export-oriented foreign direct investment that has driven the growth of many East Asian economies. Mauritius is the major exception, apart from some recent (fairly small) investment in apparel production for the US market taking advantage of quota and tariff privileges offered by the African Growth Opportunities

Act (AGOA). The long term impact of AGOA is not clear; it is possible that the investors, mainly from East Asia, will leave when the trade privileges end in 2008.

World trade has shifted from resource-based to medium and high technology-based products (Lall 2001). However, SSA is not sharing in this trend. With the exception of South Africa and Mauritius, SSA has not altered its traditional specialisation in unprocessed primary products, the slowest growing segment of world trade and also the one that offers least by way of technological learning, skill creation and beneficial externalities. Table 1 shows manufactured export performance by the case study countries and selected comparators.

Table 1 Manufactured Exports, SSA and Selected Developing Countries (US$ mill.)

	1980/81					1996/97				
	Total	RB	LT	MT	HT	Total	RB	LT	MT	HT
Kenya	706.7	606.0	58.0	31.0	11.8	913.1	519.5	257.0	103.5	33.2
Tanzania	56.7	38.4	14.5	1.9	1.9	99.1	71.1	19.1	2.2	6.7
Uganda	12.0	9.9	0.1	1.8	0.2	29.4	5.1	12.7	9.0	2.6
Ghana	144.3	135.4	3.0	2.6	3.2	N/A	N/A	N/A	N/A	N/A
Zimbabwe (85/98)	360.5	97.2	84.4	173.6	5.3	873.6	336.7	229.2	290.3	17.4
South Africa	6,490.4	4,059.6	1,096.0	1,224.3	110.5	15,907.7	7,930.2	2,730.8	4,294.2	952.5
India	4,901.9	1,431.3	2,489.9	779.5	201.1	27,178.4	8,201.1	13,227.5	3,956.2	1,793.6
China	N/A	N/A	N/A	N/A	N/A	164,209.3	17,979.6	84,998.2	32,593.3	28,638.1
Korea	16,314.5	2,156.7	8,124.0	4,286.8	1,746.9	126,053.3	13,798.7	25,568.9	49,111.0	37,574.8
Malaysia	6,121.3	3,943.5	432.0	462.8	1,283.0	68,995.2	12,393.9	7,693.0	13,718.3	35,189.9
Thailand	2,258.4	944.5	709.7	564.6	39.7	47,190.4	9,127.9	11,961.5	9,662.8	16,438.2
Distribution (%)										
Kenya	100	85.8	8.2	4.4	1.7	100	56.9	28.1	11.3	3.6
Tanzania	100	67.7	25.7	3.3	3.3	100	71.7	19.3	2.2	6.8
Uganda	100	82.7	1.1	14.7	1.6	100	17,3	43,2	30,6	8,8
Ghana	100	93.9	2.0	1.8	2.2	N/A	N/A	N/A	N/A	N/A
Zimbabwe (85/98)	100	27.0	23.4	48.1	1.5	100	38.5	26.2	33.2	2.0
South Africa	100	62.5	16.9	18.9	1.7	100	49.9	17.2	27.0	6.0
India	100	29.2	50.8	15.9	4.1	100	30.2	48.7	14.6	6.6
China	N/A	N/A	N/A	N/A	N/A	100	10.9	51.8	19.8	17.4
Korea	100	13.2	49.8	26.3	10.7	100	10.9	20.3	39.0	29.8
Malaysia	100	64.4	7.1	7.6	21.0	100	18.0	11.2	19.9	51.0
Thailand	100	41.8	31.4	25.0	1.8	100	19.3	25.3	20.5	34.8
Memo item: distribution by regions (%)										
World	100	25.4	18.8	41.9	13.9	100	18.4	18.6	39.0	24.1
Industrialised	100	22.6	17.8	44.6	15.0	100	17.2	16.1	43.0	23.7
All Developing	100	40.9	32.5	17.0	9.5	100	17.8	27.6	25.7	28.9
SSA *	100	89.3	6.3	3.0	1.4	100	40.8	44.2	13.0	1.9
East Asia	100	30.5	37.7	19.1	12.8	100	13.1	28.2	23.9	34.7
Latin America	100	71.9	15.6	10.2	2.2	100	27.6	18.7	37.3	16.5

Note: * SSA excluding South Africa but including Mauritius. RB= resource-based, LT, MT, HT = low, medium and high-technology.
Source: Calculated from COMTRADE Database, and national sources for Uganda.

National Technology Systems and Developing Countries

This section deals briefly with the analytical setting for this discussion. Much of the conventional development literature assumes away the need for capabilities as a distinct input into industrial development. It assumes that developing countries can choose and import technologies from advanced countries and use them in production at 'best practice' levels without further effort, cost or risk. If technology were transferable like a physical product (that is, if they were fully embodied in equipment, patents and blueprints), then indeed no further learning or capabilities would be called for – getting prices right would ensure that developing countries optimised their technological choice and use. Industrial capacity (physical plant) would be the same as industrial capabilities.

A large body of empirical research on developing countries suggests that this depiction is over-simplified and often misleading (Lall 1992; Pietrobelli 1997). Based on the evolutionary theories of Nelson and Winter (1982), it argues that firms do not operate on a typical neoclassical production function. There is no well defined and complete set of alternative techniques of which they have full and clear knowledge. Finding suitable technology at the right price involves cost and risk. Using it efficiently involves further cost and effort: search, experimentation, introduction of new information and learning. Adapting the technology to different scales, new input and skill conditions and different product demands further effort. Keeping up with technical change is another set of demands on local learning. Technologies have large 'tacit' elements that have to be mastered by the recipient and cannot be sold by the technology supplier like a physical product. Without additional effort to learn different aspects of the technology, no enterprise can reach best practice levels of efficiency; in a liberalising world, this is the level needed for enterprises to survive and grow.

As technologies grow more complex and involve new skills and larger scales of production, formal research and development (R&D) often becomes necessary to monitor, understand and absorb it. Much of enterprise R&D, even in developed countries, is to keep track of, copy and adapt innovations from outside the firm (Cohen and Levinthal 1989). In developing countries, the main function of R&D is to master, adapt and improve imported technologies; only at some relatively mature stage does it become truly innovative.

The way in which knowledge is used differs by level of development. In mature industrial countries, the competitive use of technology is largely a matter of *innovation* – the ability to create new products and processes. In developing countries, it is more a matter of building the ability to use existing technologies at competitive levels of cost and quality. How difficult this is and how long it takes depends on the country and the technology, but learning is always necessary. Even routine capabilities, say for quality or process optimisation, take years to build in industrial newcomers. More advanced capabilities, for modifying, improving or generating technologies, can take longer to build. The pattern of industrial success in the developing world reflects to a large extent the effectiveness with which countries have undertaken learning (Lall 1996; Pietrobelli 1998). Some have reached the frontiers of advanced technologies, others, as in Africa, have not been able to build even the basic operational capabilities needed to compete internationally in simple technologies.

The rise of globalised production under the aegis of TNCs reduces to some extent the need for building domestic capabilities. TNCs provide affiliates with intangible assets (skills, technology, production expertise, training and so on), so that the host economy needs to offer correspondingly less 'ready-made' capabilities and invest less in subsequent absorption. Considerable industrial and export growth has taken place on this basis in countries with relatively low local technological capabilities. The growth of global production systems does not, however, do away with the need for (complementary) local capabilities (Guerrieri, Iammarino, and Pietrobelli 2001; Pietrobelli and Sverrisson 2003). In later stages as more advanced technologies have to be deployed and more efficient local suppliers needed, there is again a need for local capabilities.

Firms do not learn or innovate on their own but in intense interaction with other firms, factor markets, support institutions, and governments. They respond to rules on trade, competition, employment, intellectual property or the environment, and they behave in ways fashioned by their history, culture and environment. The interaction of economic, social and political factors provides the *system* within which firms learn and innovate, and so compete in global markets. As noted, such systemic factors also apply to developing countries, where technological effort is embedded in the specific economic, policy and institutional context of each country.

Figure 1 A Developing Country's National Technology System

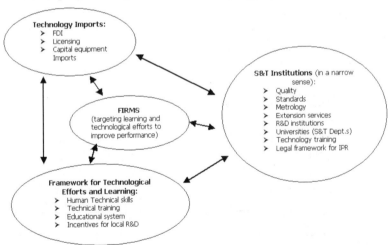

Our focus here is on two aspects of national technology systems: *technology policies* in the narrow sense and *technology institutions*. Technology policies cover such areas as technology import by licensing and FDI, incentives for local R&D and for training. Technology institutions refer to bodies such as quality, standards, metrology, technical extension, R&D and technology training. They may be government run, started by the government but run autonomously, or started and managed by industry associations or private interests.

Many services provided by these institutions are the essential 'public goods' of technological effort, difficult to price in market terms. Public research institutes and universities undertake basic research that does not yield commercial results in the short term, but provides the long-term base of knowledge for enterprise effort. Quality, standards and metrology institutions provide the basic framework for firms to communicate on technology and keep the basic measurement standards to which industry can refer. Extension services help overcome the informational, technical, equipment and other handicaps that Small and Medium-sized Enterprises (SMEs) tend to suffer. The provision of these services faces market failures of the sort that every government, regardless of its level of development, has to remedy.

Technology Imports

The main forms in which technologies are imported formally are capital goods, licensing agreements and foreign direct investment.[4]

Capital Goods Imports

Table 2 Recent Equipment Imports from Selected SSA Countries and Comparators (US$ millions and Percentages)

Country	Year	Non-electrical equipment	Electronic & electrical equipment	Total equipment imports	Total Imports	Equipment as % of total imports	Machinery imports per capita ($)	Electronics imports per capita ($)	Total equipment imports per capita ($)
Kenya	1998	427.8	218.9	646.7	3,301.8	19.6	23.8	12.2	35.9
Tanzania	1998	215.4	69.1	284.5	1,416.3	20.1	8.0	2.6	10.5
Uganda	1988	76.6	14.5	91.1	907.0	10.0	2.6	0.5	3.0
Ghana	1992	367.0	98.4	465.4	2,145.4	21.7	18.4	4.9	23.3
Zimbabwe	1998	641.5	240.1	881.6	3,157.8	27.9	16.9	6.3	23.2
South Africa	1998	4,884.0	4,343.2	9,227.2	26,624.1	34.7	444.0	394.8	838.8
India	1998	4,674.1	2,252.8	6,926.9	42,491.9	16.3	4.9	2.4	7.3
China	1998	24,371.8	27,821.8	52,193.6	140,236.8	37.2	20.1	22.9	43.0
Korea	1998	11,000.5	19,147.2	30,147.7	93,280.9	32.3	171.9	299.2	471.1
Malaysia	1998	9,700.6	24,375.8	34,076.5	57,759.4	59.0	461.9	1,160.8	1,622.7
Thailand	1998	8,562.4	9,100.4	17,662.8	42,684.1	41.4	142.7	151.7	294.4

Source: Computed from UN Comtrade Database, National Sources, Lall and Pietrobelli 2002.

The five countries import relatively little embodied technology in the form of new capital goods, either as a share of total imports or on a per capita basis (Table 2). The only other developing country in the table that imports less is India, which still had a relatively highly protected capital goods industry. East Asia largely relies on capital equipment imports. The low level of equipment imports into the African countries may seem surprising in view of the fact that none of them now imposes any restrictions on such imports, with low or zero tariffs on equipment and has subjected enterprises to import competition. The slack suggests not the lack of a suitable incentive framework for technology upgrading but the absence of capabilities to use new technologies at competitive levels. In other words, firms invest little in new embodied technology because they realise that they do not have, and cannot build in a reasonable period, the capabilities needed to use it in open markets.

Foreign Direct Investment

FDI is one of the most important sources of technology transfer to many developing countries, and its importance is rising with the globalisation of production. Table 3 shows FDI inflows by region and into African countries.

Table 3 FDI Inflows, 1988-2001

	INFLOWS (US$ mill.)		INFLOWS (shares)	
	1988-93 ann. Average	2001	1988-93	2001
World	*190,629*	*735,146*	*100.0*	*100.0*
Developed countries	*140,088*	*503,144*	*73.5*	*68.4*
Developing countries	*46,919*	*204,801*	*24.6*	*27.9*
North Africa	1,388	5,323	0.7	0.7
Sub-Saharan Africa	*2,084*	*11,841*	*1.1*	*1.6*
Latin America, Caribbean	13,136	85,373	6.9	11.6
South and East Asia	27,113	94,365	14.2	12.8
Least Developed (43)	1,361	3,838	0.7	0.5
African LDCs	*822*	*3,798*	*0.4*	*0.3*

Source: UNCTAD, World Investment Report 2002.

There has been a gradual increase in inflows into SSA but the region's shares remain very small. As noted below, FDI in Africa is also highly concentrated in a few resource rich countries (South Africa, Angola and Nigeria) and, apart from South Africa, relatively little goes into manufacturing. Of the five case study countries, Uganda has the largest recent value of and increase in FDI, followed by Tanzania and Zimbabwe (though the latter is down from 1991-94) (Pigato 1999; Lall and Pietrobelli 2002). Uganda has relied increasingly on this channel of technology transfer, to become one of the largest recipients (in relative terms) in Africa. UNCTAD qualified it as a 'frontrunner' among African countries in attracting FDI in 1992-96, along with

Botswana, Equatorial Guinea, Ghana, Mozambique, Namibia and Tunisia (UNCTAD 1998). Ghana suffers a decline after a rise in the earlier period.

What do the inflows signify for technology inflows? Unfortunately not very much, in that much of the FDI is 'either in the primary sector, particularly petroleum, or in infrastructure. And, with the exception of South Africa, other SSA countries have seen *very little inflows in the manufacturing sector* in recent years' (Pigato 1999, emphasis added). While FDI into primary and infrastructure activities is desirable and economically beneficial, in terms of transfer of technology *it does not add much to industrial capabilities or efficiency.*

Technology Licensing

As far as licence payments are concerned, patchy data from UNCTAD show that SSA excluding South Africa paid US$84 million in 1997 for imported technology, a tiny 1.5 per cent of the amount spent by the developing world. Of this amount, Kenya accounted for US$39 million and Swaziland for another $39 million, and South Africa alone spent US$258 million. In the same year, by comparison, Thailand spent US$813 million, India US$150 million and China US$543 million. Thus, *licensing is clearly not a major channel of foreign technology inflow into SSA.*

The Skill Base

Skills in general, and technical skills in particular, are the base on which technological capabilities are built. With the rapid pace of technical change, the spread of information technologies and intensifying global competition, skill needs are growing and changing (Lall 1999). While it is not possible to capture the complex nature of the skill base with national data, table 4 shows two available measures. They are enrolments at the tertiary level in all subjects and in technical subjects (science, mathematics and computing, and engineering). Enrolment data are not optimal for assessing the national skill base,[5] but they are the only data available on a comparative basis.

The dispersion in skill creation is much wider for technical subjects than for general enrolments. The leading 3 countries in terms of total technical enrolments – China (18 per cent), India (16 per cent) and Korea (11 per cent) – account for 44 per cent of the developing world's technical enrolments, the top ten for 76 per cent. SSA, with about 12 per cent of the developing world population, accounts for 4.4 per cent of its total tertiary, 3.1 per cent of technical tertiary, and 1.7 per cent of engineering, enrolments. The total number of engineers enrolled in the whole of SSA (about 70,000) is only 12 per cent of the numbers enrolled in Korea (577,000).

Table 4 Tertiary Enrolments in Total and Technical Subjects, 1995

	3 level enrolment		Technical Enrolments at 3 level							
	No. students thousands	% of population	Natural Science		Math's, computing		Engineering		Total Tech. Subjects	
			numbers	%	numbers	%	numbers	%	numbers	%
Sub-Saharan Africa										
Ghana	9,600	0.055	1,200	0.007	200	0.001	700	0.004	2,100	0.012
Kenya	31,300	0.115	3,600	0.013			1,000	0.004	4,600	0.017
Tanzania	12,800	0.043	800	0.003	100	0.000	2,700	0.009	3,600	0.012
Uganda	27,600	0.140	800	0.004	300	0.002	1,500	0.008	2,600	0.013
Zimbabwe	45,600	0.408	2,200	0.020	800	0.007	6,700	0.060	9,700	0.087
South Africa	617,900	1.490	21,700	0.052	30,500	0.074	20,000	0.048	72,200	0.174
Sub-Saharan Africa	1,542,700	0.28	111,500	0.02	39,330	0.01	69,830	0.01	220,660	0.04
Comparators										
Developing countries	*35,345,800*	*0.82*	*2,046,566*	*0.05*	*780,930*	*0.02*	*4,194,433*	*0.10*	*7,021,929*	*0.16*
Argentina	1,069,600	3.076	69,700	0.200			92,600	0.266	162,300	0.467
Chile	367,100	2.583	8,800	0.062			94,300	0.664	103,100	0.726
India	5,582,300	0.601	869,100	0.094			216,800	0.023	1,085,900	0.117
Korea	2,225,100	4.955	163,700	0.365			577,400	1.286	741,100	1.650
Taiwan	625,000	2.910	16,800	0.078	32,800	0.153	179,100	0.834	228,700	1.065
Malaysia	191,300	0.950	8,800	0.044	4,600	0.023	12,700	0.063	26,100	0.130
Sri Lanka	63,700	0.355	8,100	0.045	300	0.002	6,800	0.038	15,200	0.085
Developed countries	*33,774,800*	*4.06*	*1,509,334*	*0.18*	*1,053,913*	*0.13*	*3 191 172*	*0.38*	*5,754,419*	*0.69*

Source: UNESCO (1997), national sources.

Technological Effort

Technological effort is essential to building capabilities. Much of the effort is informal, and is impossible to measure and compare across countries. What is available and used commonly for this purpose is formal R&D. There is some justification for using this measure: R&D become important for technology absorption and adaptation in industrializing countries, even if they do not innovate as they industrialize. This is also true at the enterprise level, where a substantial part of R&D is for monitoring and absorption rather than frontier innovation (Cohen and Levinthal 1989).

Table 5 shows comparative spending on R&D and scientists and engineers employed in R&D for various regions. SSA performs poorly, particularly for R&D most directly relevant to industrial technology – R&D financed by the productive sector. The available data suggest that by this measure none of the five case study countries spend anything on technological activity. This is not surprising, given the

275

recent history of industrialisation in SSA and its specialisation in natural resource-based and low-technology activities.

Table 5 R&D Propensities and Manpower in Major Country Groups (Simple, Averages, Latest Year Available)

Countries and regions (a)	Scientists/engineers in R&D		Total R&D	Sector of performance (%)		Source of Financing (% distribution)		Source of financing (% of GNP)	
	Per mill. popula-tion	Numbers	(% of GNP)	Produc-tive sector	Higher edu-cation	Produc-tive enter-prises	Govern-ment	Produc-tive enter-prises	Productive sector
Industrialised market economies (b)	1,102	2,704,205	1.94	53.7	22.9	53.5	38.0	1.037	1.043
Developing economies (c)	514	1,034,333	0.39	13.7	22.2	10.5	55.0	0.041	0.054
Sub-Saharan Africa (exc. S Africa)	83	3,193	0.28	0.0	38.7	0.6	60.9	0.002	0.000
North Africa	423	29,675	0.40	N/A	N/A	N/A	N/A	N/A	N/A
Latin America & Caribbean	339	107,508	0.45	18.2	23.4	9.0	78.0	0.041	0.082
Asia (excluding Japan)	783	893,957	0.72	32.1	25.8	33.9	57.9	0.244	0.231
Mature NIEs (d)	2,121	189,212	1.50	50.1	36.6	51.2	45.8	0.768	0.751
New NIEs (e)	121	18,492	0.20	27.7	15.0	38.7	46.5	0.077	0.055
World (79-84 countries)	1,304	4,684,700	0.92	36.6	24.7	34.5	53.2	0.318	0.337

Source: Computed from UNESCO *Statistical Yearbook 1997*. Notes: (a) Only including countries with data, and with over 1 million inhabitants in 1995. (b) USA, Canada, West Europe, Japan, Australia and N Zealand. (c) Including Middle East oil states, Turkey, Israel, South Africa, and formerly socialist economies in Asia. (d) Hong Kong, Korea, Singapore, Taiwan Province. (e) Indonesia, Malaysia, Thailand, Philippines.

Evidence on informal technical effort and technological capabilities in SSA is provided by some recent studies.[6] In general, the findings suggest that external sources of information and learning are poor, with firms forced to rely almost exclusively on internal efforts to build their technological capabilities. This is not by itself a problem, as internal efforts are often the most important source of technological capabilities among successful small-scale exporters in Asia and Latin America.[7] However, the problem in Africa is that internal technical efforts – however measured – are weak, inadequate and sporadic (Biggs, Shah, and Srivastava 1995). These efforts are not supported by the S&T system, considered below.

Science and Technology Institutions in SSA

During the colonial period and even after independence, there was little attempt to develop an explicit science and technology (S&T) strategy in most African countries. S&T policy was pursued implicitly by technical government departments (e.g. medical services, agriculture, mines, geological surveys, industry and education). Sometimes, the organisation of research was handled by inter-territorial research institutions set up by the colonial administrators to cater to the needs of the whole regions. This was the case of West as well as East Africa. Let us briefly consider some of the main technology institutions, extensively analysed elsewhere (Lall and Pietrobelli 2002).

Institutions for Metrology, Standards, Testing and Quality (MSTQ)

MSTQ institutions provide the basic infrastructure of technological activity in any country. Standards are a set of technical specifications used as rules or guidelines to describe the characteristics of a product, a service, a process or a material. The use of recognized standards and their certification by internationally accredited bodies is increasingly demanded in world trade. This reduces transactions costs, information asymmetries and uncertainties between the seller and the buyer with respect to quality and technical characteristics. Metrology provides the measurement accuracy and calibration without which standards cannot be applied. The application of standards and the certification of products necessarily imply (accredited) testing and quality control services.

The importance of industrial standards has risen because of the fast pace of technical progress, the growing complexity of new products and the increasing multiple use of technologies. Therefore, standards importantly contribute to the diffusion of technology within and across industries. Most importantly, in a developing country a standards institution can disseminate best practice in an industry by encouraging and helping firms to understand and apply new standards. Redundant experimentation with new technologies is reduced, and enterprises are forced to use a common language that is also shared by the international market. In turn, this reduces the complexity of inter-firm technical linkages and collaboration.

The International Standards Organisation (ISO) has introduced the best known quality management (not technical) standards in use today: the ISO 9000 series. ISO 9000 certification is becoming an absolute must for potential exporters, signalling quality and reliability to foreign buyers, retailers as well as transnational corporations seeking local partners and subcontractors. In the whole of Africa (including Northern Africa) only 23 such institutions were operating at the end of 2000 (www.iso9000.org).

Evidence on standards institutions in the case study countries was collected, and is reported and assessed in all details elsewhere.[8] The Ghana Standards Board (GSB) is the main organisation in Ghana for ensuring industrial quality, through standards, metrology, testing and quality assurance, and has some of the typical features of similar institutions in SSA. Thus, a major shortcoming is its low funding, and especially the share of the budget devoted to activities oriented to the internal development of the Board and its linkages to local industry. Salaries account for an increasing and disproportionate share of the budget, and no funding is available for any kind of R&D (Table 6). Total revenues amounted to about US$2.2 million, twice as much as in the analogous institution in Uganda, but much less than in other SSA countries. In most instances, the share of self-financing by selling services to local firms is invariably very low in this as well as in other similar institutions in SSA.

The Standards Association of Zimbabwe (SAZ) was set up in 1957 as a non-government and non-profit making body, and had developed some capability in the ISO 9000 area by 1997, with internal assessors who certified around 20 companies. However, SAZ lacks the ability to accredit private testing laboratories, and is also handicapped by not having a metrology facility: most metrology work for Zimbabwean enterprises is done in South Africa and some (for mining equipment) in Zambia.

Table 6 Summary Financial Indicators of the Ghana Standards Board

	1994/95	1998/99
Revenues		
Revenues in US$ mill. *	1.46	2.25
Sources of Revenues (%)		
From government scheduled	90	82
From services sold	10	18
Expenditures (%)		
Salaries	60	77
Materials & buildings	4.5	6.6
Training	1	2
Equipment	10	5
R&D	-	-
Others	24.5	9.4
Total	100.0	100.0

Note: * Approximate figures due to variable exchange rate.
Source: Interviews to GSB Staff during UNCTAD field mission /Jan.2000).

The Kenya Bureau of Standards (KEBS), set up in 1974, is the most active and efficient of the five standards bodies studied here. KEBS is funded by a standards levy on all manufacturers (0.2 per cent of ex-factory sales up to a ceiling of US$4,000 per annum), import quality inspection fees, annual government grants and services sold to industry, such as training. Other institutions studied include the Tanzania Bureau of Standards and the Uganda National Bureau of Standards (UNBS).[9] They both complained of the extremely low quality consciousness in their countries, a problem shared by most SSA countries.[10]

R&D Institutions

The largest and most active public R&D institutions in most African countries are involved in agriculture rather than manufacturing. As private sector R&D in industry is virtually absent (apart from South Africa, see UNIDO 2002), public institutions have a vital role to play in local efforts to absorb, adapt and improve imported technologies. The R&D institutions analysed in details (Lall and Pietrobelli 2002) are Uganda's National Agricultural Research Organisation (NARO), always with the largest annual budget (around US$10 million), Ghana's Food Research Institute (FRI),

Uganda's Industrial Research Institute (UIRI), conceived in the 1970s by the East African Community (EAC) as a regional project to promote research in industry,[11] Tanzania's Industrial Research and Development Organisation (TIRDO), Kenya's Industrial Research and Development Institute (KIRDI), and Ghana's Industrial Research Institute (IRI).

In Zimbabwe, despite its large manufacturing sector and reasonable base of industrial capabilities (Lall 1999)[12], there were no public R&D institutions in manufacturing technology till the end of the 1990s. The only bodies that could do R&D for industry were the engineering departments at the university, but these had few links with enterprises. In response, in 1997 the government launched an ambitious programme of building seven R&D institutes under the Scientific and Industrial Research and Development Centre (SIRDC), placing the Centre directly in the President's office. The subsequent political turmoil has inevitably negatively affected the programme.

In sum, the most active (and well-funded) R&D institutions in the five countries have so far focused on agriculture and not manufacturing. Industrial R&D institutes have tended to perform poorly, failing to offset (and to some extent reflecting) the paucity of technological activity in industrial enterprises.

There are several common threads running through the industrial R&D institutions in the region. They generally lacked the facilities (physical and human) to provide meaningful support to industrial enterprises. Their personnel tended to be poorly paid and motivated, with little incentive to reach out to and interact with their prospective clients. They had no means of assessing the technological needs of industrial enterprises or of diffusing to them the technologies they had created (or, more commonly, adapted). As a result, the institutions carried little credibility with the private sector and had very few continuous linkages with it apart from providing routine testing services. And they were not (unlike similar institutions in more advanced countries that also failed to link up to industry) conducting advanced research for publication in international journals.

Their poor performance reflected not just internal constraints, but also technological apathy in much of local industry. Most enterprises were not technologically active and aware; few had responded to liberalization by mounting technology-based upgrading strategies. In the absence of technological activity in enterprises, however, it is difficult for R&D institutions to provide effective assistance (Rush et al. 1996). Governments had not given much priority to promoting industrial research in these countries, in private or public institutions. This reinforced the general feeling of marginalization and discouragement in most institutions.

To sum up on S&T institutions, the picture of national technology systems emerging from this sample is discouraging; what is worse is that it is (South Africa excepted) likely to be representative of the whole SSA region. The main elements of the system are weak. The technology infrastructure is small, passive and largely ineffective. It is often poorly funded and motivated and tends to be de-linked from industry. Its ability to develop, adapt and disseminate industrial technologies is weak.

It has little awareness of the needs of local industry, even less of how new technologies can be introduced to potential users. Enterprises, for their part, conduct little formal technology activity and generally lack awareness of the need to do so to cope with the severe challenges posed by import liberalization. The government is largely indifferent to industrial technology and provides little support to inherited technology institutions. Nor does it do much to promote a more active technology culture in industry.

Not only is each element weak, but there is also *little systemic interaction* between them to support industrial technology development.[13] An additional dimension of the problem, not discussed in this paper, is the similar absence of linkages between industry and *educational institutions*. Very few firms collaborate with universities or polytechnics, despite the reservoir of theoretical and engineering knowledge there, nor Universities and high schools plan their programmes and syllabus on the needs of industry.

All countries that have industrialized successfully, as in Asia or Latin America, have developed strong public technology infrastructure institutions to support technological development in industry. More recently, they have undertaken reforms and new measures to strengthen their linkages with industry (Amsden 2000; Lall 1996). Private enterprises in some newly-industrializing economies are acquiring a technology culture – they undertake meaningful R&D in-house and contract R&D to other institutions. If African countries are not able to mount a similar reform, it is difficult to see how their industrial enterprises will become dynamic competitors in world markets.

Conclusion

Sub-Saharan Africa's recent industrial and technological performance is disappointing. The manufacturing sector is tiny in most countries and has been losing shares in world markets despite some years of liberalization and opening up to globalised production. Enterprises are smaller, less efficient and less innovative than counterparts in other developing countries. Quite apart from the political and governance problems affecting the region, there are binding structural constraints on industry (UNIDO 2002; Lall and Wangwe 1998; Collier and Gunning 1999). The supply of modern skills is inadequate and the physical infrastructure is weak and often deteriorating. In addition, this paper has noted the inadequacies of the technology system that underlies industrial competence and dynamism. This aspect has been unduly neglected in the ample literature on African economic problems, but is of vital significance to long-term development.

This paper suggests that despite liberalization and structural adjustment in much of the region, the manufacturing sector is lagging in international competitiveness – a far less optimistic picture than portrayed by the World Bank in the early nineties (World Bank 1994). Unlike many other industrializing countries in East Asia, there has been little attention given to the technology system. Even the most advanced industrial economy in the region after South Africa – Zimbabwe – suffers from a weak and slothful technology system. In general, the MSTQ infrastructure is weak, R&D support is minimal and linkages between public institutions and universities, on the one hand, and industrial enterprises, on the other, are negligible.

The strengthening of the national technology system is necessarily a long-term process. It entails the gradual building of institutions, changing of attitudes, creation of new links and networks and, inevitably, substantial resources over a lengthy period. Needless to say, it also needs a conducive social, political and economic setting in which enterprises, governments and institutions can plan and implement long-term strategies. It is beyond the scope of this paper to discuss the array of policies needed to do all this in Africa (but see Lall and Pietrobelli (2002), for specific recommendations on technology development drawing upon the experience of other industrialising countries). However, we conclude by noting two priorities for policy: technology strategy formulation and co-ordinating and planning the technology system.

Technology strategy formulation is particularly weak in Sub-Saharan Africa. In Kenya, for instance, there is no institutional mechanism for evaluating and setting S&T priorities. In Ghana the strategy still consists largely of statements of good intent and over-ambitious plans. Overall S&T policy exists largely on paper, and comes very low in the pecking order of government priorities. This differs greatly from the dynamic Asian developing countries (Amsden 2000; Lall 1996), where technology upgrading and strategy have become important policy priorities. The most fundamental policy gap in Africa is perhaps the lack of official appreciation of the importance of technology development to manufacturing growth and competitiveness – without such appreciation clearly no effective strategies can be formulated or implemented. Governments in the region pay little attention to technological needs in industry or to the promotion of technological activity within firms or in support institutions. Not only does industry lack a technology culture, so does the government. No national technology system can function effectively unless such a culture is created.

Coordinating and planning the technology system is another area of policy concern, in turn reflecting the low priority attached to technology. In most of Africa, technology policy formulation is uncoordinated and spread over a number of different ministries and departments. Where institutions exist to formulate S&T policy (COSTECH in Tanzania, CSIR in Uganda or MEST in Ghana), they tend to be too weak to affect other ministries and to coordinate their efforts. Government agencies generally guard their turf jealously, unwilling to part with the information, functions and resources that a coordinated effort would need.

Fragmentation means that partial objectives are pursued without reference to national goals. What is more, the private sector is rarely involved in the design and implementation of a technology strategy. Private sector business associations do not, for their part, formulate technology strategies for their sectors or members, and do not attempt to influence government policy in this respect; most tend to stick to their traditional role of seeking government favours and extending protection. However, no

281

technology development strategy can succeed unless the private sector is convinced of its need and is willing to play its part. The most effective technology strategies in East Asia, for instance, have involved private sector collaboration and resources. R&D linkages have generally been stimulated by schemes where private firms financed half the cost.

Ultimately, and not surprisingly, the development of strong technology systems in Africa needs a *systemic change* in all elements. The institutions themselves cannot accomplish much unless the government and the private sector also commit themselves to technology development. At this time, the possibility of such a change appears rather remote. To the extent that technology upgrading is a necessary element of industrial development in a liberal and globalised economy, this is a matter of grave concern.

Notes

[1] This paper draws on the empirical findings of Lall and Pietrobelli (2002). The authors are grateful to UNCTAD and the Commonwealth Secretariat for funding the research on which this paper draws.

[2] Of the large and growing literature on this subject see Freeman (1997), Lundvall (1992), Metcalfe (1995), Nelson (1993), Edquist (1997), and Edquist and McKelvey (2000).

[3] See, for instance, Lall (2001), Pietrobelli (1997) and UNIDO (2002).

[4] There are, of course, also many informal forms of technology import like copying, migration, trade fairs, journals and the like, but these are difficult to measure and so are not considered here

[5] Data on educational enrolments may be misleading because they do not take account of the quality (and drop-out rates) of the education or its relevance for local industry.

[6] See Biggs et al. (1995), Lall et al. 1994, Lall (ed.) 1999, and Wangwe 1995.

[7] Berry and Escandon 1994; Levy et al. 1994; Pietrobelli 1998; Wignaraja 1998.

[8] This evidence was collected by the authors in fieldwork funded by UNCTAD, the Commonwealth Secretariat, the European Commission and the World Bank during 2002 and 2001. See Lall and Pietrobelli (2002) for further details.

[9] In conjunction with the Kenyan and Tanzanian standards bureaux UNBS is also involved in the elaboration of East African harmonised standards within the framework of East African Co-operation.

[10] By the year 2000 only two firms had obtained ISO 9000 certification in Tanzania.

[11] Initially based in Nairobi, in 1974/75 the Research Council of the EAC decided to decentralise industrial research in the three partner states (Uganda, Kenya and Tanzania) on the basis of local raw materials and resources. Kenya set up the Kenya Industrial Research and Development Institute (KIRDI) and Tanzania the Tanzania Industrial Research and Development Organisation (TIRDO).

[12] The comparison of technological capabilities in Zimbabwe with those in Kenya and Tanzania suggested that its industrial enterprises were technologically in advance of its neighbours (Lall 1999). This was also the conclusion of the total factor productivity analysis in a World Bank study (Biggs et al. 1995), showing that average technical efficiency was higher in Zimbabwe than in Kenya or Ghana. However, Lall (1999) argued that capabilities in Zimbabwe were well below levels reached in other developing countries, and that this was being manifested in the competitive difficulties facing enterprises being exposed to direct import competition.

[13] See Lall and Pietrobelli (2002), Biggs et al. (1995), Enos (1995), Lall and Wignaraja (1998), Latsch and Robinson (1999), Wignaraja and Ikiara (1999) and Wangwe and Diyamett (1998).

References

Amsden, Alice. H. (2000), *The Rise of the Rest: Late Industrialization Outside the North Atlantic Region*, New York: Oxford University Press.

Berry, Albert and Jose Escandon (1994), *Colombia's Small and Medium-Sized Exporters and Their Technology Systems*, Policy Research Department Working Paper, No. 1401, Washington, D.C.: The World Bank.

Biggs, Tyler, Manju Shah, and Pradeep Srivastava (1995), *Technological Capabilities and Learning in African Enterprises*, World Bank Technical Paper, No. 288, Washington, D.C.: World Bank.

Cohen, Wesley M. and Daniel A. Levinthal (1989), 'Innovation and Learning: The Two Faces of R&D', *Economic Journal*, Vol. 99, No. 4: 569-96.

Collier, Paul and Jan W. Gunning (1999), 'Why Has Africa Grown Slowly?', *Journal of Economic Perspectives*, Vol. 13, No. 2: 3-22.

Edquist, Charles (ed.) (1997), *Systems of Innovation: Technologies, Institutions and Organisations*, London: Pinter Publishers.

Edquist, Charles and Maureen McKelvey (eds.) (2000), *Systems of Innovation: Growth, Competitiveness and Employment*, two volumes, Cheltenham: Edward Elgar.

Enos, John L. (1995), *The Pursuit of Science and Technology in Sub-Saharan Africa Under Structural Adjustment*, London: Routledge.

Freeman, Christopher (1997), 'The 'National System of Innovation', in Historical Perspective', in Daniele Archibugi and Jonathan Michie (eds.), *Technology, Globalisation and Economic Performance*, Cambridge: Cambridge University Press.

Guerrieri, Paolo, Simona Iammarino, and Carlo Pietrobelli (2001), *The Global Challenge to Industrial Districts: SMEs in Italy and Taiwan*, Cheltenham: Edward Elgar.

ISO (2000), *The ISO Survey of ISO 9000 and ISO 14000 Certificates: Ninth Cycle*, The International Organization for Standardization, Geneva: www.iso.ch.

Lall, Sanjaya (1986), *Learning From the Asian Tigers: Studies in Technology and Industrial Policy*, London: Macmillan.

Lall, Sanjaya (1992), 'Technological Capabilities and Industrialization', *World Development*, Vol. 20, No. 2: 165-186.

Lall, Sanjaya (1998), 'Exports of Manufactures by Developing Countries: Emerging Patterns of Trade and Location', *Oxford Review of Economic Policy*, Vol. 11, No. 2: 54-73.

Lall, Sanjaya (1999), *Competing with Labour: Skills and Competitiveness in Developing Countries*, Working Paper, No. 31, Issues in Development, Geneva: ILO.

Lall, Sanjaya (2000), 'The Technological Structure and Performance of Developing Country Manufactured Exports, 1985-98', *Oxford Development Studies*, Vol. 28, No. 3: 337-69.

Lall, Sanjaya (2001), *Competitiveness, Technology and Skills*, Cheltenham: Edward Elgar.

Lall, Sanjaya (ed.) (1999), *The Technological Response to Import Liberalization in Sub-Saharan Africa*, London: Macmillan.

Lall, Sanjaya, Giorgio Barba-Navaretti, Simon Teitel, and Ganeshan Wignaraja (1994), *Technology and Enterprise Development: Ghana under Structural Adjustment*, London: Macmillan.

Lall, Sanjaya and Carlo Pietrobelli (2002), *Failing to Compete: Technology Development and Technology Systems in Africa*, Cheltenham: Edward Elgar.

Lall, Sanjaya and Ganeshan Wignaraja (1998), *Mauritius: Dynamising Export Competitiveness*, Economic Paper, No. 33, London: Commonwealth Secretariat.

Lall, Sanjaya, Peter Robinson, and Ganeshan Wignaraja (1998), *Zimbabwe: Enhancing Export Competitiveness*, Study for the Ministry of Industry, Zimbabwe, and the Commonwealth Secretariat, London.

Lall, Sanjaya and Samuel Wangwe (1998), 'Industrial Policy and Industrialisation in Sub-Saharan Africa', *Journal of African Economies*, Vol. 7, Supplement 1: 70-107.

Latsch, Wolfram W. and Peter Robinson (1999), 'Technology and Responses of Firms to Adjustment in Zimbabwe', in Sanjaya Lall (ed.), *The Technological Response to Import Liberalization in Sub-Saharan Africa*, London: Macmillan.

Levy, Brian, Albert Berry, Motoshige Itoh, Linsu Kim, Jeffrey Nugent, and Shujiro Urata (1994), *Technical and Marketing Support Systems for Successful Small and Medium-Sized Enterprises in Four Countries*, Policy Research Department Working Paper, No. 1400, Washington, D.C.: World Bank.

Lundvall, Bengt-Åke (ed.) (1992), *National Systems of Innovation: Towards a Theory of Innovation and Interactive Learning*, London: Pinter Publishers.

Metcalfe, J. Stanley (1995), 'Technology Systems and Technology Policy in an Evolutionary Framework', *Cambridge Journal of Economics*, Vol. 19, No. 1: 25-46.

Nelson, Richard R. (ed.) (1993), *National Innovation Systems: A Comparative Analysis*, New York: Oxford University Press.

Nelson, Richard R. and Sidney G. Winter (1982), *An Evolutionary Theory of Economic Change*, Cambridge: Harvard University Press.

Pietrobelli, Carlo (1998), *Industry, Competitiveness and Technological Capabilities in Chile: A New Tiger from Latin America?*, London: Macmillan.

Pietrobelli, Carlo (1997), 'On the Theory of Technological Capabilities and Developing Countries: Dynamic Comparative Advantage in Manufactures', *Rivista Internazionale di Scienze Economiche e Commerciali*, Vol. 44, No. 2.

Pietrobelli, Carlo and Arni Sverrisson (eds.) (2003), *Linking Local and Global Economies: The Ties that Bind*, London: Routledge.

Pigato, Miria (1999), *Foreign Direct Investment in Africa: Old Tales and New Evidence*, Africa Region: World Bank.

Rush, Howard, Mike Hobday, John Bessant, Erik Arnold, and Robin Murray (1996), *Technology Institutes: Strategies for Best Practice*, London: International Thomson Business Press.

UNCTAD (1998), *Foreign Direct Investment in Africa: Performance and Potential*, UNCTAD/ITE/Misc.5, Geneva: UNCTAD.

UNCTAD (Various Years), *World Investment Report*, Geneva: United Nations.

UNIDO (2002), *Industrial Development Report 2002-2003*, Vienna: United Nations Industrial Development Organization.

Wangwe, Samuel M. (1995), *Exporting Africa: Technology, Trade and Industrialisation in Sub-Saharan Africa*, London: Routledge.

Wangwe, Samuel M. and Bitrina Diyamett (1998), 'Cooperation Between R&D Institutions and Enterprises: The Case of the United Republic of Tanzania', in UNCTAD, *New Approaches to Science and Technology Cooperation and Capacity Building*, Geneva: 193-210.

Wignaraja, Ganeshan (1998), *Technological Capabilities and Industrialization: The Case of Sri Lanka*, London: Macmillan.

Wignaraja, Ganeshan and Gerrishion Ikiara (1999), 'Adjustment, Technological Capabilities and Enterprise Dynamics in Kenya', in Sanjaya Lall (ed.), *The Technological Response to Import Liberalization in Sub-Saharan Africa*, London: Macmillan: 57-111.

World Bank (1994), Adjustment in Africa: Reforms, Results, and the Road Ahead, Washington, D.C.: The World Bank.

18

Barriers to and Opportunities for Innovation in

Developing Countries

The Case of Ghana

Olav Jull Sørensen

Innovation and Development

Some of the popular catchwords for the present state of affairs in the economy are 'global economy', 'network economy', and 'innovation and technological driven economy' (Dicken 2003; Dunning 1997; Sørensen 2000). Building on these concepts, the aim of this analysis is to study the innovation capability of the Ghanaian companies and economy at large. In this study, a distinction is made between an invention, an innovation and the implementation or commercialisation of the innovation. This conceptual trilogy corresponds, on the human level, broadly speaking to the scientist, the innovator and the entrepreneur – a distinction between roles often used in the literature (Swedberg 2000).

Innovation obviously takes on different forms in different types of companies. In micro-enterprises, innovations normally emerge out of practice and experience, for example, as a consequence of a specific challenging task from a customer. A customer may have seen a certain design of a chair and asks the wayside furniture maker if he can make it. Thus, innovation processes in micro-enterprises (the informal sector) are primarily customer driven.

In entrepreneurial or small-sized firms, innovative activity is mostly associated with the manager-owner, but it may also involve other employees. The process may be either customer or company driven, i.e. a customer may approach the entrepreneur with a problem in his productions or the manager may spot a market opportunity.

Organisationally, companies may have a quality control unit, which they also use to test new ideas, or an equipment maintenance unit, which has the task of improving the production process. More advanced companies may have a small innovation unit and perhaps even a research and development lab. Furthermore, much innovation takes place in the relations between partners in a network or strategic alliance involving producers, users, suppliers and special research units such as technology labs and universities.

Turning to the literature on developing countries, it seems that the concept of innovation is just beginning to be taken seriously. Conventional concepts such as 'technology transfer', 'appropriate technology', 'intermediate technology', and 'indigenous technology/development' have been very 'sticky'. This is unfortunate for three reasons:

1. The concept of technology transfer assumes that the technological trajectory is given. There is only one best path and one best technology, normally the one developed by Western countries.
2. The concept also assumes a 'catching up' way of thinking, i.e. a stages theory both at the macro and the company level. Lall (1992) provides a typology for the catching-up process at company level. It has four steps, i.e. learn to (1) operate the transferred technology, (2) repair the transferred technology, (3) improve the transferred technology and (4) in the end learn to innovate (R&D).
3. The concept of technology transfer views innovation as a formal learning process and thus rejects experiential and tacit knowledge generated locally.

The concepts of appropriate, intermediate and indigenous technology leave a little more room for local innovative activities, but then these concepts signal that the absorptive capacity for change is relatively low in developing countries. Furthermore, all the conventional concepts are related to technology. Innovation is a much broader concept, which implies change not just in technology but also in the organisational forms, the economic thinking and in institutional arrangements. However, the transfer of management know-how, marketing know-how and other concepts related to the running of economic units have either not been developed or have been embedded in the concept of transfer of technology. Technical assistance could be accepted, not organisational and institutional assistance.

For business development thinking, it is 'unhealthy' to stick to these concepts and notably the concept of technology transfer. It is similarly constraining to build on the concepts of 'intermediate and appropriate technology'. Although any marketing book would agree that a product or a technology should be adapted to the specific context in which it is to be used, the concept in the development literature became synonymous with 'locally developed' technology under the import substitution development regime.

Returning to the present movements towards a network economy and the generation of innovations through relationships rather than internally in the company, it is obvious that the concepts of 'technology transfer' and 'intermediate technology' need to be replaced by more dynamic concepts and the institutions to assist companies in developing countries to be more focused on learning and knowledge generation processes compared to technology per se. In other words, they must turn the companies innovative.

Before such a change in the orientation of support institutions, consultants, etc. can take place, we need to know more about the present innovation activities in the developing countries – in this case, in Ghana. To what extent do companies today innovate; what are the barriers to innovation; are there any so-called innovative industrial districts or clusters; to what extent is there a support system for the innovation activities in the companies and to what extent does the system constitute a system of innovation?

The present study aims to contribute to answering these questions. The framework for discussing the issue of innovation in business development is a simple three-concepts model of a market economy (Sørensen et al. 1998), comprising the behaviour of entrepreneurs and managers, the structure of the industries and markets, and the governmental regulative and promotional policies, see Figure 1.

Figure 1 Business Systems in Global Context

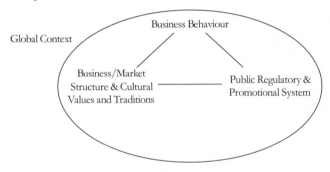

Barriers to Innovation

Business Behaviour and Innovation

The Ghanaian business system is configured by three main types of companies: Ghana has a few large- and medium-sized companies; the backbone of the economy comprises many small-scale companies, owned and managed by entrepreneurs who often own several small undertakings. Finally, as the third layer, Ghana has numerous micro businesses notably in trade and other services but also in production, e.g. furniture, carvings, metal works, etc. The common names for this last category is the informal sector and for the trading part of it, due to the dominant position of women, the Mammy Trading System (Sørensen 1978). Peasant farming dominates the agricultural sector.

The larger companies have generally speaking a functional organisation structure with specialised managers. Combined owner-managers run the small-scale companies. The micro-businesses are in most cases one-person undertakings with a few family hands to help out. In the following, the first category will be termed larger companies, the second entrepreneurs and the third micro-business/traders. The chapter deals primarily with entrepreneurs, but a number of the issues discussed are equally relevant for all three categories of businesses.

Given the composition of companies in Ghana, the entrepreneurial firm is and is expected to be an important building block of the economy. The two core characteristics of an entrepreneur are that he is the owner and manager, i.e. the centre of control and involved in almost all decisions. Furthermore, he is also the centre of excellence and action. In entrepreneurial companies, innovation is not associated with any department or position. It is 'floating around' in the company and the manager-owner is responsible for the organising of the innovation activities and he himself is mostly part of them.

Related to innovation, five entrepreneurial and managerial issues appear of importance in the Ghanaian context: (1) the financial orientation of entrepreneurs; (2) the superiority of formal education; (3) the low level of trust; (4) the inclinations to imitate; and (5) Internet and innovation.

THE FINANCIAL ORIENTATION OF ENTREPRENEURS

Studies have shown that a number of entrepreneurs in Ghana are relatively well educated and they have often studied abroad (Kuada and Sørensen 2000). These entrepreneurs have thus not followed the conventional path from apprenticeship to the employment as a craftsman before finally becoming an entrepreneur. Furthermore, most have a general academic or commercial background rather than an engineering background.

Due to this background, the entrepreneurs are more oriented towards the creating of businesses and the financial side of the business than the production/product side and they have little insight into the technical aspects of production and the product. What product they produce does not matter. They have invested little love and commitment in the product itself; only in its financial potential. This orientation has two implications:

A. It may partly explain why Ghanaian entrepreneurs are diversification oriented rather than specialisation oriented. Through diversification, the entrepreneurs try to spread their financial risks. Furthermore, a finance-oriented entrepreneur may also be more inclined to create businesses rather than perfect production and products. Thus, the finance oriented entrepreneur focuses on the spotting of new businesses. This is possible, as, in general, you will find more market gaps in an emerging economy compared to the mature market economies, where the density of entrepreneurs has vacuum cleaned the market for opportunities.[1]

B. It may also explain why Ghanaian products often are of low and inconsistent quality. To develop a quality product, it is necessary to work with the product and be devoted to it. An entrepreneur, holding an MBA-degree, is most likely committed to the cash flow, not the product.

Other explanations for the low and inconsistent quality in Ghana are, that the market cannot pay for quality or the technology in use is not refined enough to produce high quality products. Combining the quality orientation with a maintenance orientation, we may explain poor quality from the point of view of *a natural circle orientation*. In Ghana, things are continuously used until they break down and then they are abandoned, seemingly not complaining much over the fact that the air conditioner does not work; the telephone is out of order, etc. Apparently, things have a natural course or circle: invest and operate; breakdown and abandon; raise funds again and run, etc. In a later section, the natural circle will be combined with how resources are allocated in the Ghanaian context.

In conclusion, qua the background of managers, entrepreneurship is on the agenda, not innovation. There is no big need for the entrepreneur to be innovative, as the market in an emerging economy has many gaps waiting for the entrepreneur to exploit them. Furthermore, the quality consciousness of buyers, consumers and industrial buyers alike, is relatively low. There are no qualified buyers to push the producers to innovate (Lundvall 1988, 1992). However, this comfortable situation came to a halt, when it was decided to open the economy and thus expose companies to international competition on the home market and face quality conscious buyers on export markets.

THE SUPERIORITY OF FORMAL EDUCATION

Under this heading, an attempt is made to integrate a number of issues, including the role of education, the power distance between managers and employees, and the need for direct supervision. Together these issues have severe implications for the innovative spirit and capacity of the firm.

Many visits to Ghanaian companies have shown that employees, including middle managers, often sit idle. Why is it so and does it have implications for innovation?

It seems to begin with formal education. In Ghana, education is important. There is still the belief that if you are well educated, you are safe (and can support the family). Related to many development issues, Ghanaians often say that the issue would be solved if people were more educated. Education normally means scholarly education. It does not include skilled and unskilled personnel. What they get is training and experience from their work. This understanding of education has important consequences for the management of the company and the relations between managers and personnel.

As mentioned earlier, entrepreneurs are often well educated and, as a consequence of the high social status of education, they are almost per definition 'born to lead' and assumed to have insight into everything, also insight into technical matters.

In general, this understanding of education and know-how manifests itself into a gap and a distance between the managers and the employees. As the manager is not just the centre of excellence but also of power, the employees come to play a passive role. They are not trusted with much responsibility and hardly ever consulted. Their experiences from the floor are rarely accumulated. In the end, the employees become passive and direct supervision is required to make them perform. It is a kind of *vicious circle* that ends in distrust and which has the implications that the innovative potential embedded in experiential and tacit knowledge is under-utilised.

THE LACK OF TRUST

Although the Theory of the Firm and Markets (Ferguson and Ferguson 1998) does not incorporate the concept of trust, it is nevertheless the glue that makes a market economy work. 'Trust is a key component of any market economy because of transactional uncertainties […] Without minimal levels of trust and confidence that commitments will be honoured, markets cannot function' (North 1990: 34-35; Whitley 1992: 20).

A distinction between personal and systemic trust is useful in this context. By system's trust is meant the trust you have in established non-personal routines, institutions, etc., e.g., the legal system and the traditions established over many years in mature industries. Obviously, in an emerging economy, where such routines and institutions are under establishment, the systemic trust will be low and business has to be conducted by relying on personal relations and the building of personal trust.

You do not need many interviews in Ghana before you realise that Ghanaian businessmen do not trust each other to any large extent. How can this trust-free Ghanaian economy be explained and what are the implications for innovation?

One implication could be that in a market economy without trust, companies will tend to be small, as the owner has to directly oversee all activities himself. As shown in Sørensen and Nyanteng (2000), a Ghanaian trader and her commodities are inseparable from the time of the very act of buying foodstuffs in the rural areas to the selling of the commodities in the urban markets. The trader is 'stapled to the produce', so to speak. If she is not, 'an incidence on the road may happen', as it is often expressed in Ghana, hinting that drivers, police at road barriers, etc. have assisted in diverting the goods.

The general distrust does not prevent economic exchange, but it increases transaction costs and it prevents a company from growing, let alone from becoming innovative. Only by using family members and other people whom the entrepreneur can trust or control is it possible to enlarge the business activity. But even family members are not always trusted (see below) and entrepreneurs try to avoid employing them, as they cannot easily dismiss a relative even if he or she has misconducted.

Distrust means that the economy can only be a transaction economy and not a network economy, dominated by long-term relations. The source of innovation is the company itself and perhaps even only the entrepreneur as he or she may distrust the employees. The consequence is that the needed resources for innovation cannot be accessed through the building of networks. An economy lacking both personal and systemic trust will face difficulties in building innovative capabilities.

THE INCLINATIONS TO IMITATE

Ghanaians seem to be more inclined to imitate than innovate. When an entrepreneur has successfully established a production of an item, others will copy him or her. This inclination gives rise to a 'more-of-the-same-thesis' rather than the treading of new paths, an issue we shall return to. A good example of imitation is the many followers when one company, Astek, came out with locally bottled water – the product in itself an imitation of the foreign brands on the market. Several companies followed and the first-mover company was out-competed and had to close down.

Imitation is part and parcel of a market economy and essential for creating competitive conditions. On the other hand, too much imitation and too many 'band wagon-entrepreneurs', may destroy the market as no one can accumulate capital for investment and take upon them innovative projects.

There is no doubt that certain sectors in Ghana suffer from overpopulation of entrepreneurs and the situation is made worse by none or very minor differentiation between the products offered to the market.

INTERNET AND INNOVATION

A recent study of the use of the Internet by Ghanaian companies revealed that the necessary IT-infrastructure is poor and that companies have reached only the e-mail stage. A number of companies have a homepage but it is not used much as a promotion tool. The Internet is hardly used for active information search, including search for partners and know-how. Lack of training and costs were mentioned as the two main barriers for the use of the Internet (Sørensen and Buatsi 2001).

The Internet has been mentioned as yet another panacea for the catching up by the developing countries. Due to the net's ability to reach all corners of the world at a low cost, the millions of web pages, containing much of the existing know-how of the world, could be downloaded anywhere and used to upgrade production activities. However, access to information is only one step in a three-stepped development formula: the generation of knowledge, the dissemination of knowledge, and the use or implementation of knowledge. What rules the world today is the generation of knowledge. The strength of the Internet as of today is the provision of information and the spread of existing knowledge. However, with the Internet's potential for interactivity, it is believed that it will emerge as a strong means for the generation of knowledge (Sørensen and Buatsi 2001).

Africa is already far behind in terms of infrastructure and access to the Internet. Thus, the first catch-up issue is related to the establishing of the very basic means, that is the prerequisite for catching up with the advanced market economies. In the meantime, as the heavy users of the Internet acquire experience, they also become the best innovators, and the risk is, that the gap between developed and developing countries will widen rather than be closed.

While the Internet is no panacea for catching up, it will be and is already to some extent an important tool in linking business in the North and the South. India's position in the software development industry is a case in point. However, in case of

Ghana we are far from a situation where the Internet plays any role in business development in general and in innovation in particular.

The Economic Structure and Culture as Barriers to Innovation

From the theory of industrial districts and clusters, it is well known that innovation has its contingencies. Under certain structural conditions, innovations will mushroom.

In Ghana, it has been observed that there are both economic structures and cultural traits that constrain an innovative capability to emerge. Three of them will be discussed:

THE NON-ACCUMULATING ECONOMIC STRUCTURE

From microeconomics, it is well known that profits and capital accumulation is low under certain market conditions and as innovation requires investment resources, the competitive conditions become even more important.

Parts of the Ghanaian economy, notably the Mammy Trading System, come close to the perfectly competitive market. Capital accumulation is negligible and especially petty traders are vulnerable to even small misfortunes such as smaller fluctuations in demand and supply, illness, accidents etc. No funds are available for innovative undertakings. Innovations will only come about by customers asking and paying directly for new services.

Entrepreneurs are less vulnerable as they have diversified their business. This may make them survive. However, they may have spread their resources so thinly, both the financial and the managerial ones, that they will not be able to make any of the businesses grow and be successful.

While individual businesses may face problems growing and accumulating funds for innovations, the Ghanaian economy is very dynamic in terms of an ability to expand and contract according to demand and supply. In Sørensen (1999) it is noted that many sectors in Ghana, and especially the retailing sector, are good at 'creating more of the same'. For example, the retailing sector can expand quickly by adding more retailers, who transact the exact same way as the present ones. Within this framework transaction costs are low and so is profit for expansion and innovation.

RECIPROCITY AND RESOURCE LEVERAGE

Turning to cultural issues with economic implications reciprocity seems to be one of the more important ones for the accumulation of funds for innovation.

Reciprocity is a social principle for the allocation of resources. It states that 'if I am in a needy situation, I have a right, or at least an expectation, to receive resources from more resource rich members of the community or extended family'. Furthermore, 'I do not have to repay what I received, but I have an obligation to help the benefactor if he or she later has a need for support and I am now in a position to help'. The principle requires or assumes some agreement as to what constitutes a needy situation. Illness is the most obvious one, but educational needs and the needs to establish a small business for a living are popular ones as well.

Living according to this principle influences the orientation of people, including the orientation of businessmen. The principle creates a tendency to look for resources outside one-self, i.e. instead of developing ones own resources attempts are made to leverage the resources from others. Furthermore, in the pure version of the principle, you do not stipulate any conditions. If you do, they are soft and the conditions are easily relaxed and mostly negotiable. This invites to the culturing of excuses and forgiveness (not apologies) for not repaying.

The extended family is an excellent forum for such a resource orientation as it has a large number of members to search among. Furthermore, in business and especially business associations, the international donor community is perfectly designed for this resource attitude and orientation.[2] An association lives well with nice brochures, excellent equipment, etc. when the money flows from the donor. When the grants dry up, the nice newsletters stop, the equipment breaks down and the organisation comes almost to a halt. Reaching the bottom, an appeal is again made to the donor community and someone may come to rescue the association (Sørensen 1999).

Perhaps the most devastating implication of the reciprocity principle is that it implies an allocation orientation and not a growth orientation. As stated above, the question for a needy person is not how can I develop resources. The question is how can resources be (re)allocated so that I get a share.

Figure 2 illustrates how the reciprocity principle together with the previously discussed circle orientation works within the productive areas.

Figure 2 The Reciprocity Based Resource Circle in Ghana

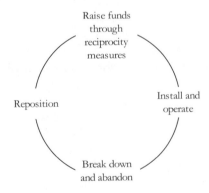

The reciprocity-based circle may be looked at from three angels:

1. Economic effectiveness: the Circle is effective in as much as the objectives are fulfilled, although the fulfilment process may be interrupted from time to time.
2. Economic efficiency: the Circle may be less efficient than a more linear model, where resources are maintained continuously.

3. Social acceptability: the Circle is in accordance with the prevailing social norms and practices.

The Circle must thus be accepted as an economic principle and tensions will arise between systems where the circle-model and the linear model meet. However, the Circle principle may be less efficient and effective in case of innovations. Although it destroys resources, it is not a creative destruction as requested by Schumpeter. Furthermore, the reciprocity principle makes people think in terms of allocation (how can I put myself into a needy position to request resources), not in terms of creation.

Opportunities for Innovation

In this section of the chapter, an attempt is made to sum up the innovative potential in Ghana and activities related to innovation. The picture painted in the previous section was relatively negative in the sense that the innovative system is thought to be weak and the innovative capabilities in the business system are not developed. However, there are also some positive traits, which will be summarised in the following.

Innovation Through Integration Into the Global Economy

Reading the literature on multinationals does certainly not provide you with much optimism related to an increase in the innovative capability and activities of Ghanaian companies. As Dicken (2003) clearly shows, the R&D activities of multinationals are normally located close to their headquarters. However, some changes can be observed. Broadly speaking, multinational companies, on the sourcing side, are on a global search mission, looking for either static efficiency or dynamic efficiency or both. To capture dynamic efficiency, multinationals increasingly establish R&D links to dynamic 'valleys', clusters, technical corridors, etc. around the globe (Dunning 1997).

However, such dynamic valleys are rarely located in developing countries. In these countries, the multinationals are primarily looking for static efficiency, i.e. low cost production and an attractive FDI promotional package from the government. Penang in Malaysia is a good example. The major part of the world production of motherboards is produced in this area where most of the global players have a production unit to take advantage of low labour costs and, by now, a well trained labour force. Penang is now at a stage where it makes an attempt to move away from static efficiency and use the know-how accumulated to convert the area into a valley.

In Ghana, neither dynamic valleys nor static efficiency have, in spite of competitive and attractive FDI-packages, offered by the government, attracted many multinationals and global contract production, for example, in the textile industry, hardly exist in Ghana. However, two observations leave some hopes, one of them being the emergence of clusters, the other, an increase in the value creation through upstream internationalisation. The former will be discussed in the next section.

A study by Kuada and Sørensen (2000) revealed that the internationalisation theories of the West are inadequate to explain the internationalisation process in emerging economies. While the Western theories focus on downstream internationalisation in order to exploit an existing competitive advantage, many companies in emerging market economies practise upstream internationalisation, i.e. they source new technology and modes of management internationally as a means to build a competitive advantage. This upstream internationalisation may take different forms from a simple transfer of technology through a technical assistance project, to a licence, and further on to a strategic alliance and joint venture. Obviously, by moving from a simple transfer of technology to a strategic alliance, the interaction between the foreign and the Ghanaian partner becomes more intensive and will, *ceteris paribus*, thereby give rise to mutual learning.

The building of competitive advantages through up-stream internationalisation is one of the main aims of the Danish Private Sector Development Programme in Ghana. Presently, more than 50 long-term alliances and joint ventures have been established between Danish and Ghanaian companies since 1994. A recent evaluation of the programme found the success rate to be modest when measured at individual alliance level and the impact at industry and sector level very small. Many factors account for this result, including cultural factors, aid and lack business thinking, and the fact that the alliances are not seen as long-term learning platforms, but more short term instruments for technology transfer through training (Danida 2001).

Cultural Synergies

Most literature on inter-cultural relations and inter-cultural management is barrier oriented, i.e. cultural issues create barriers for business relations to develop across cultures. This general conclusion is logical in the sense that the very definition of culture emphasises a coherency in the set of values, norms and behaviour constituting a culture. Doing business across cultures, thus, involves confronting two such coherencies and clashes are to be expected. In spite of this, we venture the idea that there might as well be synergies between cultures, synergies that may generate competitive advantages. Child and Faulkner (1998) have a similar line of thinking. They find that culture can be a resource rather than a barrier.

Two examples from observations in Ghana will illustrate the cultural synergy thesis: Ghanaians do not trust each other. A Ghanaian finds it easier to trust a 'white' businessman. Even if the trust is not high, business develops more easily when black and white join forces. The second example relates to an interview with a Chinese manager of a textile factory in Ghana. Looking for an example of contract manufacturing in Ghana, we were advised to contact a certain textile factory. It appeared that the contract manufacturing was for the mother company in Hong Kong and that the factory also produced its own brands. While not finding much genuine contract manufacturing, the Chinese manager did illuminate that the factory was moving towards being as efficient as similar ones in Asia. Reflecting upon this statement and having the pre-understanding in mind that Ghanaians are individualists, we concluded that we may have observed a cultural synergy in the sense that Ghanaians – in this case women – can work as disciplined and efficient as others if they are under non-Ghanaian management, in this case, the work discipline for which Chinese are well known.

Imitation and Clustering

In an earlier section it was concluded that Ghanaians are good imitators but not innovators. At face value, this is considered to be detriment to development, especially in this era with emphasis on knowledge generation, innovation and competence building. However, this is not the whole truth. A market economy rests primarily on an efficient or at least a workable competition. Competition comes about by entrepreneurs imitating other entrepreneurs, not necessarily by producing the exact same types of goods and services as others, but by differentiation, segmentation and other methods so richly presented in any basic marketing textbook.

Imitation leads to competition and, depending on the ability to develop the market and differentiate the products, it may lead to financial difficulties for all companies due to intensive price competition. However, imitation also leads to having a labour force with experience in the field, suppliers who know the needs of the producers, etc. Furthermore, as Easton (1991) has pointed out, there are many flows between competitors and these flows, together with relations to support institutions, may eventually turn into a cluster where price competition is replaced by innovation competition.

With reference to the discussion of the structure of the economy, Ghana has two main types of agglomerations of companies: the traditional agglomerations of micro-enterprises, for example, of wood-carvers, mechanics, traders, etc. and the more recent ones within pineapple processing, plastics, and aluminium utensils. While the agglomerations of micro-enterprises are based on a simple reproduction formula and able to expand and contract according to supply and demand, the more recent agglomerations are entrepreneurial based and may turn into clusters. The pineapple industry comprises the value chain from pineapple cultivation to the processing of juice; the plastics industry with a diversification into different product areas from simple plastic bags to plastic water tanks for reservoirs in houses and factories, and the aluminium industry focusing on kitchen utensils. These industries can be considered as potential clusters at a very early stage, struggling, for now, with operational problems at the same time as they try to improve products and processes based on experiential learning and partly formal knowledge.

Discussion and a New Perspective

To provide an overview of the barriers and opportunities for innovation, a summary is shown in Table 1. Based on the discussions in the previous sections, this final section will venture into how Ghana and Ghanaian companies may be more innovative.

Table 1 Barriers/Opportunities and Implications for Innovation: a Summary

			Barriers
A.	Financial orientation	→	Too little knowledge of and too low commitment to the product and product quality.
B.	Superiority of formal education	→	Leads to under-utilisation of experiential, including tacit knowledge.
C.	Lack of trust	→	Prevents the establishing of long-term relations and thus getting access to resources and capabilities controlled by others.
D.	Inclinations to imitate	→	May prevent accumulation of financial means or eventually lead to dynamic clusters.
E.	Low level of Internet access and use	→	Insufficient access to information and little chance of being discovered by potential foreign partners.
F.	Reproduction oriented economic structure	→	Leads to more-of-the-same types of enterprises and low level of accumulation of resources.
G.	Resource leverage by reciprocity measures	→	Leads to (non-creative) destruction of resources and resource allocation rather than creation thinking.
H.	The Palaver economy	→	Linguistically elegant but little substance, analysis and action.
			Opportunities
I.	International linkages	→	Long-term links to foreign companies makes it possible to transfer know-how and participate in common development projects.
J.	Cultural synergies	→	Distrust and other barriers may be overcome through cultural synergies.
K.	Cluster formation	→	Imitation, experiential learning, competent work-force, qualified customers, competent suppliers, foreign participants, support institutions, information flows between competitors, etc. may gradually lead to the establishing of a dynamic, innovation driven cluster.

INTEGRATING FORMAL AND EXPERIENTIAL KNOWLEDGE

The important role, played by experiential knowledge and especially tacit knowledge, has over the last ten years become very clear. This is perhaps of no surprise considering the outsourcing trend and the global location of production facilities, both requiring the transfer of know-how. Formal knowledge is relatively easy to transfer, for example, by means of handbooks, blueprints, and structured training courses, while the experiential and tacit knowledge is much more difficult and, at times, impossible to transfer when it is culturally bound.

In Ghana, the problem is one of discovering the value of experiential and tacit knowledge embedded in the work force. The micro-business in Ghana relies solely on tacit knowledge and its development and transfer from one generation to another. Very little formal knowledge is acquired and at use in the informal sector. Among entrepreneurs, it would be wrong to say, that almost the opposite is the case, but formal knowledge dominates over experiential knowledge and tacit knowledge is not appreciated.

Innovative processes can easily begin in Ghanaian companies, if it is possible to create a dialogue between employees with formal knowledge and those possessing tacit knowledge and thus involve the work force in the technical development and organisation of production. This dialogue would also have the effect that parts of the tacit knowledge would become explicit and thus easier to communicate and transfer.

A study in Ghana of the know-how and the ideas for improvement of production among the skilled and unskilled labour in the production sections could be helpful in finding out to what extent an increase in the interplay between formal and experiential knowledge can be productive.

INNOVATION THROUGH INTER-FIRM RELATIONS

Being a major competitive weapon, innovation has mostly been viewed as an in-house activity, taking place in a lab, well sealed off, even from other company employees. Being in the midst of the era of the network economy, innovative activities are now part and parcel of inter-firm relations, whether the relations are organised as a joint venture, a strategic alliance, etc.

Many studies have shown that new ideas are often not given birth in the labs of the producer but come from the user, who has experienced problems in his production. The first thing to do for Ghanaian companies is, therefore, to establish close contact to one or more advanced customers who can give feedback and with whom close co-operation can be achieved. The objective is for the producer to move away from simple transaction relations to become the innovative lab for the customer.

At the other end, the producer may establish close links to suppliers and challenge them in order for the supplier to improve the inputs to the company in question. This may include links to equipment suppliers who are able to upgrade the production process technologically.

To successfully establish such close links requires that the Ghanaian entrepreneur first of all becomes aware and convinced that he will be able to improve his products and processes; secondly, that he is willing to let the innovations partly be developed through his business relations rather than in-house, and thirdly, that the relations can be balanced and mutual, so that opportunism cannot unfold. This mutuality in the benefits (win-win) derived from the links to customers and suppliers is a prerequisite for the building of trust. There is no doubt that the trust problem will be the hardest to overcome and facilitators may be needed to make sure that mutuality prevails.

INNOVATION AND DEVELOPMENT – NEW PERSPECTIVES

To recapitulate: mainstream thinking on innovation and development has centred on the concept of technology transfer from the developed to the developing economies.

The dominant formula has focused on transfer of production processes according to the following formula:

Learn to operate → *Learn to maintain and repair* → *Learn to modify* → *Learn to innovate*

As indicated by the very concept of 'transfer', the formula assumes a once and for all transfer – no further interaction between the transferor and the transferee is expected. Mainstream thinking also assumes a diffusion process, a trigle down process – within the developing countries; i.e. other companies would imitate the company, that in the fist place adopted the new technology. This mainstream thinking needs to be revised in several respects: first of all, the basic transfer formula focuses on process technology. A similar formula for the transfer of product technology should be added. It would look as follows:

Learn to produce → *Learn to repair and service* → *Learn to differentiate* → *Learn to develop new products*

Rasiah (1996) developed a similar formula for Malaysian firms focusing on the gradual change in the output capabilities of the firms: original equipment manufacturer (OEM); original design manufacturing (ODM), and ending with original brand manufacturing (OBM).

Secondly, a transfer of technology is not a one-time event. The transferring company keeps improving the technology and a licence contract will therefore often contain the option of concurrent upgrading of the originally transferred technology.

Thirdly, while the concept of transfer of technology was born and useful when the developing countries were closed, import substitution, and central planning oriented, the concept has become blurred in the era of open economies with many more options for relations between companies in different countries. Rather than a one-time event, relations are, under the free trade and investment regime, continuous and embedded in more flexible organisational forms such as strategic alliances, joint ventures, etc.

This implies mutual interaction rather than a one way transfer and although one should leave no doubts that the developed economy-based firms have and control the lion's share of innovative activity, the affiliates and partners in developing countries will be participants in the processes in a more organic way than under the closed economy regime.

Fourthly, and summing up the previous points, the innovation and development formula under the open market regime may be divided into three stages:

The Entrepreneurial Stage: In the open market oriented economy entrepreneurs will spot and exploit market opportunities. As the economy is open, the entrepreneurial phase includes linkages to foreign companies. These linkages may be based on simple import or export, licences (technology transfer), strategic alliances, etc. This stage also includes foreign firms establishing affiliates in the country to capture the market or to exploit static efficiency. The entrepreneurial stage includes imitation; i.e. other entrepreneurs 'spot' the same market opportunity. Innovation is often absent due to the fact that the density of opportunities in the economy exceeds by far the density of entrepreneurs. There is little need to innovate to carve a business niche in the market. The resulting competition may have two implications: it will force the competitors to differentiate their market offers. If they do not differentiate, they will destroy the market in the sense of incurring keen price competition and the lowering of profit.

The Entrepreneurial-Innovation Stage: The entrepreneurial stage gradually created different industries and thus the accumulation of experiential knowledge at all levels from top management to a pool of workers with the basic skills in production. Links to suppliers, customers and various institutions have been developed and so has links to governmental institutions. A promotion programme may also be in place to the benefit of the industry. The seedlings for a small industrial district have been created and a dynamic innovative process may take off. International links, primarily upstream, have also been established.

The Cluster Stage: The final stage, the cluster stage, will take off when formal knowledge and experiential knowledge are being integrated and formal knowledge institutions such as science labs and universities are being established and co-operate with the companies. This three stages-model is a cumulative stage model rather than distinct stages. Innovation is added to the basic entrepreneurial stage, etc. It is a process model based on learning-experience and the institutionalisation of the process (routines and cultures) in companies, organisations and institutions.

Ghana is in the entrepreneurial stage, struggling to establish viable companies. Imitation is dominant but differentiation of products is poor. Seedlings for industrial districts exist in a few areas. However, the lack of differentiation combined with the lack of trust form barriers for the seedlings to develop into innovative and dynamic industrial districts. There is still some way to go to reach the innovation-driven economy: 'In the innovation-driven economy, competitiveness of firms increasingly depends on the ability to create and efficiently to organise the use of distinctive core competencies in one or other lines of their business' (Dunning 1997: 128).

Notes

[1] Low density of entrepreneurial activity can also be found in developed economies. The IT/Internet wave, for example, created in the beginning a situation with many opportunities and a low density of entrepreneurs with know-how on IT/e-markets.

[2] In this case, the difference between the reciprocity principle and rent seeking is not always easy to establish.

References

Child, John and David Faulkner (1998), *Strategies of Co-operation: Managing Alliances, Networks, and Joint Ventures*, Oxford: Oxford University Press.

Danida (2001): Evaluation – Private Sector Development Programme – 2001/1., www.um.dk/danida/evalueringer.

Dicken, Peter (2003), *Global Shift*, 3rd edition, London: Paul Chapman Publishing House.

Dunning, John H. (1997), 'A Business Analytic Approach to Government and Globalisation', in John H. Dunning (ed.), *Governments, Globalisation, and International Business*, Oxford: Oxford University Press.

Easton, Geoff (1991), 'Competition and Marketing Strategy', *European Journal of Marketing*, Vol. 22, No. 2.

Ferguson Paul R. and Glenys J. Ferguson (1998), *Industrial Economics*, 2nd edition, London: Macmillan.

Kuada, John and Olav J. Sørensen (2000), *Internationalisation of Companies from Developing Countries*, London: International Business Press.

Lall, Sanjaya (1992), 'Technological Capabilities and Industrialisation', *World Development*, Vol. 20, No. 2: 165-86.

Lundvall, Bengt-Åke (1988), 'Innovation as an Interactive Process: From User-Producer Interaction to National System of Innovation', in Giovanni Dosi, Richard R. Nelson, and Christopher Freeman (eds.), *Technical Change and Economic Theory*, London: Pinter Publishers.

Lundvall, Bengt-Åke (ed.) (1992), *National Systems of Innovation: Towards a Theory of Innovation and Learning*, London: Pinter Publishers.

North, Douglass C. (1990), *Institutions, Institutional Change and Economic Performance*, Cambridge: Cambridge University Press.

Rasiah, Rajah (1996), *Malaysia's National Innovation System*, IKMAS Working Paper, No. 4, Institute for Malaysian and International Studies, National University of Malaysia.

Sørensen, Olav J. (1978), *The Mammy Trading System in Ghana*, School of Administration, University of Ghana.

Sørensen, Olav J. (1999), *Business Development through Networking. An Explorative Study: The Case of Ghana*, Paper presented at the International Conference on Business and Development, Copenhagen Business School and University of Aalborg, November 18-19.

Sørensen, Olav J., P. Wad, J. Kuada, H. Schaumburg-Müller, J.E. Thorp (1998), *Business in Development: In Search of a Conceptional Framework*, Paper presented at Research Workshop on Business in Development, October 1-2, Copenhagen: Copenhagen Business School.

Sørensen, Olav J. and Seth Buatsi (2001), 'Internet and Exporting: The Case of Ghana', *Journal of Business and Industrial Marketing*, (forthcoming).

Sørensen, Olav J. and Victor K. Nyanteng (2000), 'Ghana's Export to Neighbouring Countries', *The Journal of Management Studies*, Vol. 15, No. 1: 52-75.

Swedberg, Richard (2000), *Entrepreneurship: The Social Science View*, Oxford: Oxford University Press.

Whitley, Richard (ed.) (1992), European Business Systems: Firms and Markets in their National Contexts, London: SAGE Publications.

19

Do Services Matter for African Economic Development?

An Empirical Exploration

Mark Tomlinson and Tidings P. Ndhlovu

Introduction

This chapter begins to examine the extent to which non-industrial factors such as information and knowledge infrastructures and specific types of service sector activity can enhance economic development in Africa. Following a discussion about recent debates on knowledge-intensive business services (KIBS) and industrial cluster approaches to innovation and how these can enhance understanding of the learning economy, data are analysed from a variety of sub-Saharan African countries in this context. The chapter uses some statistical techniques to explore the impact of knowledge infrastructures and knowledge-intensive services on performance. It will be argued that observing and maintaining traditional distinctions between manufacturing and services is no longer a useful way to approach economic problems of growth, innovation, competition or productivity and that a systemic approach is necessary to understand developments in the African context. Several hypotheses are tested within the framework proposed using data from 1998 for the whole of sub-Saharan Africa and using input-output data for several years from South Africa.

The chapter concludes that knowledge-intensive services and knowledge infrastructures in general have a key role to play in the overall economic development in African countries. This contrasts with many traditional theories that see services as 'tertiary' and unproductive. Services and manufacturing sectors, for example, are better understood as existing in a symbiotic relationship with each other, rather than as oppositional or separate entities. The cluster approach to economic development is proposed as a useful way forward – with the proviso that services need to be incorporated to a greater extent in the analysis.

Conceptual Framework

The mainstream view of economic development has often neglected the service sector. This, in part, stems from the legacy of the Fisher-Clark model of economic development, which placed services in a kind of residual, or 'tertiary' sector, after

305

agriculture, extractive industries and manufacturing had been accounted for (Clark 1940; Fisher 1935). Statistics are still routinely and uncritically separated into these categories.

This picture of the service sector as a laggard, non-innovative and burdensome sector is being re-examined in the developed world not least because services now account for more than two thirds of the OECD economies – whether in terms of production or employment. Some recent research has shown that producer services are also now considered to be as dynamic and innovative as manufacturing (Miles and Boden 2000; Tomlinson 1999, 2000; Windrum and Tomlinson 1999). Clearly, a reassessment of the significance of the service industry in the promotion of economic development is called for: 'Rather than seeing services as a separate and peripheral economic factor mainly producing superfluous consumer products, emphasis is beginning to shift towards looking at services as an integral part of a potentially dynamic economic system' (Tomlinson 2000: 37). These new debates focus on a more systemic and radical view, which sees services as crucial components of economic networks or clusters (see, for example, OECD (1999), Hauknes (1998)). Recent macroeconomic evidence also points to significant linkages between knowledge-intensive business services (KIBS) and all sectors of the economy (Antonelli 1998, 2000; Tomlinson 2000; Katsoulacos and Tsounis 2000).

As economic development proceeds, one of the key features of the economic transitions and changes is the mediation between old and new structures. This mediation is, in the main, dependent on the development of new services that allow the transformations caused by periodic innovation and growth to proceed. This was one of the key ideas behind Marshall's economic districts. The old structures are constantly transformed into new ones. These transformations are necessarily accompanied by technological and social change, which require new and more sophisticated services.

The Cluster Approach to Innovation and National Systems of Innovation

One major advance in recent years has been a focus on clusters and industrial networks in developing countries (van Dijk and Rabellotti 1997; Nadvi 1997; Nadvi and Schmitz 1994; Humphrey and Schmitz 1995; Ceglie and Dini 1999; Mytelka and Farinelli 2000). Rather than a focus on individual firms or enterprises the unit of analysis becomes a cluster of organisations (often closely located geographically although this does not have to be the case).[1] The cluster approach can also be extended within a National System of Innovation (e.g. Lundvall (1992)) approach to include other economic and non-economic factors and institutions (such as local labour market conditions, governmental support agencies, supply chains, transport structure and efficiency, etc.). Clusters can also be examined at different levels (macro, meso, micro) and using a variety of empirical techniques (see Roelandt and den Hertog (1999)). Although the concept is not specifically national it can readily be incorporated in a NSI framework. The 'National System of Innovation' can be thought of as including a set of overlapping clusters within a national boundary and embedded within a context of overarching national policy and specifically national conditions. This does not mean that localised conditions are not important, but that some features of economic systems are bounded and constrained by institutional frameworks that are developed and implemented at national level. These often include such areas as education, training and health systems, which cannot be ignored even when the cluster approach is adopted. Even if a cluster crosses a national border there will be considerations on both sides of the state line that are a consequence of nationally defined conditions.

There have been several case studies of clusters in developing countries. McCormick (1998) provides a useful review of some of the literature and a detailed description of eight clusters in Africa accompanied by comparisons with clusters in other developing countries such as Brazil and Pakistan – which are substantially different. The eight clusters studied in Africa also reveal a variety of different markets, external economies and types of linkage. This is summarised in Table 1. As one moves down the table the more sophisticated the clusters described. There is great diversity in the size and extent of these clusters ranging from small groups of micro-enterprises to large agglomerations of firms.

We can see that in terms of 'learning effects', co-operation is a key factor in many clusters. There is a tendency to sub-contract and to share information when there is a problem that cannot be met by an individual firm or when there is excess demand. As the clusters get more sophisticated there is a tendency for them to have more training and apprenticeship systems in place thus keeping and generating knowledge in the area. There are also usually associations of producers formed to represent the interests of the cluster firms with local and national government etc. In terms of service activity it is clear that there is a transition from the importance of basic services like security and dealing in scrap metal in the more basic clusters to more sophisticated retailing, marketing and business services as the clusters become bigger and more powerful. The South African cluster of garment makers has sophisticated computerised design and marketing activities associated with it. There is a relative lack of analysis of this service aspect of clusters in the literature and this is a problem that we now address.

Services in the 'Knowledge Economy'

In recent decades we have witnessed several shifts in the economic foundations of manufacturing industry. First, there is the increasing reliance and percolation in the economy of non-material production. This is usually referred to by such terms as knowledge-based or new economy. Second, there is the rapid development and industrialisation of large parts of the developing world. Thirdly, there is also the increased overall globalisation of production. This includes the increasing propensity for countries and firms to engage in cross-border operations, and the increased volume of trade between nations. These three aspects of economic development are entwined and feed off one another to some extent, making for a more complex and constantly changing world. Some production clusters have become rejuvenated due to these changing global conditions (Mytelka and Farinelli 2000). Some are also buckling under more intense global competition.

Table 1 The Character of the Eight African Clusters in (McCormick 1998)

Cluster	No. of producer firms	Average size of firm	Products	Learning effects and co-operation	Developments in the 'service economy'
Garments in Nairobi	600	3.5	Clothing		Basic services emerged such as retailing/suppliers/distribution
Kamukunji	2000	1-2	Metal products	Formation of an association Collaboration in marketing	Development of security
Thika	337	1.6	Vehicle repair	Extensive subcontracting and co-operation Improved manufacture of spare parts	
INDUSTRIALISING CLUSTERS:					
Suame metal	3-4000	5-6	Metal products	Technology Consultancy Centre Apprenticeship system	Trade in scrap metal
Suame vehicle repair	4-5000	6.5	Vehicle repair	Extensive subcontracting Locally made spare parts Apprenticeship system Spillover effects from engineering workshops	
Ziwani	506	1.5	Vehicle repair	Extensive subcontracting Training Sharing of knowledge	'Legal services': The association enforces contracts Security services enhanced by the association
DIVERSIFIED INDUSTRIAL CLUSTERS:					
Western Cape SA	538	126	Clothing	Co-operation and sharing of information Subcontracting	Marketing agents/market research Extensive retailing sector Computerised textile technology Design firms etc. Co-operation between retail and manufacturing sectors
Uhanya Beach Kenya	560	Fishing 3-15 Trading 1-50 Processing 35-300	Fresh and processed fish	Strong vertical links between fishing and processing plants, supply chain	Marketing and financial services Good transport infrastructure Outsourcing of repair and transport services

Source: Adapted from (McCormick 1998), Table 3.1 and text.

Insofar as innovation and economic performance is inextricably linked with 'the ability to learn (and forget) rather than the stock of knowledge' (Johnson and Lundvall 2000: 3), it can also be argued that increased competition will result in further change which, in turn, generates pressures for a more rapid rate of innovation. While knowledge-intensification becomes thus self-reinforcing, the question is whether learning and innovation are promoted by either market selection and competition (with their attendant inequalities) or by co-operation, social cohesion and/or social networks, that is civil society and institutional arrangements predicated on social norms such as loyalty, voluntary co-operation and social and mutual trust which induces collective action (Cameron and Ndhlovu 2000: 240-243) (see also (Ernst and Lundvall 2000: 16, 23-25; Freeman 1992: 186-87; Johnson and Lundvall 2000: 10, 15-24, 26)). There are also increased pressures with respect to services such as education, health care and services for the elderly. For example, the need to institute cleaner technologies and safer working conditions, find more effective new treatments, etc. (Persson 2001). Clearly, economic, social and ecological aspects can best be understood when couched within a structure that integrates technical, organisational and institutional frameworks (Chudnovsky 1999: 160; Johnson and Lundvall 2000: 6-10).

The increasing impact of what has been termed the 'knowledge-based economy' suggests that there are new forms of economic activity, which, in turn, require the development of new service structures. Knowledge-based services are becoming increasingly dominant in western economies and are increasingly changing the way manufacturing firms operate. We also saw above that services play an increasingly sophisticated role in some African clusters.

In the west at least, these processes are to some extent a result of the outsourcing of services by manufacturing firms during the last few decades as enterprises moved towards a more 'core competence' model of organisation (Prahalad and Hamel 1990). Many firms tended to shed their service elements, such as software teams, marketing departments, catering, etc. which became separate entities in their own right and hence appeared in the national accounts for the first time as 'producer services' instead of being hidden within 'manufacturing'. There has also been an increase in consumer services as the proportion of personal income spent on manufactured goods has declined with the shift towards more sophisticated 'non-material' products (telecommunications, cable TV, internet, etc.). These developments have also led to new theoretical insights about the role of services and especially knowledge-intensive services (KIS) in the economy.

There are at least three distinct ways in which knowledge-intensive services can promote innovation and learning:

1. Co-production: quite often with these knowledge-based services, innovation is a 'co-production' generated by the influence of the service provider interacting with the consumer (Tomlinson 2001; Tomlinson and Miles 1999). Client specific problems generate new solutions, which can lead to innovations on both sides. This can also lead to re-innovation with other firms.

2. Re-innovation: the service firm can maintain competitive advantage by reapplying ideas generated with one client and applying them to a set of similar client firms. These ideas are then accumulated to generate new forms of solution and break into new markets. Thus, the level of interconnectedness, complementarities and externalities (spillovers across clusters of companies, industries and supporting or associated institutions) are linked with the rate of innovation and competitiveness. There is no reason to assume that these types of activity are not equally relevant to developing countries.

3. Re-penetration: newly organised services often re-penetrate and transform older structures. This is especially the case where the impact of new ICTs has transformed the way businesses operate. We could say knowledge-intensive services in modern economies are allowing older structures to change in the face of innovation. That is new service industries can re-penetrate older structures (from which they themselves emerged) and ultimately transform them.

In order for co-production, re-innovation and re-penetration to be effective, there must exist a suitable knowledge infrastructure. If there are problems with this then knowledge and information will not diffuse effectively throughout the clusters and sub-systems that make up the National System of Innovation and service innovation and service led growth will be less effective.

The Industrialisation and Service Sector Growth of the Developing World

Along with the major shifts that have taken place in the developed world with respect to services, the last quarter of the 20th century has seen the increase in production of manufactured goods for developed economies by less developed countries (LDCs). There has been an increase between 1960 and 1990 for LDCs from 15.6 per cent to 20.7 per cent of world GDP from manufacturing and a corresponding decrease in developed countries from 28.4 per cent to 21.4 per cent (Williams 1997). The changes in the LDCs are most dramatic in Asia, which now accounts for a large share of global manufacturing, even when excluding Japan.

On the other hand, there has been a corresponding rise in service sector share of GDP in *all* regions (including developing ones). This indicates that there may be much more to the story than just the offloading of manufacturing from the first to the third world. This lends further weight to the idea that we are entering a new phase of economic development, even at the global level, with respect to the services-manufacturing debate. 'Tertiarisation' is a global phenomenon, but there are important national differences and trajectories to consider.

Data and Hypotheses

What hypotheses might be pertinent to the above discussion in the African context? Firstly, we would expect countries with larger service sectors to be wealthier and those with larger agricultural sectors to be poorer. If services are necessary for economic growth then we will see a consequently large service sector accompanying a large industrial sector where there is higher output.

Secondly, we would also expect that, in an increasingly knowledge-oriented economy influenced by increased service sector activity, education and literacy would be positively correlated with wealth production. KIS require a certain degree of personal competence and often involve the transfer of information and communication rather than material production. This requires competences that will be enhanced by the general level of education that can be provided within the system.

Thirdly, following from the previous point, we would expect that the knowledge and information infrastructure, such as the extent of telephony or computers, for example, would be positively correlated with wealth production. If knowledge-intensive services are to provide efficient inputs there must be a decent level of this type of infrastructure for it to be effective.

Finally, we would also expect that we can show that knowledge-based services are positive drivers of output throughout the whole economy especially in the more affluent African countries.

In the following we have used data published by UNDP (2000) for all sub-Saharan African countries. This data is cross-sectional, mainly from the year 1998, but with some slight variations for some countries. To test the final hypothesis, we have used various input-output tables for South Africa from several years (see STATS SA[2] various years).

The following is a list of the infrastructural variables used from the UNDP data:

- No. of main telephone lines/1000 people
- No. of PCs/1000 people
- No. of internet hosts/1000 people
- Adult literacy rate
- Education expenditure as a proportion of GNP
- Agriculture as a proportion of GDP
- Industry as a proportion of GDP
- Services as a proportion of GDP

From the input-output tables the following variables are defined at sectoral level:

- Q is gross sectoral output
- M is sectoral material and energy inputs (i.e. non-agrarian and non-service inputs)
- S is knowledge based service inputs
- L is sectoral labour costs

Following (Windrum and Tomlinson 1999) and (Tomlinson 2001) we estimate a production function of the form:

$$Q/L = A(M/L)^a (S/L)^b$$

Therefore $\log(Q/L) = \log(A) + a \cdot \log(M/L) + b \cdot \log(S/L)$, which can be estimated by OLS regression. This assumes that sectoral output per unit labour cost is a function of material and energy inputs and knowledge-intensive service inputs (again per unit labour cost). This relationship has been demonstrated for developed countries such as the UK, Germany, Japan, Netherlands, etc.

Results

HYPOTHESIS 1: THE RELATION BETWEEN THE SECTORAL ACTIVITY AND GDP PER CAPITA
These graphs (Figures 1-3) suggest that the level of development in terms of GDP per capita in sub-Saharan Africa is dependent on the advanced progress not only of industrial sectors, but also the service sector. Clearly the most backward countries are also the most agrarian ones. The most advanced countries in terms of per capita GDP have high service sector development such as Seychelles, Mauritius, and South Africa. Unfortunately we cannot break down the type of service sector activities with this database, which may give us a better idea of the knowledge-based services relationship with output.

Figure 1 GDP per Capita versus Level of Agriculture

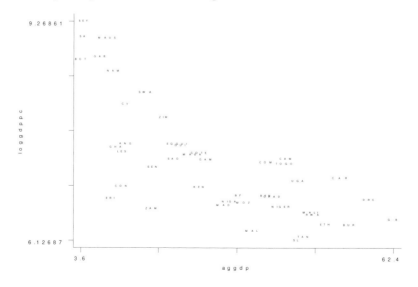

Figure 2 GDP per Capita versus Level of Industry

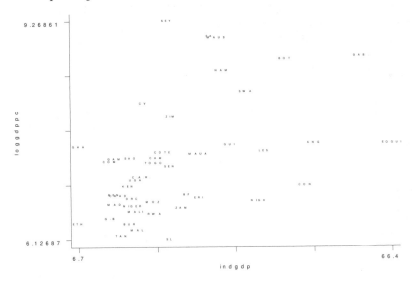

Figure 3 GDP per Capita versus Level of Services

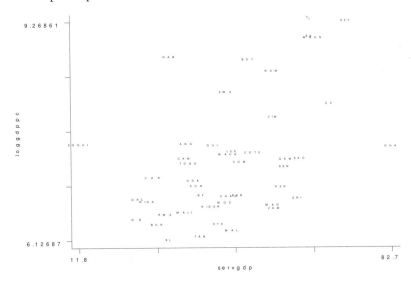

HYPOTHESES 2 AND 3: EDUCATION AND LITERACY; KNOWLEDGE INFRASTRUCTURE

Furthermore, there are also positive correlations between the basic education variables (literacy and educational expenditure – Figures 4 and 5) and the basic information infrastructure variables (PCs and telephony – Figures 6-7). The higher GDP per capita then requires an appropriate level of service development and an information and knowledge based infrastructure in order to prosper. These variables (in Figures 4-7) are also positively correlated with the overall level of service sector activity within the economy thus confirming that economic development appears to proceed hand in hand with service sector development and infrastructural advances.

Figure 4 GDP per Capita versus Level of Educational Provision

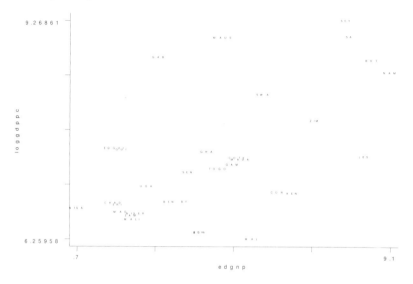

Figure 5 GDP per Capita versus Level of Literacy

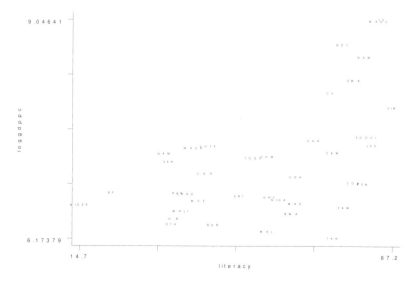

Figure 6 GDP per Capita versus Level of PC Ownership

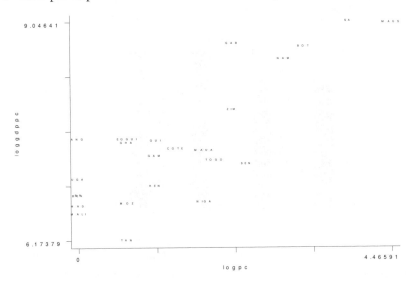

Figure 7 GDP per Capita versus Level of Telephony

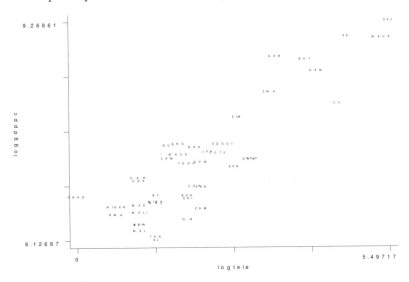

HYPOTHESIS 4: SERVICES AND OUTPUT

Figure 8 Levels of Agriculture, Industry and Services by GDP per Capita Quartiles, All Sub-Saharan African Countries, 1998

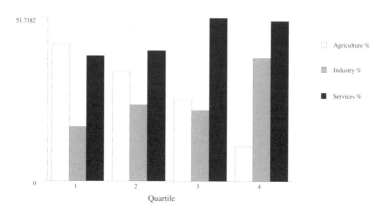

Finally, we turn to the service sector itself and its impact on overall economic development. Figure 8 breaks down the economic structure in 1998 of sub Saharan Africa by four quartiles based upon GDP per capita. The lowest two quartiles have high levels of agrarian activity, while the upper two quartiles have the highest levels of service development. However, we also need to compare the third with the fourth quartile from the perspective of systems of innovation requiring a developed service sector. The top quartile has about the same level of service sector development as the third, but has a much *more highly developed* industrial sector. This suggests that development in Africa must take into account the development of services more generally. They must form part of the explanation for economic development rather than just being a consequence of it. Clearly services are necessary for growth, but so is industry and hence the relationships between services and industry need to be examined in more detail.

To finally show that services do matter, we analysed the input-output data from South Africa going back to 1971. Table 3 shows results from 1971, 1981 and 1993.

Table 2 OLS Regression (beta coefficients) on Input-Output Data, South Africa

	1971	1981	1993
M/L	0.531**	0.272**	0.116*
S/L	0.360**	0.257**	0.852**
R square	.488	.152	.767
F	41.010**	8.222**	141.530**

Note: Dependent variable Q/L.
* Significant at 5 per cent; ** significant at 1 per cent.

What this suggests is that, as output per unit of labour increases, so do manufactured inputs as well as service inputs (per unit labour cost). All the coefficients were statistically significant. Furthermore, the beta coefficient on the service sector variable is much higher in the 1993 data than in earlier years, suggesting that it is increasingly important. These results look similar to results obtained for other developed countries (see (Tomlinson 2001)) and confirm that services appear to provide essential positive inputs into the South African economy as a whole.

Conclusion

We have argued that in order to understand economic systems in Africa, a broader overarching approach is required. This could be provided by the literature on National Innovation Systems (e.g. Lundvall (1992)).

We argue that this approach is equally valid for understanding developing nations such as African countries. It provides a framework for understanding economic development in terms of the interactions of several institutional and innovative structures within countries. This includes knowledge infrastructures, information economies and service industries (which are generally neglected in explanations of innovation and growth). The cluster approach may prove to be of benefit for this type of analysis.

With respect to services, the old models of economic development, which place industrial/manufacturing development at the core, are being replaced by theories, which see producer services as highly innovative and dynamic drivers of growth rather than unproductive laggards. This needs to be explored in the context of LDCs.

We have demonstrated that the information and knowledge infrastructure is relevant and important in the study of sub-Saharan African economies and that there are high correlations between the levels of such infrastructures and economic performance.

We have also demonstrated that knowledge-intensive services in South Africa do appear to have a significant economic impact even as long ago as 1971. More input-output data from other African countries is needed to see whether this applies to poorer African countries.

Notes

[1] There is no universally accepted definition of a cluster in the literature. We use the term here to refer to a broader concept of economic unit that extends beyond individual firms.

[2] The analysis and interpretation of these tables is the authors' and not that of STATS SA that merely provide the basic data for researchers to use.

References

Antonelli, Cristiano (1998), 'Localized Technological Change, New Information Technology and the Knowledge-Based Economy: The European Evidence', *Journal of Evolutionary Economics*, Vol. 8, No. 2: 177-198.

Antonelli, Cristiano (2000), 'New Information Technology and Localized Technological Change in the Knowledge-Based Economy', in Ian Miles and Mark Boden (eds.), *Services and the Knowledge-Based Economy*, London: Continuum.

Cameron, John and Tidings P. Ndhlovu (2000), 'Development Economics: An Institutional Bastion', *The Journal of Interdisciplinary Economics*, Vol. 11: 237-53.

Ceglie, Giovanna and Marco Dini (1999), *SME Cluster and Network Development in Developing Countries: The Experience of UNIDO*, Working Paper, No. 2, PSD Technical Working Paper Series, UNIDO.

Chudnovsky, Daniel (1999), 'Science and Technology Policy and the National Innovation System in Argentina', *CEPAL Review*, No. 67.

Clark, Colin (1940), *The Conditions of Economic Progress*, London: Macmillan.

Ernst, Dieter and Bengt-Åke Lundvall (2000), *Information Technology in the Learning Economy: Challenges for Developing Countries*, mimeo.

Fisher, Allan G. (1935), *The Clash of Progress and Security*, London: Macmillan.

Freeman, Christopher (1992), 'Formal Scientific and Technical Institutions in the National Systems of Innovation', in Bengt-Åke Lundvall (ed.), *National Systems of Innovation: Towards a Theory of Innovation and Interactive Learning*, London: Pinter Publishers.

Hauknes, Johan (1998), *Services in Innovation, Innovation in Services*, report for the TSER project SI4S, Oslo: STEP Group, http://www.step.no/

Humphrey, John and Hubert Schmitz (1995), *Principles for Promoting Clusters and Networks of SMEs*, UNIDO, SME programme discussion paper, No.1.

Johnson, Björn and Bengt-Åke Lundvall (2000), *Promoting Innovation Systems as a Response to the Globalising Learning Economy*, mimeo.

Katsoulacos, Yannis and N. Tsounis (2000), 'Knowledge-intensive Business Services and Productivity Growth: The Greek Evidence', in Ian Miles and Mark Boden (eds.), *Services and the Knowledge-Based Economy*, London: Continuum.

Lundvall, Bengt-Åke (ed.) (1992), *National Innovation Systems: Towards a Theory of Innovation and Interactive Learning*, London: Pinter Publishers.

Lundvall, Bengt-Åke (1998), 'The Learning Economy: Challenges to Economic Theory and Policy', in Klaus Nielsen and Björn Johnson (eds.), *Institutions and Economic Change: New Perspectives on Markets, Firms and Technology*, Cheltenham: Edward Elgar.

McCormick, Dorothy (1998), *Enterprise Clusters in Africa: On the Way to Industrialisation?*, IDS Discussion Paper, No. 366, Sussex: Institute of Development Studies.

Miles, Ian and Mark Boden (eds.) (2000), *Services and the Knowledge-Based Economy*, London: Continuum.

Mytelka, Lynn and Fulvia Farinelli (2000), *Local Clusters, Innovation Systems and Sustained Competitiveness*, INTECH Discussion Paper, No. 2005, Maastricht: The United Nations University.

Nadvi, Khalid (1997), *The Cutting Edge: Collective Efficiency and International Competitiveness in Pakistan*, IDS Discussion Paper, No. 360, Sussex: Institute of Development Studies.

Nadvi, Khalid and Hubert Schmitz (1994), *Industrial Clusters in Less Developed Countries: Review of Research Experiences and Research Agenda*, IDS Discussion Paper, No. 339, Sussex: Institute of Development Studies.

OECD (1999), *Boosting Innovation: The Cluster Approach*, Paris: OECD.

Persson, Göran (2001), 'How to Make Europe Work', *Financial Times*, February 8.

Prahalad, C. K. and Gary Hamel (1990), 'The Core Competence of the Corporation', *Harvard Business Review*, May-June.

Roelandt, Theo J. A. and den Hertog, Pim (1999, 'Cluster Analysis and Cluster-Based Policy Making in OECD Countries: An Introduction to the Theme', in OECD, *Boosting Innovation: The Cluster Approach*, Paris: OECD.

STATS SA (various years), *Input-Output Tables*.

Tomlinson, Mark (1999), 'The Learning Economy and Embodied Knowledge Flows in Great Britain', *Journal of Evolutionary Economics*, Vol. 9, No. 4: 431-51.

Tomlinson, Mark (2000), 'The Contribution of Knowledge-intensive Services to the Manufacturing Industry', in Birgitte Andersen, Jeremy Howells, Richard Hull, Ian Miles, and Joanne Roberts (eds.), *Knowledge and Innovation in the New Service Economy*, Cheltenham: Edward Elgar.

Tomlinson, Mark (2001), 'A New Role for Business Services in Economic Growth', in Daniele Archibugi and Bengt-Åke Lundvall (eds.), *The Globalizing Learning Economy: Major Socio-Economic Trends and European Innovation Policy*, Oxford: Oxford University Press.

UNDP (2000), *Human Development Report 2000*, New York: UNDP/Oxford University Press.

Van Dijk, Meine Peter and Roberta Rabellotti (eds.) (1997), *Enterprise Clusters and Networks in Developing Countries*, London: Frank Cass.

Williams, Colin C. (1997), *Consumer Services and Economic Development*, London: Routledge.

Windrum, Paul and Mark Tomlinson (1999), 'Knowledge-intensive Services and International Competitiveness: A Four Country Comparison, *Technology Analysis and Strategic Management*, Vol. 2, No. 3.

20

No Cause for Afro-pessimism

A Study of Technical Innovation in Sengerema District,

Tanzania

Mona Dahms

Introduction

The main questions addressed in this chapter concern the existence of technological knowledge and technological innovation in rural areas of Africa. The research has been carried out in connection with a so-called Multipurpose Community Telecentre (MCT) in Sengerema, Tanzania. The MCT is part of a larger programme, funded by the International Telecommunication Union (ITU), United Nations Educational, Scientific and Cultural Organisation (UNESCO) and the International Development Research Centre (IDRC) of Canada. The main aim of the MCT programme is to demonstrate the catalytic impact of easy access to modern information and communication technologies (read: computers with Internet access) on community development in rural areas of Africa.

Community Development, Information and Communication

In order to understand the great emphasis presently placed on (access to) information and communication technologies by development organisations it might be useful to take a closer look at some of the most important concepts, such as community development, information, communication and knowledge. These concepts and (some of) the relationships between them are the topics of this section. First, a definition of the concept of community development is presented. Second is presented the 'value chain of information', a useful tool for analysing differences between data, information, knowledge and wisdom. The value adding processes from data to information and from information to knowledge are described, and the concept of knowledge, including tacit knowledge is analysed in more detail.

Community Development

The dominant (but most often implicit) perception of 'development' even today is 'modernisation', i.e. that the so-called 'developing' countries must become like the so-called 'developed' countries, although the modernisation paradigm in development studies has been challenged as being too simplistic several decades ago. Like the modernisation paradigm, the 'digital divide' paradigm in discourses about information- and communication technologies (ICT) and development is based on technological determinism, while overlooking complex social factors influencing underdevelopment and poverty, such as power imbalances and inequalities at global, national and local levels (Wilson 2001).

In this chapter I will define 'community development' as processes of cultural change which – when assessed by the community, using assessment criteria based on shared values, interests and goals – are seen as change for the better, i.e. as change leading to improved living conditions for the members of the community (Hastrup 1990; Gullestrup 1992; Gyekye 1996). The change is brought about by community members engaging in autonomous and critical learning processes (Nyerere 1973; Wenger 1998), which expand the knowledge and capabilities of individuals – and thereby of communities. When combined with increased freedom to be in control of one's own life, these processes will lead to change (Freire 1993; Sen 1999). Conflicts of interest at the community level have to be dealt with in a constructive way in order to establish co-operation and mutual respect (Guijt and Shah 1999).

The 'Value Chain of Information'

In phrases like 'Information Society', 'Information Highway', etc. the word 'information' is used in the special sense in which it is used in the engineering discipline called 'Information Theory', i.e. as a denomination of a physical quantity which can be stored, processed and transmitted via different forms of information and communication technologies. A more precise word for this physical quantity is 'data' and use of the word 'information' should be reserved to its everyday usage, i.e. to denote meaning (Jensen and Skovsmose 1986). The qualitative difference between 'data' and 'information' is brought out in 'the value chain of information' (Fuchs 1997):

Data -> Information -> Knowledge -> Wisdom

where value is added from one step to the next. The process of adding value between the first two steps can be illustrated by the '4 As' model in Figure 1.

Figure 1 The '4 As' Model

Source: Heeks (1999: 7)

Data must be accessible, physically as well as mentally. Accessible data must be assessed and found useful in the given context and for the given purpose. Once data has been accessed, assessed and found useful, it needs to go through a process of adaptation and/or application in order to be transformed into 'information'. Once the information is available the person can begin to act based on the information received. Seen in this perspective it is rather obvious that what is found on the 'Information Highway' is 'data' rather than 'information'.

From Information to Knowledge

The value adding from 'information' to 'knowledge' takes place via a wide range of learning processes. According to the experiential learning theory of Kolb (1984), learning progresses through a learning cycle as shown in Figure 2.

Figure 2 The Experiential Learning Cycle of Kolb

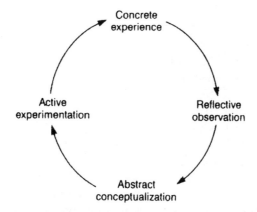

Source: Dixon (1999: 40)

Information enters the learning cycle in the form of concrete experience gained from actions performed in/on the real world. The process of reflective observation transforms the concrete experience into abstract concepts, which are then brought to trial in active experimentation that eventually leads to new concrete experience.

323

The Kolb cycle is mainly applicable to individual learning processes; when focus is on community development, however, collective learning processes are equally important. A theory of learning dealing with collective learning is the theory of 'situated learning' in which learning is viewed as an integral aspect of all social activity. Thus, learning takes place through legitimate peripheral participation in communities of practice, i.e. the novice learns through direct – although in the beginning peripheral – involvement in the practical activities taking place within the community of practice. The fact that some of the empirical studies underlying the theory of situated learning have been carried out in West Africa serves to indicate its relevance in an African context (Eskildsen 1988; Lave and Wenger 1991; Reagan 1996).

In the interaction between the individual members of the communities of practice, communication becomes crucial. In this context communication is seen as mainly fulfilling an informative function, i.e. the purposeful exchange between members of a community, of information acquired from sources of data within or outside the community (Holgaard 1999; Windahl, Signitzer, and Olsen 1992). Thus, information and communication are indeed crucial for community development. However, it is the information exchange and the interactive communication processes which are important, not the technical means of accessing and producing information or of mediating communication.

Knowledge

A useful taxonomy of knowledge is the following (Ernst and Lundvall 1997):

- Know-what (factual knowledge)
- Know-why (theoretical knowledge)
- Know-how (practical knowledge)
- Know-who (social knowledge)

'Know-what' refers to knowledge about facts and data, which might be called factual knowledge. 'Know-why' refers to knowledge about principles and laws of motion in nature, in the human mind and in society. It is most often science-based and might also be termed theoretical knowledge. 'Know-how' has to do with practical skills, i.e. the capability to do something. This is experience-based, practical knowledge. 'Know-who' is related to social competence and the capability to establish relationships to individuals and to groups of people who are important as business partners, customers or in other ways. This could be called social knowledge.

'Know-what' and 'know-why' are explicit types of knowledge, i.e. knowledge which can be codified into 'data' and thus be easily stored, processed and transmitted via any kind of information technology, from person to person and from organisation to organisation, independent upon the given context. Once codified into 'data' these types of knowledge can even be sold as a commodity in the market of the 'learning economy'. In the Western culture these two types of knowledge are highly valued and strongly emphasised in the development of formal education, not only in the Western world but also in training and education activities planned and implemented by development organisations and governments in the South (Müller and Bertelsen 2001). The learning processes that lead from 'information' to 'know-what' and 'know-why' are the ones found in formal education, i.e. reading books, attending lectures, doing exercises, performing laboratory experiments, surfing on the Internet, etc. These learning processes can (to a very large extent) be undertaken by the individual as an individual enterprise, at a time and a place suitable for the individual, and thus they lead to individualised knowledge.

The other two types of knowledge, 'know-how' and 'know-who' are tacit forms of knowledge, i.e. knowledge which is characterised by being non-systematised and non-verbal. These types of knowledge cannot easily be codified and therefore cannot be stored, processed and transmitted from person to person in an explicit form. Instead they can be transmitted in communities of practice where knowledge is transferred from the master(s) to the apprentice(s) via learning processes based on observation and imitation, rather than on oral or written instructions (Eskildsen 1988; Polanyi 1966; Reagan 1996; Rolf 1991). Practical knowledge in the form of the tacit knowledge forms 'know-how' and 'know-who' is more highly valued in African society than is theoretical knowledge (Gyekye 1996). But by its very nature tacit knowledge is excluded from the ongoing discourses on the 'knowledge-based' economy. This exclusion of tacit knowledge is unfortunate in a phase of development where the importance of tacit knowledge in processes of learning and development is stressed by several authors (Baumard 1999; Ernst and Lundvall 1997):

> [...] there is simply no other way to reduce poverty but to place learning and knowledge creation at the centre of development strategy. [...] [T]acit knowledge is as important as or even more important than formal, codified, structured and explicit knowledge. (Ernst and Lundvall 1997: 21)

According to Ernst and Lundvall the countries of the South are faced with two challenges in technological learning: acquisition of codified technological knowledge and development of tacit technological knowledge. I will come back to this statement in the later section discussing the research findings.

'Know-how' and 'know-who' are forms of knowledge which in the 'value chain of information' seem closer to the highest form of human information, 'wisdom', than to 'information' or 'data'. While theories of learning and knowledge are abundant in the West, no theories known to this author deal explicitly with wisdom, therefore the value adding from knowledge to wisdom will not be further discussed here.

Technology

Given that focus in the Multipurpose Community Telecentre project is on information and communication technologies and given that the theme of this anthology is systems of technological innovation, a section on technology theory

325

seems relevant. In the first subsection is presented the definition of technology used in this chapter and this definition is applied to information and communication technology. The next subsection deals with technological knowledge and the last subsection discusses technological innovation.

Technology – a Holistic Definition

In this chapter I will use the holistic technology theory introduced by Müller (1980) according to which technology is defined as follows:

> Technology is one of the means by which mankind reproduces and expands its living conditions. Technology embraces a combination of four constituents: Technique, Knowledge, Organisation and Product. (Müller and Bertelsen 2001: 7)

The definition of technology is symbolically illustrated in Figure 4 as four pieces of a puzzle, thus indicating the close relationship between all four constituents:

> A qualitative change in any one of the components will eventually result in supplementary, compensatory and/or retaliatory changes in the others. (Müller and Bertelsen 2001: 7)

Figure 3 The Holistic Technology Concept

Source: Müller and Bertelsen (2001: 7).

This definition of technology distinguishes itself from most other definitions by insisting upon the inclusion of the 'product' as part of technology, whereby it is possible to avoid the oft-encountered pitfall: '*Here is the technical solution – where is the problem?*' – a pitfall which is especially well known in technology transfer from the North to the South and which has often led to the creation of 'white elephants' in countries of the South.

The definition of technology indicates how the structure of a given technology can be described through a description of the four components. To the structural description should be added a process perspective, i.e. processes of constant adaptation and transformation between the four components. Analyses of the structure and the processes of adaptation together constitute a technology analysis.

Information and communication technology can be seen as a physical infrastructure technology, which can be described using the above definition. Most important is the realisation that the 'product' is the information and communication services supplied to the user and that the 'technique' component may be any kind of technique which can deliver these services, for example: computer providing access to the Internet; telephone, telefax and telegram; radio, television, and recorded audio and video material on tape or on CD-ROMs; books, newspapers and other printed material; indigenous folklore, popular theatre and proverbs; production, processing and transmission of information taking place within and between human beings (without ever being recorded explicitly). This last 'technique' may well be the most important information and communication system in many rural communities of the South. The two technology components 'knowledge' and 'organisation' will of course depend upon the specific form of the 'technique' component.

Technological Knowledge

The concept of technological knowledge can be described using the above technology definition, with focus on the human aspect. This aspect may not be immediately visible from the technology definition – it is, however, included in all the four components and can be depicted as overlaying the four components. The different forms of technological knowledge embedded in the four components are described in the following.

In the 'product' component the main human aspect is the user – female or male – of the product. Seen from a producer point of view, knowledge about the users, their needs and demands and their preferences is important in order to be able to satisfy user demands. This knowledge is part of the tacit knowledge termed 'know-who'.

In the 'technique' component the human labour force constitutes the human aspect. Only under the most extreme of Fordist factory production systems can this labour force be seen as separated from the human knowledge, know-how, and skills.

The 'knowledge' component includes human knowledge in the forms of 'know-what', 'know-why' and 'know-how'. The codifiable knowledge ('know-what' and 'know-why') may in advanced industrial production be embedded in the machines and thereby become a part of the 'technique' rather than of the 'knowledge' component.

In the 'organisation' component the human knowledge element includes knowledge about suppliers, other producers, market organisation, etc. as well as social skills in establishing interpersonal relations between people in different positions as employees and employers, i.e. 'know-who' knowledge.

Technological knowledge thus consists of a combination of different forms of knowledge, most of which are tacit, while others are explicit and can be codified. The knowledge is generated via different types of learning processes, as discussed in the previous section.

Technological Innovation

Based on the above definition of technology, a broad definition of technological innovation could be:

> A qualitative change in any of the 4 elements of technology that effectively leads to transformative move and thus change in the other elements we denote a technological innovation. (Müller and Bertelsen 2001: 8)

Such qualitative changes will be caused by and/or have an impact on the human knowledge element in any one of the four components. Thus, technological innovation is closely connected with technological knowledge which is mostly generated via interactive, socially embedded learning processes.

In summary, two of the most important value-based concepts in this chapter and in ongoing development discourses, namely Community Development and Technological Innovation are characterised by being processes of qualitative change brought about by interactive, socially embedded learning processes, taking place via participation in communities of practice. In the next section the results from the field work will be presented.

Technological Knowledge and Innovation in Sengerema: A Case Study

This section deals with the fieldwork carried out in connection with the study of technological knowledge and innovation and of indigenous systems of information and communication. The first subsection describes shortly the methodology for the fieldwork. The second subsection presents the factual findings, while the next two subsections discuss the findings in relation to the concepts of technological knowledge and technological change and innovation. The last subsection discusses the findings concerning indigenous information and communication systems.

Methodology

The field work was carried out in Sengerema District, Mwanza Region, Tanzania in February 2001. A total of 11 informal sector groups, active within a range of different areas, from agriculture over handicraft to blacksmithing and house construction, were interviewed using a semi-structured interview guide. The interviews were carried out with the assistance of a research assistant, Mr. Joseph S. Shigulu who translated to/from the local languages Kiswahili and/or Kisukuma. The interviews were taped and simultaneously notes were taken. The notes were summarised in a group report, one for each group. Mr. Shigulu compared the group reports with the tapes, in order to include any missing information in the report, after which the reports were translated into Kiswahili. The translated reports were given to the groups for consolidation and verification of the information summarised. Also, the groups were asked for permission to distribute the reports to other groups as well as to the telecentre manager. An overview of the groups is shown in Table 1.

A limitation of this method of data collection is that an interview reveals only what the interviewee wants to reveal. Another limitation lies in the short period of time for field work and a final limitation is caused by the language problem – nuances in the discussion will be lost in translation from Kiswahili to English.

Basic Facts

The Sengerema district covers 8,817 km^2 of which 3,335 km^2 are dry land and 5,482 km^2 are covered with water, i.e. Lake Nyanza (formerly Lake Victoria). It has a total population of approximately 375,000 people, with 45,000 living in Sengerema town and the rest living in 125 registered villages (Tanzania 1999). Main sources of income in the district are agriculture, livestock herding, fishing, softwood and hardwood timber production, mining and to a lesser extent industry and trade.

The 11 groups interviewed embrace a total of 187 members, of whom 107 are women and 80 are men. Three groups are women's groups, four are men's groups (although, interestingly, these groups do not call themselves 'men's groups' while the women's group are called so by both women and men) and four groups are mixed. Members are aged between 14 and 61 years. Eight of the groups are located in or very close to Sengerema town while three groups are located in villages at a distance ranging from 6 to 35 km from Sengerema town.

Technological Knowledge

A research hypothesis concerning technological knowledge is as follows:

> Many members of the community possess sophisticated technological knowledge; this knowledge is, however, tacit. If made explicit the knowledge can be communicated to other members of the community and can form the basis for social learning processes.

Based on the group interviews it can be concluded that the first part of this hypothesis is indeed confirmed as far as 'know-how' – practical knowledge – is concerned. Most of the groups actively pursue a strategy of diversification, i.e. they engage in several different types of activities, depending upon needs and seasonal variation and most of the members are capable of performing all the different types of activities, which the group is engaged in. Thus, many people have practical technical knowledge and skills in different areas.

Table 1 Informal Sector Groups interviewed in Sengerema District, February 2001

Name of group	Location	Activities	Time spent in group	No. of members F/M	Age Min/max	Registration	Date of Inter-view
Sengerema Afforestation & Nursery Supplies for Environmental Protection (SANSEP)	Sengerema	Production of tree and flower seedlings; Prevention of soil erosion; Food processing and selling; Savings and credit scheme	2 hours/day, seasonal variations	17/23	20/60	Community Based Organisation SISC	7th Feb.
Sengerema Works Group	Sengerema	Blacksmithing and metal work of any kind incl. repairs	Full time	1/11	19/35	YADEC; SISC; LAZOBCA	7th Feb.
Zana za Kilimo na Ufundi (ZAKIU)	Nyampulukano	Vegetable cultivation; fruit growing; food production; Low cost housing	Seasonal variations	0/5	40/61	YADEC; SISC	7th Feb.
Nguvukazi Partners	Nyampulukano	Metalwork, incl. tool production; carpentry; Low cost housing, incl. Production of building material; Pottery; basketry; weaving; tailoring; Trade training programmes and evening classes in basic literacy, English, mathematics	Full time	3/10	20/47	YADEC; SISC; LAZOBCA; VETA;	8th Feb.
Tupendani Wanawake Bomani	Sengerema	Savings and credit scheme; Kindergarten (under construction) (Pottery, basketry, food processing and selling, café etc. on individual basis)	Part time	40/0	25/55	SISC	8th Feb.
Wazazi Group	Sengerema	Basketry; pottery; weaving; tailoring; wood carving; Savings and credit scheme;	Twice/week	30/0	35/50	SISC	16th Feb.
Benjamin Furniture Workshop	Sengerema	Production of furniture and other wooden items	Full time	0/6	??/??	YADEC; SISC	17th Feb.
Malindima Fanuel Group	Kamanga	Boat building ; Carpentry	Full time	0/6	14/42	-	18th Feb.
Kikundi cha Wanawake Maendeleo Tabaruka	Tabaruka	Vegetable gardening; farming of cash crop; Pottery, incl. Stoves; basketry; broidery; crochet	Once/week (r) Twice/week (d)	14/0	29/55	SISC	20th Feb.
Vumilia Workshop	Sengerema	Production of knives and pangas	5 hrs/day (r) Full time (d)	0/15	18/49	SISC	20th Feb.
Nyamililo Tool Makers Group	Nyamililo	Blacksmithing and metal work, incl. tool production; Carpentry; Tailoring	Full time	2/4 2/6 app.	18/50	SIDO; VETA; LAZOBCA	21st Feb.

This knowledge is, however, not tacit in the sense that it cannot be expressed in words. Rather, the knowledge is non-articulated because there has so far been no need for articulation – until some outsider (such as the researcher) starts asking questions about the activities which they pursue. Possibly there is a tacit element of the technological know-how, which will only be revealed in a practise situation, but to a large extent the group, members are able to describe precisely what they do when performing the activities undertaken by the group.

Also, members of the groups possess 'know-who' – social knowledge. The chair of one of the metal working groups expressed it in this way: 'all the members of the group have sufficient self-confidence, can introduce themselves and talk to customers'. Another indication is that from several of the groups members have split off and have established their own workshops, drawing upon the 'know-how' and the 'know-who' acquired in the group.

Although most of the persons interviewed have had no formal training beyond primary school all groups say that they are looking for opportunities to learn more, in the form of short formal courses, study tours to other groups and/or contacts to groups outside Tanzania working within the same area. Thus, acknowledgement of the value of formal training and explicit knowledge is outspoken among the groups.

The second part of the research hypothesis, that technological knowledge made explicit can form the basis for social learning processes, has not yet been documented. However, some elements of learning appeared during the process of interviewing simply by bringing information about the activities from one group to another. An example of this happened when a boat-building group complaining about the prices of imported handheld carpentry tools were told that such tools were locally made and cheaply available in a nearby village.

Also, there are indications that reflective learning processes have been set in motion simply by being asked and by answering the questions posed by the researcher. Several of the groups expressed this in their concluding comments to the researcher. Further, anecdotal information received from Mr. Shigulu approximately one year after the field work indicates that learning has happened by sharing the group reports (where the technological know-how has been made explicit, i.e. written down) between all groups whereby the (good) examples of more successful groups have been adapted by less successful groups. Thus, members from one blacksmith group went as apprentices to another blacksmith workshop, and members from one of the women's groups went to learn how to use the potter's wheel from one of the mixed groups.

In connection with this discussion of technological knowledge I would like to return to the statement by Ernst and Lundvall (1997) about the two challenges for technological learning in countries of the South: acquisition of codified technological knowledge and development of tacit technological knowledge.

Concerning acquisition of codified technological knowledge it is a fact that due to the crisis of the educational systems and weak research and development institutions prevalent in many African countries, there is a general lack of explicit knowledge, as expressed by the groups. Since this knowledge can be easily codified it could be made available to African countries and more specifically to rural areas, for example via combinations of different forms of information and communication technologies, such as local radio broadcasting, combined with user controlled searching on the Internet. One problem is that much of this codified knowledge has been turned into a commodity and therefore has to be bought at prices, which may be beyond the reach of many people and countries of Africa.

Concerning the development of tacit knowledge it seems justified to conclude that in rural Africa there is a lot of tacit technological knowledge in the form of 'know-how' and 'know-who' and that therefore there is no need to develop such knowledge but rather a need to acknowledge the existence of this knowledge, to include it as a practical element in formal education and training and to support the further development of it by enhancing the collective learning and exchange of information between the communities. For this purpose modern information and communication technologies can be very useful.

Technological Innovation

Based on the above definition of technological innovation as a qualitative change in any one of the four elements of technology it can be concluded that all the groups interviewed have been making technological change and/or innovation during the period they have existed, although some groups more so than others.

'Responding to varying needs of customers' is a statement often mentioned as a main cause of innovation as far as changes in 'product' is concerned. This is true for the boat builders who changed from softwood boats to hardwood boats in response to customer demands for more durable boats that could travel longer distances. It is also true for the women who introduced colour patterns in their basketry; they did it because – as opposed to their mothers – they produce for a market where there is a demand for more beautiful products. Similarly, the vegetable growers introduce new types of vegetables, depending upon customer demands.

Changes in 'technique' are often caused by the search for more efficient ways of doing things or for time saving methods. This is the case for the knife makers who bought a manual metal cutting machine from one of the metal workshop groups as a replacement for the chisel, which had hitherto been used for cutting metal. It is also true for the carpenter group, which has bought some handheld electric tools. The vegetable growers have purchased a water pump from one of the metal working groups because they assume that it will help increase production.

The groups interviewed are all informal sector groups with only a few of them registered. Changes in the 'organisation' component of the technology are few and mainly consist either of establishing a group as a replacement of an individual activity or of having the group registered with one or another organisation/association.

Concerning the 'knowledge' component, a characteristic of all groups is that the group members have acquired practical knowledge ('know-how') and social knowledge ('know-who') through learning processes which have taken place via participation in a community of practise, constituted either by the group itself or by another group with similar activities. Thus, all groups mention that they either 'learned from the chair' or 'are learning from each other', i.e. they emphasise the interactive and collective learning processes.

The boat builder learned the trade as an apprentice when a young man. He in turn has passed on the skills to his apprentices by working with them in the workshop. The apprentices explicitly state that they learn purely through doing practical work, there is no theoretical training. In the blacksmithing group the chair, born into a family of blacksmiths, was an apprentice in his grandfather's workshop from the age of 7 to the age of 10, while the other members of the group have learned the skills from him. In his own workshop, in the training of apprentices, practical training is integrated with theoretical training. This is done by having some schoolteachers being members of the group. The women in the women's groups have learned the skills of pottery and basketry from their mothers and grandmothers and are passing on these skills to their daughters.

This said, it should be added that the most innovative groups are characterised by the fact that at some point in time there has been an input of 'exogenous knowledge' in the form of a member participating in some kind of formalised training, leading to the acquisition of 'know-what' and 'know-why'. Examples of such input are: a two year course on handicraft and domestic science in Rubya Home Craft School, Bukoba; a formal education as agricultural extension officer; a two year carpentry course at the National Vocational Training Centre in Mwanza; a two year blacksmithing course at Mwanza Technical College. These 'exogenous' inputs have given rise to technological innovation in the form of new products, new tools and/or new methods of production.

An exception to this rule is the chair of one of the metal work groups who, having received nothing but primary school education, constructed an aeroplane out of scrap metal and made it fly for 100 m before it collapsed. He also constructed an automatic rifle the possession of which led to a police arrest for illegal weapon possession. Ingenious persons are found everywhere, also in rural Africa!

Indigenous Systems of Information and Communication

A research hypothesis concerning indigenous[1] systems of information and communication is:

> The indigenous systems of information and communication are based upon 'organic' technology, i.e. they rely on human beings as the main media for storing, processing and transmitting information between members of the community and between the community and the outside world.

All of the groups interviewed state that their main source of information is other people and the main means of communication is face-to-face communication, with suppliers, customers and others. Most of the groups regularly send members travelling to bigger towns such as Mwanza, Geita or even Dar es Salaam to obtain information about materials and tools and to buy if prices are right.

Only four of the groups mention letter writing as a means of communication, four mention telephones (which are available in Sengerema town although the outside plant at the time of interview was old and the services within the town were of low quality) and one group uses e-mail which is available from the SIDO office in Mwanza. There is no mention of newspapers, newsletters, etc. for finding information, which is maybe not surprising since there is no regular sale of newspapers in Sengerema. One group does, however, mention that they themselves use newspapers and newsletters for advertising.

Thus, the hypothesis about 'organic' technology as the basis of indigenous information and communication systems seems to be confirmed, at least for the time being. It should be mentioned, however, that all the groups interviewed complained about the lack of good communication facilities, including transport facilities.

An interesting observation concerning telecommunications was made in a study, which examined the impact of infrastructural constraints on entrepreneurship in Northwest Tanzania:

> [...] reliable telecommunications remain of rather minor importance to entrepreneurs. If telecommunications were more reliable, perhaps the entrepreneurs would trust them more and extend their roles in various operations. However, today, telecommunications are not dependable. (Trulsson 1997: 133)

The explanation offered by Trulsson for this apparent lack of concern with telecommunication infrastructure is that the personal encounter is what matters in business relationships. Maybe this could explain why one of the groups interviewed expressed a certain scepticism towards the use of telephones: 'Sometimes you do not get the right information via the telephone'.

According to Tanzania Telecommunication Company Limited (TTCL) an upgrading of the total telephone network in Sengerema is planned for the first half of 2001, in connection with the establishing of the permanent Multipurpose Community Telecentre. An interesting question for future research therefore is to what extent the use of 'organic' communication technology will be replaced by digitalised technologies, such as telephone and e-mail services offered by the telecentre in Sengerema town. Similarly, it will be interesting to see to which degree information will be sought via the Internet, once there is easy access and people have learned how to use it.

Conclusion

Based on the fieldwork it can be concluded that many people possess technological knowledge in the form of 'know-how' and 'know-who' and that technological innovation is happening continuously. It can also be concluded that there is an unsatisfied quest for 'exogenous' knowledge in the form of short courses, study tours, etc. Further, it can be concluded that technological innovation in rural Africa is a process of qualitative change brought about by interactive, socially embedded learning processes, taking place via participation in communities of practice.

The title of this chapter is based on the findings about technological knowledge and innovation: there is no reason to be overly pessimistic about the future technological development of Africa – provided that the indigenous technological knowledge and the existing capacity for technological innovation is being recognised, acknowledged and enhanced. One way of doing this could be to support and promote the exchange of information between the informal sector groups and to supply them with inputs of 'exogenous' knowledge in the form of short courses, study tours, etc. The use of a wide range of information and communication technologies could enhance such processes.

Notes

[1] I use the same definition of the word 'indigenous' as is given in (Müller and Bertelsen 2001: 4), i.e. indigenous systems include traditional systems but also all recent and contemporary exogenous systems which have been indigenised.

References

Baumard, Philippe (1999), *Tacit Knowledge in Organisations*, London: SAGE Publications.

Dixon, Nancy (1999), *The Organisational Learning Cycle: How We Can Learn Collectively*, 2nd ed., London: McGraw-Hill Book Company.

Ernst, Dieter and Bengt-Åke Lundvall (1997), *Information Technology in The Learning Economy: Challenges for Developing Countries*, DRUID Working Paper, No. 97-12.

Eskildsen, Jørn (1988), *Autoritetens dilemma: i u-landspædagogik og solidaritetsarbejde*, Copenhagen: Mellemfolkeligt Samvirke.

Freire, Paulo (1993), *Pedagogy of the Oppressed*, London: Penguin Books.

Fuchs, Richard (1997), *If you Have a Lemon, Make Lemonade: A Guide to the Start-up of the African Multipurpose Community Telecentre Pilot Projects*, Paper submitted to the International Development Research Centre, http://www.idrc.ca/acacia/outputs-/lemonade/lemon.html.

Guijt, Irene and Meera K. Shah (eds.) (1999), *The Myth of Community. Gender Issues in Participatory Development*, London: Intermediate Technology Publications.

Gullestrup, Hans (1992), *Kultur, kulturanalyse og kulturetik: eller hvad adskiller og hvad forener os?*, Copenhagen: Akademisk Forlag.

Gyekye, Kwame (1996), *African Cultural Values*, Accra: Sankofa Publishing Company.

Hastrup, Kirsten (1990), 'Udvikling eller historie: antropologiens bidrag til en ny verden', *Den ny verden*, Vol. 23, No. 1, Copenhagen: Centre for Development Research.

Heeks, Richard (1999), *Information and Communication Technologies, Poverty and Development*, Development Informatics, Working Paper Series, Institute for Development Policy and Management, Manchester: University of Manchester, http://www.man.ac.uk-/idpm/.

Holgaard, Jette E. (1999), *Miljøkommunikation i produktionskæden*, Ph.D.-project proposal submitted to the European Doctoral School of Technology and Science, Aalborg: Aalborg University.

Jensen, Hans Siggaard and Ole Skovsmose (1986), *Teknologikritik: et teknologifilosofisk essay*, Copenhagen: Forlaget Systime.

Kolb, David A. (1984), *Experiential Learning: Experience as the Source of Learning and Development*, London: Prentice-Hall.

Lave, Jean and Etienne Wenger (1991), *Situated Learning. Legitimate Peripheral Participation*, New York: Cambridge University Press.

Müller, Jens (1980), *Liquidation or Consolidation of Indigenous Technology*, Aalborg: Aalborg University Press.

Müller, Jens and Pernille Bertelsen (2001), *Changing the Outlook: Reconciling the Indigenous and the Exogenous Systems of Innovation in Tanzania*, Paper presented in the International Workshop on African Inovation Systems and Competence-building in the Era of Globalisation, March, Aalborg: Aalborg University.

Nyerere, Julius K. (1973), *Freedom and Development: Uhuru na Maendeleo*, Dar es Salaam: Oxford University Press.

Polanyi, Michael (1966), *The Tacit Dimension*, Cloucester: Peter Smith.

Reagan, Timothy (1996), *Non-Western Educational Traditions: Alternative Approaches to Educational Thought and Practice*, Mahwah: Lawrence Erlbaum Associates.

Rolf, Bertil (1991), *Profession, tradition og tyst kunnskap: En studie i Michael Polanyis teori om den professionella kunnskapens tysta dimension*, Lund: Bokförlaget Nya Doxa.

Sen, Amartya (1999), *Development as Freedom*, Oxford: Oxford University Press.

Tanzania (1999), *MCT Project Document on the Establishment of Pilot Multipurpose Community Telecentre (MCT) at Sengerema District*, Mwanza, Tanzania.

Trulsson, Per (1997), *Strategies of Entrepreneurship: Understanding Industrial Entrepreneurship and Structural Change in Northwest Tanzania*, Tema Research, Linköping: Linköping University.

Wenger, Etienne (1998), *Communities of Practice: Learning, Meaning and Identity*, New York: Cambridge University Press.

Wilson, Merridy (2001), *Information and Communication Technology, Development and the Production of 'Information Poverty'*, Unpublished thesis submitted for the Degree of Master of Philosophy in Development Studies, Oxford: Oxford University.

Windahl, Sven, Benno Signitzer, and Jean T. Olsen (1992), *Using Communication Theory: An Introduction to Planned Communication*, London: SAGE Publications.

21

African Systems of Innovation

What Can We Learn From Telecommunications Firms

Gillian M. Marcelle

Introduction

Investigating how developing country firms build technological capabilities can yield interesting and controversial insights into the performance, and value of innovation systems. The lens used to analyse the empirical research findings is an original conceptual framework – the Technology Capability Building (TCB) system approach – which focuses on firm level innovation and capability development routines of developing country firms (Marcelle 2002). This study sets out to explain and account for firm level variation in the innovation and capability development performance in developing countries. The study found that a firm's relative effectiveness in innovation and technological capability building derived from implementing a systematic approach, in which simultaneous and proportional effort was expended on internal processes for technological capability building, and the management of boundary relationships including between firms and the innovation system. Many insights emerged from this study which are relevant to understanding how developing country firms interact with the innovation system, and how that interaction influences their capability development processes.

The empirical setting for the research consists of four developing countries in Africa, viz. Uganda, Ghana, South Africa, and Tanzania. Primary data were collected through in-depth interviews with decision-makers in telecommunication operating companies, regulatory authorities, and S&T policy-makers, in each of the selected case study countries. This method allowed for collection and analysis of information on the key features of the TCB processes at the firm level, as well as on the context and environment facing telecoms producers. Within each country, the contextual information included review of policy instruments, major shifts in policy direction, important specific events and important illustrations of policy dynamics and overview of the national innovation system. In summary the research provides a rich source of quantitative and qualitative data at a sufficiently detailed level to improve understanding of innovation and technological capability building processes of the sample firms.

The rest of this chapter is organised in three sections, the first presents an outline of the conceptual framework developed in this research project; the second, presents the empirical evidence from the sample of twenty-six telecommunications firms and the final section presents the implications of these findings for understanding African innovation systems and producing research on this area.

The TCB System Approach

The TCB system approach argues that variations in innovation and capability development in firms cannot be fully explained by country level factors, and are likely to be influenced by developments that occur endogenously within the firm.

In the TCB system approach, the collection of firm specific assets that constituted a *technological capability (TC)* are defined to include the equipment, skills, knowledge, aptitudes and attitudes which confer the ability to operate, to understand, to change and to create production processes and products. The definition further suggests a firm's TC is likely to include elements with intensive scientific and technological content, such as disciplinary-specific technical knowledge, codified product and process specifications, and tacit knowledge about production processes, as well as elements that enhance the ability of a firm to benefit from the presence of the technical components. These non-technical elements of a firm's TC are components that support acquisition of technological knowledge and learning, both at the individual and firm-wide level. For firms to benefit from TCs they must not exist in isolation, but must be integrated across an organisation (Leonard-Barton 1995; Pettigrew and Whipp 1991; Tushman and Nadler 1996).

The process of *technological capability building (TCB)* is defined as a process of assembling or accumulating technological capabilities. TCB is considered to be an investment activity that is undertaken by productive units. Although firms undertaking TCB expect their efforts to produce economic benefits, the process is not automatic and is best characterised as a learning process in which existing levels of capability inputs are transformed over time through a process of trial and error into an improved or enhanced capability level. The impetus for undertaking these investments derive from internal factors such as recognition of technological trends and existing capability gaps, and external factors, such as changing market structures, the behaviour of competitors, actions of regulators and policy-makers, and information from suppliers and other knowledge creating institutions.

Variations in TCB expenditure by firms are not fully explained by country level factors, but are strongly influenced by endogenous factors within the firm. Therefore, within a country, one should expect to find firms that respond to changes in external factors by carrying out a range of technological capability building activities and other firms with a much slower response time and a much less developed range of technological capability building activities. By drilling down to the level of the firm, and considering intra-firm processes as explanatory variables, this approach sheds light on the innovation performance of developing country firms. The specification of the process for TCB integrates insights from organisational development theorists (Argyris 1996; Argyris and Schon 1996; Schein 1992; Senge et al. 1999; Starkey 1996; and Vaill 1996); strategic management scholars (Baden-Fuller and Stopford 1994; Leonard-Barton 1995; Pettigrew and Whipp 1991; and Senge 1992); and scholars who focus on structural enablers for organisational learning (Barney 1991; Nelson 1991; Nelson and Winter 1982; Rumelt 1984; Teece, Pisano, and Shuen 2000). By drawing on evolutionary theories of the firm, the TCB system approach uses departures from neo-classical theorising about economic growth at the macro and micro levels and methodological approaches, such as the importance of focusing on variations in firm performance (Barnett and Burgleman 1996; Dosi, Nelson, and Winter 2000; Thomson 1993; Foss 1999; Levinthal 2000; Nelson 1991 and Nelson and Winter, 1982). One of these important assumptions is that the management of boundary relationships involving exchanges of tacit information and the mastery of exchange of tacit information across organisational boundaries is considered to be an integral element of TCB and a source of competitive advantage (Winter 1987).

Insights from development studies scholars are also integral to this conceptual framework, including authors such as (Bell and Pavitt 1993, 1997; Enos and Park 1988; Ernst, Mytelka, and Ganiatsos 1998; Hobday 1995; Katz 1987; Kim 1997; Lall 1987, 1992; and Westphal 1985) who identify many features common to successful technological capability building and the weaknesses inherent in the innovation system of developing countries.[1] In Ernst et al. (1998), the following institutions are considered to be important in the learning and technological accumulation process of firms, because they increase the flow of *public knowledge*: domestic education system, university and public research institutions, regulatory bodies, standards bodies, intellectual property regimes and supplier networks.

The insights of scholars who question the surety of policy intervention, and emphasise endogenous factors and, in particular, management competencies in the capacity development process is particularly useful, for considering innovation in regions of the world where NIS systems are rudimentary. The TCB system approach maintains a sceptical viewpoint with respect to the effectiveness of policy actions, and does not assume that all policy intervention impacts beneficially on firms' investments in technological capability development. Using an approach similar to that suggested by Cooper (1980, 1991) the study undertook a more open-ended inquiry into the effects of a specific set of policy interventions. A very wide range of policies such as those to support firms in acquiring foreign technology, and to encourage investment in education, training and research, provide incentives for imitation and innovation, and provide necessary institutional support, were investigated. The TCB system approach provides a framework to question how developing country firms can use interaction with public institutions to be effective in capability development. The TCB system concept positions the productive operating companies as active respondents to signals originating from public sector institutions rather than passive agents. In some of the early literature, scholars tend to regard public sector bodies as being somewhat omniscient and better placed than decision-makers in firms to select suppliers, choose technological platforms, and adjust technological capability building efforts.

For firms to be effective in innovation and technological capability building, they must implement *TCB systems*, defined as a set of integrated processes and mechanisms incorporating five critical components; three internal processes and two boundary relationships. The internal processes – financing, management practices, culture and leadership – are assumed to be under the control of the firm, while the boundary processes managing relationship with suppliers and with the innovation system are considered to be only partially under the firm's control. The intra-firm TCB processes are embedded in institutions and practices outside the firm. Proportional, balanced and simultaneous investment in all elements, those under direct control and partial control is expected to improve effectiveness of technological capability building. TCB is considered to be path dependent process in which the direction and effectiveness is influenced by the existing stock of accumulated technological capability.

The firm's ability to manage its *relationship to the innovation system* and to access TC resources from the innovation system (national and global) is considered to be an important factor in its innovation and capability development performance. In this framework, the following institutions within the domestic innovation system are considered to be important sources of technological inputs include – knowledge creating institutions such as universities, technical vocational colleges, training institutes, national research centres etc.; policy making bodies and regulatory authorities. This is consistent with the NIS approach of Nelson (1982), Lundvall (1992), Bell and Pavitt (1997), Mytelka (1999), Kim (2000). The types of technological inputs that firms can derive from relationships with these institutions include codified knowledge, tacit knowledge, improved understanding of technological trends and patterns through regular interaction, information about sources of technological information and know-how, information on what TCB activities are permissible or feasible under existing legislative and regulatory rules, information regarding changes in legislative and regulatory rules. These institutions can also be a source of embodied skills and know-how, to the extent that the local setting can provide skills and experience required by operating companies. The domestic innovation system institutions can also improve cost efficiency in technological search activities by providing common information services to all firms, and so reducing the duplication of search costs. In this framework relationships between users and producers of knowledge in the national systems of innovation (NIS) is important (see Lundvall 1988, 1992, 1995; and Lundvall and Johnson 1994). In particular, the boundary relationship is characterised as a process of 'learning by interacting', in which producers of knowledge, users of knowledge and governments interact in a multistage process. The distinctive characteristics of this multi-stage process include: flows of qualitative information, direct co-operation, and co-ordination routines that involve not purely hierarchical relationships but mutual trust and mutually respected codes of behaviour (Lundvall 1988). The emphasis that Lundvall (1992) places on trust and respect between the relating parties as a facilitating factor for innovation by reducing uncertainty associated with technological change is pertinent. The TCB system approach also uses the argument that spatial proximity and cultural commonality are factors that tend to increase levels of trust and respect.

Lessons From African Telecommunications Firms

Analysis of evidence showed that there was considerable variation across the sample in terms of the ability to manage relationships with the innovation system. Firms that had features in common with the ideal-system for innovation and capacity building, presented in the last section were able to benefit from the innovation system, while firms with less well developed TCB system did not benefit or interact effectively with the innovation system. It is important to note that there were intra-country variation, and this demonstrates the value of including firm-level variables in the explanation of innovative performance in developing countries. In the empirical research, detailed analysis was made of interaction with policy and regulatory bodies, universities, training colleges and research centres as the representative institutions of the innovation system.

Variation in Interaction With Innovation System Across Firms

In the empirical study on which this chapter is based, the TCB routines of twenty-six telecommunication operating firms were examined in detail. The sample firms were found to make use of 61 individual TCB mechanisms, which were arrayed in seven functional categories. The empirical research also confirmed that the sample firms used a combination of internally and externally orientated TCB mechanisms, and varied in their level of TCB system development. Three categories of TCB system development were established corresponding to the absolute number and breadth of TCB routines in use. The analysis of data also included the calibration of an intensity of usage score; this metric attempts to capture and indicate the extent to which firms in a particular category of TCB system development made use of a type or group of TCB mechanisms. The minimum score is zero, representing the event where no firms in that range category use any of the mechanisms in that group. The maximum score is one, corresponding to an event where the total number of firms in that range category use all the mechanisms in the group. These data and analytical metrics were used to explore how the behaviour of firms varied across firms with different levels of TCB system development. It was found that the level of intensity of usage of different types of TCB routines varied according to the level of TCB system development

Analysis of the data confirms that firms with more developed TCB systems were three times more likely to interact with innovation systems, than firms with fairly well developed TCB systems (see intensity scores for Group VI). Statistical techniques were used to confirm that the indicated variation across different levels of TCB development was significant[2]. Summary data on patterns of use of TCB mechanisms for firms across the different categories of development of TCB systems are presented in Table 1.

Table 1 Patterns of Usage of Groups of TCB Mechanisms versus Level of TCB System Development

TCB Group Number and Function	Intensity of Usage Scores		
	High	Medium	Low
1. Increasing people skill base	0.73	0.80	0.27
2. Organisational development	0.94	0.42	0.15
3. Technological Search	0.75	0.38	0.07
4. Acquiring complementary knowledge from industry	1.00	0.68	0.04
5. Acquiring expatriate people skills	0.75	0.54	0.20
6. Interaction with innovation systems	0.63	0.23	0.00
7. Funding TCB	0.75	0.85	0.40

Note: N=26

Innovation System as a Source of Technological Capability Input – the Ideal versus Reality

The study compared firm's desired or *ideal* support from innovation system organisations and the current reality. As shown in Table 2, the sample firms were in favour of innovation system institutions supporting their capability development activities.

Table 2 Desired Public Sector Support for TCB in Firms

	Operators % in Favour	Public Officials % in Favour	All Respondents % in Favour
Direct expenditure	92	100	94
Support activities	92	67	87
Direct funding of innovation institutions	75	80	76
Operator performance requirements	67	83	72
Supplier performance requirements	50	83	63
Public sector subsidies for TCB effort in firms	42	20	35

Source: Author's analysis of primary empirical data

The definitions of these categories of support are provided here: subsidies – financial payments made to incentivise TCB activity including partial funding of TCB investment through indirect mechanisms, where firms would apply for state sponsored write-offs for TCB activities; direct expenditure by line ministries and regulatory bodies on innovation activities, for example cross-industry technological training; direct funding of public research institutions or other knowledge creating and dissemination institutions to carry out TCB activities that would benefit firms; support activities, such as organising technological information search service, setting up technical resource information centres, providing supplier evaluation services and technology assessment services to firms at a cost; setting performance requirements for TCB activities in the licences of operating companies and in their contracts with suppliers; and other specific measures that go beyond the five categories above.

The respondents from firms welcomed these areas of support from innovation system institutions:

- Organisation of formal and informal training opportunities.
- Gaining access to codified information and knowledge from technical resource centres and information services.
- Improving the ability of other institutions such as universities, training centres and research centres to provide information and knowledge inputs.
- Clear guidelines and standards for TCB activities being imposed through licence requirements or special purpose rules.

There was also high degree of variation between the views of representatives of firms and those of public sector officials representing innovation system organisations in each of the four countries. The empirical research also showed conclusively that at the time of conducting the detailed study, the reality of relationships between telecommunication firms and the innovation system was far from this ideal. The evidence suggests that the innovation system in the four countries did not work to optimise technological learning for telecommunication firms. Table 3 summarises the mechanisms that are currently used by policy and regulatory bodies to interact with telecommunications firms in their jurisdictions; these data are based on reports gathered during in-depth interviews with operating companies and public policy and regulatory officials.

The evidence presented above, supported by the detailed accounts characterise the interactions between telecommunications firms and policy and regulatory bodies as infrequent, and designed to facilitate routine and procedural objectives rather than to facilitate co-operation on strategic issues such as innovation and capability building efforts. Analysis of the data shows that interactions between firms and the innovation system, represented here by policy and regulatory bodies, varied in intensity across the lifecycle of telecommunications projects. During the bidding, licence evaluation and licensing stages, there was intensive interaction between firms and policy and regulatory bodies. In addition to this phase being the most intensive in terms of frequency of contact, it often coincides with highest levels of exchange of information and is usually associated with greatest leverage for the public sector bodies. After licensing, the intensity of interaction falls off considerably, and gives way to more formal, procedural interactions, associated with monitoring of licence conditions. When policy and regulatory rules are being reviewed and/or revised, there also tends to be intensive lobbying by firms; but these phases are infrequent.

Table 3   Interaction Between Telecom Operators and National Public Policy and Regulatory Bodies

Means of Interaction	Uganda	Ghana	Tanzania	South Africa
Regular reports on operations, financial accounts	Yes	Yes	Yes	Yes
Regulatory accounts				Yes
Communicating via government appointees to the Board	Yes			
Communicating via govt appointed Chairman of the Board	Yes			
Communicating via presidential appointee to Board		Yes		Yes
Scrutiny during pre-qualification and bidding process for licences	Yes	Yes		Yes
Scrutiny during process of licensing and agreeing licence terms			Yes	Yes
Monitoring performance against specific clauses in licences and shareholders agreements etc.	Yes	Yes		Yes
Regular private sector consultation		Yes		Yes
Specific meetings to discuss technical regulatory issues e.g. interconnection, rate filings, spectrum pricing			Yes	Yes
Accessing information subsidised international training opportunities and fellowships	Yes			
Regular communication via a specific functional team			Yes	Yes
Lobbying activities			Yes	Yes
Joint participation in international meetings e.g. ITU study groups and standards setting organisations				Yes
Interacting regularly in government organised consultations				Yes
Joint development of national testing and certification standards				Yes
Joint design of public/private funded education and training initiatives and curriculum development				Yes
Developing policy guidelines e.g. GMPCS policy				Yes
Financial contributions to Human Resource Development Fund				Yes
Providing technical advice to public sector bodies				Yes

Source: Author's analysis of primary empirical data

It was also evident that the objectives of interaction between operating firms and policy and regulatory bodies were context specific. Given the problems of low levels of penetration of telecommunication services in Africa, policy and regulatory objectives in this region have tended to focus on setting explicit criteria for network rollout and have made the assumption that if operators are successful in meeting rollout targets they would *automatically* acquire technological capabilities and improve innovative performance. This assumption has meant that during the licence bidding stage and the negotiation of licence conditions, when intensity of interaction and leverage of the state is highest, the focus has not been on technological capability of operating companies. The failure to take account of innovation or technological capability as objectives in their own right, also means that during regular and routine interactions between public sector bodies and operating companies, there is very little opportunity to influence or facilitate innovation or capability development, either at the level of a formal legal mandate or on the basis of informal best practice standard setting. For example, while the operating companies in all four countries were required to provide regular reports on operational and financial performance, only in Uganda and South Africa did operating companies have to conform to guidelines on TCB effort in firms. None of the other three *ideal* roles for policy and regulatory bodies suggested by operating companies were adopted in current practice and the specific suggested roles for removing import duties on telecommunications equipment and sponsoring large projects on public-private partnership basis were also not implemented.

This study also demonstrated that there was variation among firms in their ability to manage boundary relationships with the innovation system and to use the inputs from the innovation system to facilitate TCB and innovation. South African firms appeared to have made more progress, as shown in Table 3, the interaction between South African operating companies and policy and regulatory bodies extended beyond the procedural level and worked in such a manner to provide tangible technological inputs for the TCB efforts of firms, often working in joint projects with public sector institutions. For example, South African firms undertook learning experiment, to trial wireless access technologies, involving the operating company, suppliers and state institutions; made contributions to internal skills development programmes, to sectoral skills development and human resource development funds, and to the Human Resource Development Fund, which partially financed a skills development programme championed and organised by the line ministry. Although South African firms encouraged state activity in provision of 'public goods' such as standardised training, technological literacy, and certification of training programmes, they were resistant to state direction or intervention into their management of their TCB effort. South Africa firms were proponents of *technological neutrality* in any public rules, and were adamant that the public sector should not specify or determine technological choice. These firms regarded public sector institutions as being less up to date with changes in technologies. There were however sharp differences between firms operating in competitive segments of South African telecommunications and those in monopolistic or duopolistic segments. The latter were more pragmatic and accepting of public policy involvement, while the former group was more resistant to any public sector role and identified the government's restrictive interpretation of the 1996 Telecommunications Act as a critical barrier to TCB efforts. It is also worth noting that the most pro-active institution in the South African innovation system was a line ministry, which had extended its mandate of policy setting beyond the traditional interpretation, and had established skills training programmes in partnership with firms.

Uganda, Ghana and Tanzania also provided examples of firms that were successfully able to manage boundary relationships with regulatory and policy bodies and enhance their innovative and capability development performance. The Tanzanian public network company used appointments to the board of directors as a mechanism of accessing technical skills and knowledge and secured the services of prominent and well-respected professors from the local university using this method. In this instance the state was able to exercise influence more effectively TCB activities in the state-owned public network firm through the board of directors rather than through its policy-making or regulatory role. In Ghana, firms used informal consultation and regular communication with the policy and regulatory bodies to solicit information and know-how, thus highlighting the importance of shared social-networks and trust between public officials and decision-makers in firms as a feature of the innovation system[3]. In the three smaller African countries, the scarcity of training resources and skilled personnel in knowledge centres, meant that the resources of the individual firms was greatly superior to the domestic innovation system and opening access to the knowledge resources of the operating firms was therefore an important effort. In Uganda, the example of success derives from the role played by a single individual, who was able to ensure that the large state-owned company had a leading role in gaining access to international sources of technical information as a representative of the Ugandan state. The representation of the country at the highest levels by an executive of the operating company was unusual and accelerated technological capability development in the firm.

For all firms in the sample, those with some degree of public ownership had more interactions with policy and regulatory bodies, (the innovation system) than those companies that were 100 per cent privately owned. There was also a high level of correlation between size of the company, and the reported level of interaction with regulatory and policy bodies. Smaller firms, 100 per cent privately owned firms, and firms operating in competitive segments were more dissatisfied with the facilitating efforts of the policy and regulatory bodies than larger firms and those firms that had some degree of state-ownership and/or monopoly control. Small firms in the smaller countries, with the least developed institutional systems reported highest levels of distrust and disaffection with the innovation system and displayed least ability to compensate for the institutional failings of the national innovation system. We believe that this finding is consistent with the argument of the TCB system approach which would suggest that firms with clearly designated, professional responsibility for managing interface with public and regulatory bodies would be better able to benefit from these interactions and to compensate for institutional failures. The relative high level of disaffection on the part of small firms might also be explained by the tendency for policy and regulatory bodies to be genuinely more solicitous of large firms that exert greater economic impact and wield more political influence. The evidence presented here shows that at present the relationships between public sector bodies and telecommunication operating companies do not work to optimise technological learning in these firms. The following section reviews some of the factors that lead to the discrepancy between ideal relationships and current reality.

Telecommunication Policy and Regulatory Frameworks and Promotion of Innovation

In the four countries examined in this study, promotion of innovation and TCB was considered to be positively associated with the priorities of network expansion and growth, and was therefore not explicitly defined as distinct policy objectives. Policy makers in the African countries under investigation reported that they would find it difficult to justify placing specific emphasis on innovation and TCB promotion. There was also a common perception that inclusion of specific requirements for TCB might introduce difficulties in attracting private international capital. In a region constrained by debt burdens and capital scarcity, this effectively limited the delineation of innovation and TCB specific objectives. While policy-makers understood that the licensing of new operators offered the opportunity for the state to exercise leverage in defining innovation and TCB objectives, this was often not put into practice, because the negotiation of licensing conditions was also a time when difficult and politically sensitive decisions about the restructuring of state-owned companies that were overstaffed and poorly managed were made. In this context, policy-makers often made a trade-off between not highlighting TCB in exchange for reducing levels of retrenchment. The financial objectives of privatisation and the state requirement to maximise proceeds also often took precedence over any long-term objectives such as facilitating innovation.

The legal mandate for public policy support for innovation and TCB in firms was unclear and weakly provided for in three out of the four countries. When there were provisions, such as in South Africa, TCB was included under the rubric of skills development, or in Uganda under technology transfer. The skills development provisions in South Africa were included in the context of the partial privatisation of a state-owned company and in providing remedies for the discriminatory access to skills development in apartheid era South Africa. Unlike requirements for network rollout, the targets for skill development and technology transfer were not well defined, and often did not include measurable targets or indicators and did not specify mechanisms for enforcement or monitoring. South Africa provides some exceptions in so far as it designed explicit provisions for funding of skills development through the Human Resource development fund, and mechanisms to set up skill development and training agencies that were specialised on a sectoral basis. The line ministry responsible for telecommunications in South Africa also adopted a pro-active approach in implementing training programmes with industry.

The operating companies emphasised more the general weaknesses in the policy and regulatory framework rather than on specific measures to facilitate innovation and TCB.

TCB System Approach and African Innovation Systems

Applying the TCB systems approach, and its related empirical methodology, has generated insights that are useful for exploring African innovation systems and developing policy recommendations. This study has shown that a firm's endogenous effort to manage relationships with the innovation system is a key factor, in influencing innovation performance and may be more important than the relative performance of public sector institutions in the innovation system. The results generated by this study are quite surprising, and possibly controversial, since the theoretical literature on capability development, particularly by developing country firms would suggest that institutions within the national innovation system would be an important source of technological capabilities.

However, it is clear that for the sample firms operating in Uganda, Ghana and Tanzania, the national innovation systems did not appear to exert a positive influence of their innovation efforts. In these three countries, firms that were able to develop effective TCB systems undertook many activities that can be characterised as 'compensation mechanisms', to overcome the failings of the national innovation system. The respondents from operating companies in these countries did not perceive their TCB effort to be in response to signals or inducements from the local innovation system. Although the South African firms operated in a more favourable environment, the representatives of these firms also did not regard local public sector institutions as sources of technological inputs. There are several factors that led to this poor performance.

First, since the pace of change in the technological knowledge required by operating companies is so rapid, the public sector bodies that are not engaged in production activities were not considered to be repositories of appropriate knowledge. Telecommunication firms often require equipment specific information regarding the operating conditions and functioning of network components. Public sector bodies are unlikely to be sources of that information which is often vendor specific and proprietary. This suggests that the type of knowledge and the use to which that knowledge is to be applied have implications for the relative usefulness and importance of institutions in the national innovation system. There is an important distinction to be made between the nature of knowledge required by service organisations and by manufacturing firms. Previous studies of capability development in the telecommunication industry have been based on analysis of the requirements of aspiring producers of telecommunication equipment, and in those settings, the role of national research laboratories has been significant (Hobday 1990; Mytelka 1999). This study does not lead to similar conclusions and, conversely, suggests that public sector research centres were more dependent on telecommunication operating companies as a source of knowledge and information than vice versa. This was particularly the case in instances where the telecommunication operating companies had foreign shareholders and had undertaken restructuring of in-house learning and training facilities. This result departs from the expected relationship between innovation systems as a source of technological capability inputs and has important policy implications.

Second, the legal mandate for supporting technological capability in firms was weak in all the case-study countries. Where there were provisions to support technological capability development in firms, these focused almost exclusively on skill development and were administered by institutions with very weak enforcement and compliance capability. Importantly, the study found that when general industrial development policy instruments, such as the Joint Economic Development regulations administered by the Department of Trade and Industry in South Africa, included provisions requiring firms to undertake technological capability development, they were more successful than sector specific policies.

Third, in understanding the problems associated with national innovation systems and their role in supporting TCB efforts in firms, it is also necessary to recognise that initial conditions matter. In Uganda, Ghana and Tanzania, gaining access to individuals with appropriate levels of skills was a major problem for the firms. In these countries, there was evidence that the national university systems had, by the late 1990s, been so plagued by under funding and structural problems, that their ability to provide support to the innovation and TCB effort in firms was severely restricted. Finally, the research findings appear to indicate that the ability of firms to benefit from boundary relationship with organisations in the national innovation system varied with firm size, with the larger firms reporting greater levels of interaction with public sector bodies. Interaction between public sector bodies and telecommunication operating companies appeared to be intense during the bidding and licensing stages and trailed off after networks were operational. This pattern may have contributed to the inability of sector-specific bodies to be effectively involved in long-term strategic issues such as technological capability development.

New Directions for Innovation Policy and Firm Strategy

These results also suggest some new directions for scholarship and innovation policy. Developing country firms are in favour of public sector institutions playing a pro-active role in facilitating their innovation and TCB effort. All sample firms supported the increased provision of public-goods, including the dissemination of 'know-why' and 'know-who' by the innovation system. This evidence suggests that national governments should strengthen the ability of public sector organisations to play the role of the 'honest broker' assisting firms with vendor selection, technology search and scan, and can encourage information and skills sharing across different firms and communities of interest. The evidence also suggests that public sector institutions have to improve their outreach to smaller firms and 100 per cent privately owned firms. Officials in public sector organisations have to reach outside of their own social class and professional networks that intersected with those of publicly owned firms to ensure that they can provide services to a wider range of firms on the basis of mutual trust and respect.

This analysis also suggests that systems of innovation and the tools used to support innovation and capability development should be widened. This is particularly so in the case of the telecommunication sector that is characterised by rapidly changing technology. To improve effectiveness, the employees of public sector knowledge producing institutions should update their skills, and improve their level of familiarity with and understanding of technological trends. Public sector institutions should also make their technological training curricula more suitable to providing graduates with an understanding of technological trends and fundamental principles as well as applied skills and industry-specific technological knowledge. In their interactions with industry, public sector institutions should become more facilitating and supportive of technological knowledge exchange.

The TCB system approach confirms that firms can compensate for unfavourable environments to become effective in innovation and capability accumulation. Operating in Tanzania was found to have a statistically significant negative influence on the effectiveness of capability development. However, there were Tanzanian firms among the top ten (out of 26) in terms of effectiveness of capability development. This suggests that the explanations for 'effectiveness' in capability development lie more in the endogenous variables than in the country-level factors. Moreover, it is the Tanzanian firm that adopted a TCB system that conformed to the characteristics of the 'ideal-system' that proved to be effective in its efforts to build technological capabilities. This firm succeeding by deploying a balanced and proportional systematic approach to innovation and TCB, in which a range of learning routines were effectively implemented in an integrated manner. It is these endogenous variables that compensated for the failings of the national innovation system and allowed the firm to outperform the level of effectiveness in innovation and capability development that would be predicted by the national innovation system with its small market size and unfavourable macro-economic conditions.

Weaknesses within firms also contribute to their inability to benefit from relationships with the innovation system. Firms that lack people with technological confidence fail to encourage experimentation and support open learning, develop management practices to integrate learning across firms and align with overall business strategy;. Firms also sharing responsibility for joint learning and knowledge acquisition with suppliers and institutions in the innovation system were found to be unable to optimise their technological learning through interaction with the innovation system. These weaknesses in internal capability development processes and organisational integration weakened their ability to draw available inputs from the innovation system and to compensate for weaknesses in their external environment. The implication of these empirically supported findings and analysis is that that policy relevant research on African innovation systems should drill down to the firm level to provide more cogent explanations of variation in innovation and capability development performance.

Notes

[1] Despite this literature's focus on South East Asia, the relevant insights were used and built upon to develop the TCB system approach.
[2] Further details on statistical techniques for data analysis and tests are reported on in Marcelle (2002) (see Statistical Annex and Chapter 5, Section 5-1).

[3] This feature of Ghanaian telecommunication companies appears consistent with business practices in other sectors; see Kuada (1994).

References

Argyris, Chris (1996), 'Prologue: Towards a Comprehensive Theory of Management', in Bertrand Moingeon and Amy Edmondson (eds.), *Organizational Learning and Competitive Advantage*, London: SAGE Publications: 1-6.

Argyris, Chris and Donald Schon (1996), *Organizational Learning II: Theory, Method and Practice*, Wokingham: Addison-Wesley.

Baden-Fuller, Charles and John M. Stopford (1994), *Rejuvenating the Mature Business*, Boston: Harvard Business School Press.

Barnett, William P. and Robert Burgleman (1996), 'Evolutionary Perspectives on Strategy', *Strategic Management Journal*, Vol. 17: 5-19.

Barney, Jay (1991), 'Firm Resources and Sustained Competitive Advantage', *Journal of Management,* Vol. 17, No. 1: 99-120.

Bell, Martin and Keith Pavitt (1997), 'Technological Accumulation and Industrial Growth: Contrasts Between Developed and Developing Countries', in Daniele Archibugi and Jonathan Michie (eds.), *Technology, Globalisation and Economic Performance*, Cambridge: Cambridge University Press: 83-137.

Cooper, Charles (1980), *Policy Interventions for Technological Innovation in Developing Countries*, World Bank Staff Working Paper, No. 441, December, Washington D.C: World Bank.

Cooper, Charles (1991), *Are Innovation Studies on Industrialised Economies Relevant to Technology Policy in Developing Countries?*, UNU/INTECH Working Paper, No.3, Maastricht.

Dosi, Giovanni, Richard R. Nelson, and Sidney G. Winter (2000), *The Nature and Dynamics of Organisational Capabilities,* Oxford: Oxford University Press.

Enos, John and Woo-Hee Park (1988), *The Adoption and Diffusion of Imported Technology: the Case of Korea*, London: Routledge.

Ernst, Dieter, Lynn Mytelka, and Tom Ganiatsos (eds.) (1998), *Technological Capabilities and Export Success: Case Studies from Asia*, London: Routledge.

Foss, Nicolai J. (1999), 'Incomplete Contracts and Economic Organization: Brian Loasby and the Theory of the Firm', in Sheila C. Dow and Peter E. Earl (eds.), *Contingency, Complexity and the Theory of the Firm: Essays in Honour of Brian J. Loasby, Vol. 2*, Cheltenham: Edward Elgar.

Hobday, Michael (1990), *Telecommunications in Developing Countries: The Challenge from Brazil,* London: Routledge.

Hobday, Michael (1995), *Innovation in East Asia: The Challenge to Japan*, Cheltenham: Edward Elgar.

Katz, Jorge M. (ed.) (1987), *Technology Generation in Latin American Manufacturing Industries.* London: Macmillan.

Kim, Linsu (1997), *From Imitation to Innovation: The Dynamics of Korea's Technological Learning,* Boston: Harvard Business School Press.

Kim, Linsu and Richard R. Nelson (eds.) (2000), *Technology Learning and Innovation*, Cambridge: Cambridge University Press.

Kuada, John (1994), *Managerial Behaviour in Ghana and Kenya: a Cultural Perspective*, Aalborg: Aalborg University Press.

Lall, Sanjaya (1987), *Learning to Industrialise: The Acquisition of Technological Capability by India*, London: Macmillan.

Lall, Sanjaya (1992), 'Technological Capabilities and Industrialization', *World Development*, Vol. 20, No. 2: 165-186.

Leonard-Barton, Dorothy (1995), *Wellsprings of Knowledge: Building and Sustaining Sources of Innovation*, Boston: Harvard Business School Press.

Levinthal, Daniel (2000), 'Organizational Capabilities in Complex Worlds', in Giovanni Dosi, Richard R. Nelson, and Sidney G. Winter (eds.), *The Nature and Dynamics of Organisational Capabilities*, Oxford: Oxford University Press: 363-379.

Lundvall, Bengt-Åke (1988), 'Innovation as an Interactive Process: From User-Producer Interaction to the National System of Innovation', in Giovanni Dosi, Richard R. Nelson, and Christopher Freeman (eds.), *Technical Change and Economic Theory*, London: Pinter Publishers: 349-369.

Lundvall, Bengt-Åke (1995), *The Social Dimensions of the Learning Economy*, Inaugural Lecture, Department of Business Studies, Aalborg University, Denmark.

Lundvall, Bengt-Åke (ed.) (1992), *National Systems of Innovation: Towards a Theory of Innovation and Interactive Learning*, London: Pinter Publishers.

Lundvall, Bengt-Åke, and Björn Johnson (1994), 'The Learning Economy', *Journal of Industrial Studies*, Vol. 1, No. 2, 23-42.

Marcelle, Gillian M. (2002), *Technological Capability Building and Learning in the Developing World: The Experience of African Telecommunication Companies*, Unpublished D.Phil thesis, Brighton: University of Sussex.

Mytelka, Lynn K. (ed.) (1999), *Competition, Innovation and Competitiveness in Developing Countries*, Paris: OECD Development Centre.

Nelson, Richard R. (1991), 'Why Do Firms Differ, and How Does It Matter', *Strategic Management Journal*, Vol. 12: 61-74.

Nelson, Richard R. and Sidney G. Winter (1982), *An Evolutionary Theory of Economic Change*, Cambridge: Harvard University Press.

Pettigrew, Andrew and Richard Whipp (1991), *Managing Change for Competitive Success*, Oxford: Blackwell.

Rumelt, Richard P. (1984), 'Towards a Strategic Theory of the Firm', in Robert B. Lamb (ed.), *Competitive Strategic Management*, Englewood Cliffs: Prentice-Hall: 556-570.

Schein, Edgar H. (1992), *Organizational Culture and Leadership*, 2nd ed., San Francisco: Jossey-Bass.

Senge, Peter M., Charlotte Roberts, Rick Ross, Bryan Smith, George Roth, and Art Kleiner (1999), *The Dance of Change: The Challenges of Sustaining Momentum in Learning Organizations*, London: Brealey.

Senge, Peter M. (1992), *The Fifth Discipline: The Art and Practice of the Learning Organization*, London: Century Business.

Starkey, Ken (ed.) (1996), *How Organizations Learn*, London: International Thomson Business Press.

Teece, David J., Gary P. Pisano, and Amy Shuen (2000), 'Dynamic Capabilities and Strategic Management', in Giovanni Dosi, Richard R. Nelson, and Sidney Winter (eds.), *The Nature and Dynamics of Organisational Capabilities*, Oxford: Oxford University Press.

Thomson, Ross (ed.) (1993), *Learning and Technological Change*, London: Macmillan.

Tushman, Michael L. and David Nadler (1996), 'Organizing for Innovation', in Ken Starkey (ed.), *How Organizations Learn*, London: International Thomson Business Press: 135-155.

Vaill, Peter (1996), 'The Purposing of High-Performing Systems', in Ken Starkey (ed.), *How Organizations Learn*, London: International Thomson Business Press: 60-81.

Westphal, Larry, Linsu Kim, and Carl Dahlman (1985), 'Reflections on the Republic of Korea's Acquisition of Technological Capability', in Nathan Rosenberg and Claudio Frischtak (eds.), *International Technology Transfer*, New York: Praeger Publishers: 167-221.

Winter, Sidney G. (1987), 'Knowledge and Competence as Strategic Assets', in David J. Teece (ed.), *The Competitive Challenge: Strategies for Industrial Innovation and Renewal*, Cambridge: Ballinger: 159-184.

Post Scriptum

Mammo Muchie[1]

This edited volume has been produced to re-examine and open up issues of development and underdevelopment in Africa. The persistence of underdevelopment despite the involvement of external actors saying they are there to contribute to Africa's development is one of several anomalies. Local actors have often vied for resources from outside and seem to be engaged largely in framing the development question in Africa with the paradigm of donors. This has short-changed Africa. Forty years after decolonisation, the continent that has rich natural resources has not managed to transform the largely agrarian economic structure.

A new focusing device is needed to conceptualise the problem of structural transformation in Africa. What we have proposed collectively here is that there is some merit in using the concept of national systems of innovation in the context of Africa. One advantage of the national system of innovation perspective lies in focusing debate on learning and competence building as the major source wealth and on employing science and technology for eradicating poverty in Africa. It is not with a tinkering with ad-hoc poverty alleviation programmes that poverty will be eradicated. It is by building the very foundation of wealth. In order to do this, each nation is required to organize its own system of innovation. A systemic and national perspective is helpful in mobilizing Africa's own resources.

But the national perspective may be too narrow in Africa where nation states are weak and economic activities are fragmented. A Pan African national system of political economy is what Africa does not have or has not been allowed to forge. I believe that there is a great potential in organising the structural transformation of the fragmented African economic space at a continental scale. Building the competence of old and new stakeholders and establishing new institutions is central if Africa is to claim the 21st century as a prosperous and wealthy new partner to the industrial world.

When the World Bank asks: can Africa claim the 21st century, the simple answer is no – at least if the seesaw between neo-classical economic recipes alternate with the human and social face of the same paradigm. The experience of the last forty years attests that there is something in the very core of the ideas that frame the development question in Africa that does not work. There is nothing that indicates that the conditions for development or take-off will be created as long as African countries, as fragmented entities, engage in 'privatisation, stabilisation and liberalisation'. This has been on the agenda now for the last twenty years. It has not happened on a scale that has transformed structurally the African economic space.

The broader global economy has been undergoing privatisation, deregulation and the so-called new economy has flourished and went into crisis. The assumption is that from the prosperity following the neo-liberal recipe, economic trickle-down will take place to all parts of the global system. What in fact has been happening is that 'trickle-up' to fewer and fewer wealth builders took place at a faster rate than any trickle-down effects to uplift the conditions of the poor world.

African leaders have tried to capture the 21st century with the upbeat rhetoric of the Africa Renaissance in a neo-liberal world. What they propose suggests that they would like to break the neo-liberal rules. The renaissance suggests that Africa has its own agency already or is willing to bring its agency back to drive nascent Africa. The African renaissance and the policy recipes flowing from neo-classical economic discipline mix like oil and water. The national system of innovation idea fits much better with the acceleration and strengthening of the African Renaissance.

Competence building and innovation takes place in many places in Africa, but there is no Pan African and integrated conception of how learning, competence building, science and technology may be integrated with production and users. In practice, a common Pan African perspective on how to build innovation and competence building systems needs yet to be forged. The political context exists in the form of the African Union (AU). The economic context exists in the form of NEPAD. The motivation exists in the form of a Pan African Renaissance. We offer this volume to open minds and encourage debate and encourage more research in connecting AU/NEPAD, the African Renaissance, Pan African citizenship and Pan African systems of innovation and competence building.

Notes

[1] This book is a result of conversations I had with B.A. Lundvall since 1999. It has been a learning experience. Intuitively I was attracted by the recognition of the role of nation building implicit in the systems of innovation approach. Africa needs to create a unification-nation. And that is a project that is related to creating a fully decolonised future for Africans the world over. For myself the interest is mainly to align the national system of innovation with the AU/NEPAD and African Renaissance efforts in order to create a parallel Pan African integrated science and technology system by creating the Pan African University that will produce knowledge and research to solve African problems. As I was working on this book with my co-editors, I had to move to the University of Natal in South Africa to direct a research programme on civil society in and for African integration on a secondment from the Middlesex University Business School, London. My thanks to the Ford Foundation and colleagues at the University of Natal, Adam and Vishnu.

Index

357

About Globelics

The Global Network for the Economics of Learning, Innovation, and Competence Building Systems (Globelics) is a global network of scholars who apply the concept of 'learning, innovation, and competence building system' (LICS) as their analytical framework. The network is especially dedicated to the strengthening of LICS in countries in the South.

The basic intent is to share experiences worldwide regarding analytical results, policy relevant experiences, and methodological issues among scholars, policy makers, and development practitioners as well as doctoral students and the informed public.

There are several reasons, which motivated us to initiate this project: the first and foremost is that as research finance in the U.S., Europe and Japan increase under pressures of international competitiveness between the Triad, there is a growing gap emerging with institutional support to scholars in the Southern and Eastern Hemispheres. It is our belief and claim that European knowledge institutions in particular, as unique transnational institutions, need to take on a bigger responsibility in taking initiatives that counter this tendency.

The many contributions to the literature on 'national' systems of innovations have from this perspective been valuable in bringing to the forefront the importance of state institutions in inducing or hindering processes of national competence building in a variety of countries. There is in our view a crucial need to broaden this framework not just geographically but also content wise to the rapid rise in globalisation pressures and the lack of global governance.

The analytical approaches are inspired by different disciplines and subdisciplines such as:

- Economics of knowledge and innovation
- Development economics and economic geography
- International business studies and organisation theory
- Theories on competence building in labour markets and in education systems

International comparative analyses aiming at locating unique systemic features as well as generic good practises will be stimulated within the network. The research will aim at feeding into policy making in the fields of industrial policy, innovation policy, regional policy, labour market policy and education policy as well as informing management of knowledge and innovation at the firm level.

The interim Scientific Board of Globelics includes: Prof. Bengt Åke Lundvall, Prof. Luc Soete, Prof. Richard Nelson, Prof. Christopher Freeman, Dr. Jose Eduardo Cassiolato, Prof. Frieder Meyer-Krahmer, Dr. Jorge Niosi, Prof. David Kaplan, Dr. Manuel Heitor, and Dr. Shulin Gu.

www.globelics.org